AQUATIC INSECTS

AQUATIC INSECTS

D. Dudley Williams, PhD, DSc

and

Blair W. Feltmate, PhD

Division of Life Sciences
Scarborough Campus
University of Toronto
Canada

C·A·B *International*

C·A·B International   Tel: Wallingford (0491) 32111
Wallingford   Telex: 847964 (COMAGG G)
Oxon OX10 8DE   Telecom Gold/Dialcom: 84: CAU001
UK   Fax: (0491) 33508

A catalogue record for this book is available from the British Library

ISBN 0 85198 782 6

Printed and bound in the UK by Redwood Press Ltd., Melksham

CONTENTS

PREFACE

Study of aquatic insects has been proceeding for centuries, possibly millennia. Primitive man could hardly have failed to be aware of the adults of biting species, as early settlements were invariably close to water. Whether or not these humans made the connection between the immature stages in the water and the adults flying around them on land, however, is another matter. Claudius Aelianus (ca. 175 - 235 A.D.) recorded swarms of adult caddisflies over the River Astraeus in Macedonia and described the use of artificial flies to catch trout and grayling, a practice stemming from the time of the ancient Greeks and later documented in a more scientific manner by Dame Juliana Berners (1496).

Macan, in an early comprehensive review of the ecology of aquatic insects, in 1962, chose to organize the then known facts by insect order. This was despite it being "preferable to use principles or factors as the headings in an article on ecology" but was prudent, he felt, because "if these can be illustrated only by examples drawn from one group, discussion of them is necessarily one-sided and incomplete". To a certain extent, the imbalance in coverage of the different orders that Macan alluded to is still a reality. Taxonomically, although many groups are reasonably well known today, there are still many gaps in the study of immature and pupal stages (e.g. Coleoptera and Diptera) and still regions of the world where the faunas are poorly known and collected, e.g. central Africa and South America. Ecologically, the imbalance is just as evident even though the overall number of studies has increased almost exponentially (Fig. P.1). For a variety of reasons (personal preference, ease of study, importance to human health, funding opportunities, designated research/academic appointments, serendipity, etc.) some insect groups are still studied more than others. The Trichoptera, Ephemeroptera, Plecoptera and Diptera (especially black flies and mosquitoes) remain perennial "favourites" whereas relatively little is known of the ecology of the aquatic Coleoptera or Hemiptera.

In the past three decades, the study of aquatic insects has been revolutionized. Not only have pure research studies shown the pivotal role played by the larvae of aquatic species in the breakdown of terrestrial leaf litter and the pathways by which this plant energy is incorporated into the tissues of fishes, birds and other vertebrates but a host of more applied research has revealed the importance of aquatic insects in the spread of diseases, in the biological assessment of water quality, and in the reconstruction of past environments on earth. Most recently, aquatic insects are being used, increasingly, to test many elegant hypotheses in contemporary ecological theory.

The purpose of this book is to try to give the reader a taste of these exciting new findings in the study of aquatic entomology, together with some of the

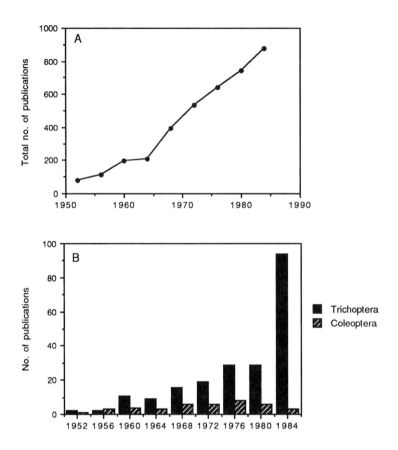

Fig. P.1 (A) Number of publications on the ecology of aquatic insects during
the period 1952 - 1984; (B) Comparison of the number of publications
on the ecology of the Trichoptera and aquatic Coleoptera over the
same period (from the authors' own reference database).

basic background information necessary to comprehend them, and to provide
references to more detailed studies in the primary literature. To these ends,
coverage begins with an introduction to insects, in general, and designation of
the aquatic groups. Chapter 2 characterizes each aquatic order by highlighting
morphology, special features of development and differences in basic biology,
and by comparing ancestry and the zoogeography of modern forms. Chapter 3
provides keys (to the ordinal level) for both mature and immature stages. Chapter
4 deals with aquatic insect habitats and the communities that they contain, and
considers insect distribution at both the macro- and micro-level. Chapter 5 relates
life history traits to habitat type and discusses the selection pressures at work.
This is followed (Chapter 6) by coverage of the problems facing insects (which,
originally, evolved on land) that live in water. Features of the population

biology of aquatic insects are discussed (Chapter 7), with emphasis on dispersal and colonization patterns, population dynamics, coexistence and competition. Two chapters are devoted to trophic relationships, the first (Chapter 8) deals with the role of insects as processors and converters in aquatic food webs, and the second (Chapter 9) concentrates on predation and its consequences, especially the effects on species fitness. Experimental design and sampling methods are covered next (Chapter 10), as these often become pitfalls during quantitative studies, and holistic (field) versus reductionist (laboratory) approaches to experimental design are discussed for similar reasons. The final chapter (11) deals with the relationships between aquatic insects and mankind and discusses the former's role as biological indicators of degraded water quality, their importance to human health, the use of fossil insects as palaeoecological tools, and the development of artificial lures for angling.

This book could not have been completed without the valued assistance of several people. Foremost among these are Professors Nancy Williams and Noel Hynes who critically examined the entire text; Dr. Maurice Lock who provided a quiet oasis where (in the company of sheep) the senior author spent a sabbatical leave, writing; and Mrs. Judith Smith for her advice on the vagaries of Microsoft Word (or our interpretations of it). We are grateful also to all the authors who gave us permission to cite their published or unpublished works, but especially to Dr. Mike Lehane (Bangor), Dr. Lena Peterson (Lund), Prof. Rich Merritt (Michigan), Prof. Cal Fremling (Minnesota) and Prof. Doug Craig (Edmonton). The Natural Sciences and Engineering Research Council of Canada contributed funds to the project.

D. Dudley Williams   Blair W. Feltmate
"Glanffrwd"   Scarborough
Lon Gernant   Toronto
Menai Bridge

1991

Illustrations by

Annette Tavares-Cromar

and

Marilyn Smith

FOREWORD

General books on aquatic insects have a long history, starting with Miall (1895). This dealt, as its title stated, with natural history, and because of that it had a great influence on me as a child thirty years after its first appearance. It was followed by a series of volumes in Europe and North America which dealt primarily with identification, and thus were of less interest to the general reader and the budding biologist. Details of these can be found in various chapters of Resh and Rosenberg (1984), where the history of the study is discussed among many other topics at a fairly high scientific level.

Then, during and after World War II, we started to get books that dealt again with natural history as well as identification, and I recall reading these avidly. I refer particularly to Wesenberg-Lund (1943) and Usinger (1956). The former dealt with the world literature, and in my enthusiasm to learn from it I acquired the ability to understand German, which my schooling had not given me. So books that inspire young people to follow scientific careers can have great value.

Since then many books and specialized treatises have appeared dealing with the aquatic fauna of various parts of the world, such as areas of Europe, Japan, New Zealand, Australia, West Africa and North America and several of them, most notably Merritt and Cummins (1978), incorporated information analogous to the natural history of Miall and Wesenberg-Lund but framed in modern ecological concepts. These are the handbooks of the professionals and although they are fascinating they are tough reading for the generalist.

In the present volume the authors have tried to produce a more general work that will give the interested reader an overall picture of what is going on in the study of aquatic insects, and how they have become important in this age of environmental concern. Indeed, aquatic insects feature nowadays in all kinds of research, be it toxicology, behaviour or general ecological theory. From being the mere interesting delight that was perceived by Miall they have become an important part of non-molecular biology, and this book serves as an introduction to them and their present-day status in science. I wish it success, and I think that Professor Miall would be happy with his intellectual descendants.

References

Merritt, R.W. and Cummins, K.W. (eds) 1978. An Introduction to the Aquatic Insects of North America. Kendal/Hunt Publishers, Dubuque, Iowa. 441 pp.

Miall, L.C. 1895. The Natural History of Aquatic Insects. Macmillan, London. 395 pp.

Resh, V.H. and Rosenberg, D.M. 1984. The Ecology of Aquatic Insects. Praeger Scientific, New York. 625 pp.

Usinger, R.L. (ed.) 1956. Aquatic Insects of California. University of California Press, Berkeley and Los Angeles. 508 pp.

Wesenberg-Lund, C. 1943. Biologie der Susswasserinsekten. Springer-Verlag, Berlin. 682 pp.

H.B.N. Hynes
Waterloo, Ontario

December 1991

1 INTRODUCTION

1.1 What is an insect?

Basically, an insect exhibits the following surface features: an external skeleton, or cuticle, made from chitin (a nitrogen-containing polysaccharide) and protein; a body divided into three sections (head, thorax and abdomen), although in the larvae of some forms the abdomen and thorax cannot be distinguished; three pairs of jointed legs attached to the underside of the thorax; a pair of antennae, usually a pair of compound eyes and up to three single eyes (ocelli) on the head; and, usually, one or two pairs of wings attached to the last two segments of the thorax. The abdomen generally lacks appendages, except for external genitalia and sensory cerci at the tip, but in the larvae of some aquatic forms there may be segmentally arranged gills. Excluding the areas around joints, the exoskeleton of most adult insects, together with that of many pupae and larvae, is hard. This hardness is due to sclerotization, a process by which the protein component of the cuticle is transformed into sclerotin, by the action of quinones. Cross-linkages are formed between the protein molecules which produce a substance of such strength that the mandibles of certain beetles are capable of biting through soft metals such as lead. Of these characteristics, probably the exoskeleton, six legs, and wings together with a crawling gait allow most lay people to recognize an insect. Scientists, however, are more precise and have painstakingly refined the study and definition of insects so that they may be distinguished from related animals such as spiders, crustaceans and centipedes.

Internally, the main body cavity is not the true coelom characteristic of other protostome coelomates (e.g. annelid worms and molluscs) but a system of large sinuses in the tissues (haemocoel) which are filled with blood. The latter contains a variety of amoebocytes and, depending on the species, the copper-based respiratory pigment haemocyanin or occasionally the iron-based pigment haemoglobin. The gut is divided into three regions. The large foregut and hindgut are lined with chitin, as they are derived from the ectoderm of the embryo, whereas the midgut is endodermal in origin and is frequently invaginated to increase its surface area for secretion and absorption. The central nervous system is concentrated in the head to form a brain (cephalization) but there are segmental ganglia located along the length of the ventral nerve chord. Respiration is accomplished by a system of air passages, the tracheae and the much smaller tracheoles, which communicate with the outside world via paired openings, or spiracles. These are located laterally, one pair per segment, on the last two thoracic and, usually, the first eight abdominal segments. Aquatic insects exhibit interesting adaptations of this system in order to allow them to

1

breathe under water. Most insects are dioecious, that is they have separate sexes, and fertilization is internal. The gonads (ovaries or testes) are located in the abdomen and ducts from these open to the outside near its tip. The male duct opens ventrally on the 9th abdominal segment and is surrounded by a sperm-transfer organ, the aedeagus. The female duct generally opens on the ventral side of the 8th abdominal segment and, associated with it, may be structures used for sperm reception and storage, provision of nutrients and protective covers for the eggs, and glands that secrete adhesive material used for attaching the eggs to some substratum as they are laid.

Insects differ both in the degree of development completed prior to hatching and in the number of stages through which they pass prior to becoming adults. Growth typically follows a geometric progression and necessitates the entire exoskeleton being shed, periodically, so that a new, larger individual can be produced. The stages in the life cycle between these "moults" are known as *instars*.

1.2   The origin of insects

The fossil history of the arthropods can be traced back to the Early Cambrian Era (540 million years Before Present; Table 1.1) and indicates that the main groups and a number of subgroups were already in existence then. This early divergence, before the first known arthropods were fossilized, means that reconstruction of their early phylogeny is difficult. It is clear, however, that the arthropods themselves arose from primitive segmented worms (polychaetes) or from some common ancestor. Modern species from both groups show homologous features in terms of body segmentation, the central nervous system, embryonic development, and segmental appendages.

The Subphylum **Uniramia**, composed of the myriapods (centipedes and millipedes) and the hexapods, is thought to have evolved along a separate line from those leading to the chelicerates and the crustaceans with the bulk of its diversification taking place on land. A marine uniramian, *Aysheaia*, has been found, however, in the Burgess Shale of the mid-Cambrian (530 million years B.P.) (Whittington, 1978). The limbs and cuticle of *Aysheaia* are similar to the most likely ancestors of the Uniramia, the **Onychophora**, small, elongate, soft-bodied animals that today are restricted to damp terrestrial habitats in the tropics and southern hemisphere. The Onychophora, in turn, seem to have been derived from marine polychaete worms. Onychophorans exhibit many features in common with both segmented worms (Annelida) and arthropods and often have been likened to a missing link between these two phyla. Their phylogenetic status continues to be debated so that some workers consider them to be a separate phylum while others place them in a superclass within the Uniramia (Manton and Anderson, 1979). With the exception of the Onychophora, the oldest terrestrial fossil uniramians are myriapod-like animals from the Devonian

Table 1.1 Geological time scale together with major events in the evolution of the insects (based on Borror *et al.*, 1981; Geological Society of America, 1983).

Era	Millions of Years B.P. (beginning of period)	Periods Epochs	Event
Coenozoic	1.6	Quaternary	
		Pleistocene	First humans
	66.4	Tertiary	
		Pliocene	
		Miocene	Rise of modern insect genera
		Oligocene	
		Eocene	
		Palaeocene	
Mesozoic	144	Cretaceous	Most modern insect orders present; first flowering plants
	208	Jurassic	
	245	Triassic	Ephemeroptera & Odonata
Palaeozoic	286	Permian	Rise of most modern insect orders
	360	Carboniferous	First winged insects, most now extinct, e.g. Palaeodictyoptera
	408	Devonian	First insects (*Rhyniognatha*)
	438	Silurian	First land arthropods (scorpions & millipedes)
	505	Ordovician	First vertebrates
	570	Cambrian	First arthropods (marine)
Precambrian			Primitive invertebrates

(380 million years B.P.).

Until recently, all arthropods with six legs were classed as insects but it is now thought that the hexapod condition may well have evolved more than once from terrestrial myriapods. Current opinion proposes four classes of hexapod: **proturans**, minute animals living in moist soil or decaying plant matter; **collembolans**, or springtails, small animals similarly found in moist plant matter, under bark and in the soil but also on the surfaces of freshwater pools and along the shores of rivers, lakes and the sea; **diplurans**, small animals again found in a variety of damp places such as in rotting wood, under bark and stones, in the soil and in caves; and **insects**. The first three classes (sometimes collectively known as the Entognatha) differ from the insects primarily in terms of head structure. In the entognaths, the mandibles and maxillae are deeply recessed in pouches within the head capsule and articulate with it at only one point. During feeding, the tips protrude and their action reduces food pieces to very small particles that are then ingested. In insects, the mandibles articulate with the head capsule at two points. This enables the mandibles to move transversely, an action suitable for biting off and grinding particles of food (Daly *et al.*, 1978). Members of the different hexapod classes also differ in the ways in which they use their three pairs of legs: the Collembola can crawl but have become specialized in jumping through use of their unique ventral, forked structure, the furcula, combined with an ability to hold their bodies rigid; this contrasts with the marked flexibility of the trunk in the Protura and Diplura; whereas the insects (except the Thysanura) exhibit trunk stability through a plantigrade tarsus and a fixed sternal pleurite, which prevent rocking and provide the coxa with a firm base, attributes that contributed to the evolution of flight and the tremendous success and adaptive radiation of the group (Manton, 1979).

The earliest known hexapod is *Rhyniella praecursor*, a collembolan from the Lower Devonian of Scotland. However, as its morphological features show, this was already an advanced member of its class indicating that the transition from myriapod to hexapod must have occurred earlier. Another hexapod, *Rhyniognatha hirsti*, was contemporary with *Rhyniella* and probably represents the first known insect proper (?Ectognatha, possibly a small early progenitor of the mandibulate orthopteroids - Tillyard, 1928; Fig. 1.1).

In the Upper Carboniferous (290 million years B.P.) are remains of the palaeopterygote *Erasipteron larischi*, the neopterygote *Stygne roemeri* (an orthopteroid) and *Metropator pusillus* which may have been an oligoneopteran. Some ten orders of insect were evident by the close of the Carboniferous. These were still extant in the Permian (286 - 245 million years B.P.) and others appeared or firmly established themselves at this time (especially the neopterygote orders Orthoptera, Psocoptera, Hemiptera, Mecoptera and Coleoptera), coincident with the rapid development of land plants. By the end of the Permian the Palaeodictyoptera, the earliest known winged insects (belonging to the infraclass Palaeopterygota) and the possible ancestors of the Neopterygota, were extinct (Fig. 1.2). It is likely that primitive mayflies (Ephemeroptera) were

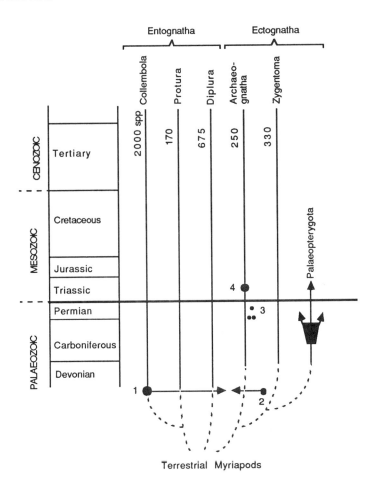

Fig. 1.1   Events in the early history of insects. Arrows from individual fossils intersect groups whose synchronous existence can be inferred from these fossils. 1, *Rhyniella praecursor*; 2, *Rhyniognatha hirsti*; 3, Monura (?Archaeognatha); 4, *Triassomachilis uralensis* (modified from Hennig, 1981).

present in the Permian (and possibly even in the Upper Carboniferous) and that by the early Mesozoic they had already separated into some of the subgroups still in existence today (Hennig, 1981).

The Mesozoic was a period (245 - 66.4 million years B.P.) of dramatic change in the insect fauna. Some 12 orders became extinct but many of those that survived underwent considerable diversification, especially the Neuroptera, Mecoptera and Diptera (Riek, 1970a). Within the Odonata (dragonflies and damselflies), a number of subgroups extant today are thought to have existed at

Palaeopterygota

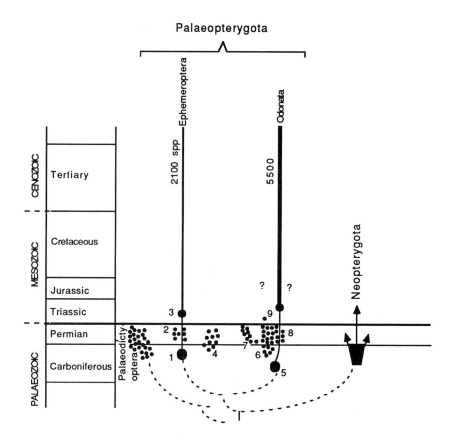

Fig. 1.2   Phylogenetic tree of the Palaeopterygota. 1,*Triplosoba pulchella*
(Protephemeroptera); 2, Permoplectoptera (Ephemeroptera); 3,
Ephemeroptera (?Heptagenioidea); 4, Archodonata (systematic
position unsure, possibly Ephemeroptera); 5, *Erasipteron larischi*
(Protodonata); 6, Protodonata; 7, Protanisoptera; 8, Protozygoptera;
9, Upper Triassic species from the stem-group of the Anisozygoptera/
Anisoptera (approximate numbers of species known, world-wide, are
shown; redrawn from Hennig, 1981).

least in the Upper Triassic and Jurassic. Archodonatans existed in the Palaeozoic
(Permain and Upper Carboniferous) but their systematic position is uncertain as
they show certain features in common with the sister-group of the Odonata, the
mayflies (Hennig, 1981; Fig. 1.2).

Tertiary insects (66.4 - 1.6 million years B.P.) are essentially modern and
the good fossil record available in Baltic amber and other beds shows that many
recent families and genera have been in existence since that time (Riek, 1970a).

Details of the early history of the aquatic insect orders will be given in Chapter 2.

1.3 The spread of insects

As we have seen, the first known insects occurred in the Lower Devonian in Laurasia, which was then a tropical environment - this region roughly corresponded to what is now the Canadian Shield, plus western Europe to the Ural mountains. Gondwanaland (large in the Palaeozoic, encompassing Brazil, Africa, Arabia, Madagascar, India, Australia, New Zealand and Antarctica) became glaciated in Permo-Carboniferous times and, as it cooled many insect groups migrated from it to Laurasia, and then subsequently to Angara (present-day Siberia) as Laurasia cooled. Insect evolution was much affected by the fragmentation of Gondwanaland into some of the continents that we know today. In the mid-Cretaceous (97 million years B.P.), some groups of insects spread with the first angiosperm (flowering) plants from East Gondwanaland to Angara, colonizing the Northern Hemisphere during the Tertiary. Descendants of Mesozoic Gondwanaland insects are now to be found all over the world, in a wide range of climates, despite the fact that most of them evolved in a subtropical Gondwanaland (Gressitt, 1974). Apart from cockroaches (Orthoptera), most Laurasian groups have become extinct. In contrast, some descendants of Permian insects still live in Australia today, e.g. the tabanid fly *Exeretoneura*, alongside their fossil counterparts, and this may be a consequence of Australia being almost completely isolated from other regions throughout most of the Tertiary.

Combining the known facts on faunal distributions with inferences about phylogeny suggests that there were three major phases to the spread of animals, in general, over the globe: (1) an early period during which dispersal was almost random; (2) a middle period during which dispersal was predominantly in wide latitudinal bands roughly corresponding with present-day northern and southern land masses, with the southern band being subdivided into temperate and tropical belts; and (3) a later period in which dispersal was primarily longitudinal but dominated by Holarctic radiation with subsidiary radiation centres such as in the tropics and Australia. Transition from phase one to phase two appears to have begun in the latter part of the Palaeozoic, and from phase two to phase three in the late Mesozoic (Mackerras, 1970).

Many genera of aquatic insects (e.g. belonging to the Ephemeroptera, Odonata, Plecoptera, Megaloptera, Trichoptera and Diptera), like certain plants, have wide southern hemisphere distributions, particularly in Chile, Tasmania, Australia and/or New Zealand, and sometimes southern Africa. It has been hypothesized that most of these "fragile insects" must have spread over land that was contiguous and therefore might represent Mesozoic survivors. Some faunal elements must have reached Australia from the north, prior to the Tertiary, and endemic genera evolved during the period of isolation. Subsequent climatic fluctuation in the Pleistocene (1.6 million years - 10,000 years B.P.) promoted speciation (Evans, 1958; Mackerras, 1970).

Biogeographical studies of mayflies (Ephemeroptera) have contributed to the theory of continental drift by providing a detailed documentation of the steps involved in the breakup of Gondwanaland. The sequence, proposed by Edmunds (1972), was: (1) India drifted north; (2) Africa and Madagascar drifted north, with Africa still joined to South America in the north and Madagascar separating from from Africa much later; (3) New Zealand and New Caledonia drifted north; (4) New Caledonia separated from New Zealand; (5) Australia drifted north and Antarctica drifted south; (6) South America drifted northwest in relation to Antarctica. Whether South America was a single land mass was questioned because the mayfly faunas of Chile and Argentina are very different. The suitability of mayflies (both nymphs and adults) for use in historical biogeography lies in the fact that their dispersal ability is limited, so that few occur on oceanic islands or on isolated mountains, and that, as we have seen, they have an ancient lineage stretching back to at least the Permian.

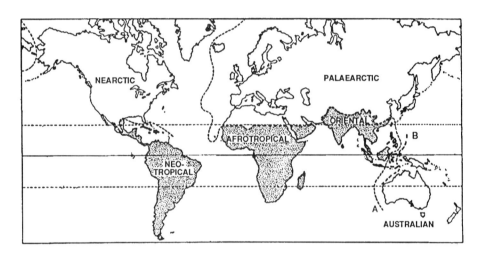

Fig. 1.3   Zoogeographical regions of the world (after Wallace, 1876). The curved broken lines A and B mark the positions of Wallace's and Weber's lines, respectively. The transitional zoogeographic zone, Wallacea, is bound by these two lines.

Past distribution processes combined with evolution have led to present day distribution patterns of insect species. For convenience, and the furtherance of zoological research, the world has been divided into various zoogeographical regions (Wallace, 1876). Briefly summarized, together with the characteristics of their faunas, they are (Fig. 1.3): the **Palaearctic** and **Nearctic** regions, the faunas of which are related and frequently combined as the Holarctic fauna. This fauna is characterized by the great diversity of many groups resulting from radiation during the Tertiary, but it also contains some older elements and some of both Afrotropical and Neotropical origin. In the **Neotropical** region, the fauna is distinctive, with a strong southern component related to southern Afrotropical and Australian forms, tropical radiations of mixed relationships, and intrusions of more recent Nearctic elements from the north. The **Afrotropical** region supports a fauna consisting of a southern component, a Palaearctic component, tropical elements derived from both, and an intrusion from the Oriental zone in the east. In the **Oriental** region the fauna results primarily from a tropical Palaearctic expansion, with relicts from Gondwanaland, and a significant Afrotropical intrusion from the west together with a smaller Australian one in the east. The Oriental fauna enters the Pacific through **Wallacea**, a wide zone of transition and filtration that is bordered by Wallace's and Weber's lines[1]. The **Australian** region, which extends south and east from Wallacea, is characterized by four faunal elements: a small archaic element of primitive animals surviving from Palaeozoic or early Mesozoic times; a major southern element consisting of a number of terrestrial and freshwater invertebrate groups, most orders of insects and some vertebrates; an older northern element related to centres of evolution in Africa or Madagascar, around the Indian Ocean, New Guinea, and the Pacific with connections with the older Neotropical fauna; and a prominent, younger northern element related to important groups in the Oriental fauna and containing the most highly evolved and frequently most conspicuous elements in most insect orders (Mackerras, 1970).

1.4   Classification of insects

Arthropods are remarkably diverse and constitute over 75% of all living animal species. More than one million species of terrestrial and freshwater insect have been described making them the most numerous of the arthropod groups. It is estimated that at least as many species again have yet to be described (Ross, 1965) and there is serious concern that in the course of current global habitat destruction, particularly in the tropics, many of these will be lost to science for ever.

---

[1] Much discussion has arisen over the significance of these two lines as biogeographers have sought the "ultimate line" separating the Asian and Australian biotas. The real significance of Wallace's line appears to be in demarcating a rich and diversified mainland Asian biota from a much poorer insular one. It divides some groups very well, for example amphibians, freshwater fishes, mammals and birds, but it is less important as a dividing line for insects and plants. Weber's line represents a line of faunal balance between Oriental and Asian elements (Keast, 1983).

Insects, in general, have become adapted to a wide range of environments, from hot to cold, from forests to grasslands, from valley bottoms to mountain tops, and from lakes, ponds and rivers to deserts. Some species feed on fresh plant material whereas others feed on decaying matter, plant or animal, while others are carnivorous or parasitic. Much of this diversity derives from the morphological and physiological adaptations that allowed their early colonization of terrestrial environments, such as: the highly chitinized exoskeleton that prevents desiccation; the small size and ability to fly which have enhanced dispersal capabilities; the short generation time which allows rapid adaptation to shifts in environmental forces; and the ability, in the more advanced orders (Oligoneoptera), to exist in two discretely functioning stages, larva and adult, which may occur in different habitats and therefore be subject to different selection processes (Danks, 1979). Faced with such diversity and sheer numbers of species, an effective classification scheme is essential to their study.

Organism diversity is readily apparent as is the discontinuity between different forms. It is this discontinuity that has been exploited to devise a scientific classification of organisms and the hierarchical nature of these observed discontinuities readily lends itself to this purpose. For convenience and convention, these discrete clusters are termed species, genera, families, orders, classes and phyla, with additional subgrouping where appropriate. The system is somewhat subjective in that the observer chooses, within limits, which cluster is to be designated a genus, family, order, etc., but the basic unit of the system is the species which can be defined quite rigidly as a population, or array of populations, between which the gene exchange is limited or prevented by reproductive isolating mechanisms. Although the system is a pragmatic and man-made one for the orderly storage and retrieval of information, it appears natural in that it identifies discernible discontinuities in the variation amongst organisms, thus the dividing lines between species, genera, etc. are made to correspond to the gaps between the discrete clusters of living forms (Dobzhansky, 1982).

The species is commonly considered not only to be the basic unit of taxonomy but also the taxonomic unit of evolution. Some scientists (e.g. Nelson, 1989) argue against this, claiming that there is no empirical difference between species and other (higher) taxonomic groups, and that species need not evolve solely through speciation of other species. This view leads to the conclusions that evolution is not a phenomenon confined to the species level, and that taxa give rise to other taxa only in the sense that ancestral taxa differentiate, in other words taxa give rise to subtaxa but all taxa evolve.

Formally, in the Linnean system of hierarchical classification, insects belong to the Class Insecta (from the Latin word *insectum,* meaning cut into sections) contained in the Superclass Hexapoda (from the Greek *hex* = six, and *pous* = foot), Subphylum Uniramia (L, *unus* = one; *ramus* = branch, referring to the unbranched nature of the appendages). The latter, together with the subphyla Crustacea, Chelicerata (spiders, scorpions, mites and their ilk) and the extinct

Table 1.2   Higher classification of living insects and related terrestrial arthropods, together with some important extinct groups* (after Scudder *et al.*, 1979; aquatic hexapods are marked •).

Superphylum **Arthropoda**
Phylum Stelechopoda - Class Tardigrada (water bears)
Phylum **Entoma**

Subphylum Trilobita *	Class **Insecta**
Subphylum Chelicerata	Subclass Archaeognatha
Class Arachnida	Order Microcoryphia
Subclass Rostrata	Subclass Zygentoma
Order Solifugae (wind scorpions)	Order Thysanura
Subclass Pectinifera	Subclass Ptilota
Order Scorpionida (scorpions)	Infraclass Palaeopterygota
Subclass Chelonethida	Order Protodonata *
Order Pseudoscorpionida	Order Palaeodictyoptera *
Subclass Labellata	Order Ephemeroptera •
Order Araneae (spiders)	Order Odonata •
Subclass Phalangida	Infraclass Neopterygota
Order Opiliones (harvestmen)	Cohort Polyneoptera
Subclass Acari (mites)	Order Plecoptera •
Order Parasitiformes	Order Dictuoptera
Order Acariformes	Order Notoptera
Class Pentastomida (tongue-worms)	Order Dermaptera
Subphylum Crustacea	Order Grylloptera •
Class Malacostraca	Order Orthoptera •
Subclass Eumalacostraca	Order Cheleutoptera
Superorder Eucarida	Cohort Paraneoptera
Order Decapoda (shrimp, crabs, etc.)	Order Psocoptera
Superorder Peracarida	Order Phthiraptera
Order Amphipoda (amphipods)	Order Hemiptera •
Order Isopoda (isopods)	Order Thysanoptera
Subphylum **Uniramia**	Cohort Oligoneoptera
Superclass Onychophora (onychophorans)	Order Megaloptera •
Superclass Myriapoda	Order Raphidioptera
Class Pauropoda (pauropods)	Order Neuroptera •
Class Diplopoda (millipedes)	Order Coleoptera •
Class Chilopoda (centipedes)	Order Mecoptera
Class Symphyla (symphylans)	Order Diptera •
Superclass **Hexapoda**	Order Siphonaptera
Class Protura (proturans)	Order Lepidoptera •
Class Collembola (springtails) •	Order Trichoptera •
Class Diplura (diplurans)	Order Hymenoptera •

Trilobita, belongs to the Phylum Entoma. The Entoma and the Phylum Stelechopoda (tardigrades) form a Superphylum of jointed-legged metazoan animals known as the Arthropoda (Gr, *arthron* = joint; *pous* = foot). This classification (Table 1.2) is a compromise and differs in certain respects from many traditional (textbook) schemes. It does, however, incorporate many of the recent concepts put forward in the literature (Scudder *et al.*, 1979).

2 THE AQUATIC INSECT ORDERS

The orders of insects may be divided into two groups based on the development of individuals. More primitive orders show **hemimetabolous** development, that is the external form of the nymph gradually approaches, through a series of instars, that of the adult; the last nymphal instar resembles the adult very closely, but is not sexually mature. In the Megaloptera and all higher insect orders, the immature stage (traditionally known as a "larva", as compared with "nymph" for hemimetabolous insects) typically does not resemble the adult so that a marked change in external appearance takes place at metamorphosis. This change, combined with an additional stage in the life cycle, between the larva and the adult (the pupa), denotes **holometabolous** development. The Holometabola are also known as endopterygote (literally, "inner wing") insects because the wing buds develop as imaginal discs beneath the larval cuticle. These buds are usually not visible until they are extruded between the discarded larval skin and the epidermis at the end of the larval stage. They thus become apparent only in the pupal stage. In the Hemimetabola the wing buds develop externally and are clearly visible throughout the nymphal stages. Hemimetabolous insects are, therefore, also known as exopterygotes (literally, "outer wings").

2.1   Collembola

It is likely that the Collembola, or springtails, diverged early on in the ancestry of the hexapods as they are entognathous and show many embryological features in common with the myriapods. Recall that the earliest known fossil hexapod is *Rhyniella praecursor*, a collembolan from the Lower Devonian. Although, strictly speaking, they are no longer considered to be insects, traditionally they have always been included in entomological texts.

Collembolans are minute to small animals, generally less than 3 mm long, and are most commonly found in moist terrestrial habitats such as humus or the soil. A moist environment is essential to their survival, as most species lack true tracheae and respire cutaneously over the entire body surface. Many are found on the margins or surfaces of freshwater ponds and streams, where their small size and water-repellent cuticle keep them afloat. Some species have more specialized habitats like caves, underground water systems, the interstitial spaces beneath streambeds, the marine intertidal area, rain pools, permanent ice fields and hygropetric habitats (where thin sheets of water flow over rock surfaces) in general.

The body may be elongate, showing distinct segmentation, or more

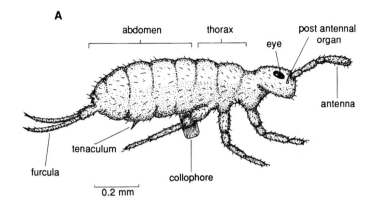

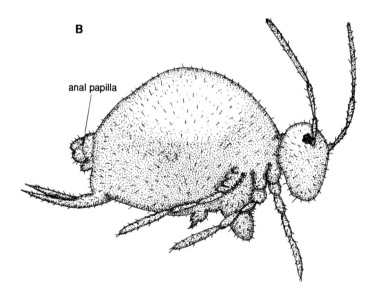

Fig. 2.1   General structure of two types of Collembola: (A) *Isotoma* sp. (Isotomidae); (B) *Katianna* sp. (Sminthuridae) (redrawn after Wallace and Mackerras, 1970).

globular, in which case segmentation is indistinct (Fig. 2.1). Scales or hairs cover the cuticle and may produce marked pigmentation or an irridescence, but subterranean forms are usually pale. In elongate species, a head, three-segmented thorax and six-segmented abdomen are clearly visible. The mouthparts are adapted either for chewing or for sucking food such as decaying plant or animal matter, fungal hyphae, bacteria, algae, spores and even insect faeces (frass). Patches of simple ocelli are present, on the head, in most species, and pseudocelli may occur on the general dorsal surface. The single pair of antennae varies in length and complexity of structure according to species. Also on the head is a postantennal organ, the function of which is unknown. The legs of many springtails have elaborate toothed claws (Scott, 1971).

Three of the six abdominal segments bear structures (the collophore, tenaculum and furcula, or spring) derived from primitive segmental appendages. The collophore, or ventral tube, the exact function of which is unsure, is a diagnostic feature of the Collembola. It was originally thought to be an adhesive or respiratory organ but now has been shown to be involved with the absorption of water from moist surfaces (Imms, 1990). The furcula and tenaculum form the basis of the jumping apparatus although they are not present in all species. The furcula consists of a pair of whiplike appendages whose bases are fused and attached to the fourth and sometimes the fifth abdominal segments. When not in use, the spring is held under the abdomen by a catch formed by the tenaculum on the third abdominal segment. When released, the furcula strikes the substratum and the whole animal is flicked upwards.

Sexual dimorphism is mostly restricted to differences in size, with males being smaller, but the aquatic genera *Sminthurides* and *Bourlettiella* show a strong dimorphism associated with an elaborate behaviour during reproduction. Some species appear to be parthenogenetic but in sexual species fertilization seems to involve the transfer of spermatophores. Eggs may be laid singly or in clumps (Richards, 1979).

Metamorphosis during development is minimal, with the only external changes between newly emerged and adult animals being progressive concentration of pigment and increased differentiation of segmental appendages. The number of moults required to reach sexual maturity is between three and 12, however moulting continues throughout the animal's life.

Springtails are very widely distributed and are one of the few groups of hexapods that are as well represented in temperate regions as in the tropics. They are often found in very harsh climates and occur both in Antarctica and on the Arctic Islands. Many species are cosmopolitan, e.g. *Isotomurus palustris* which is typically found on the surface of stagnant water, and *Sminthurides aquaticus*, found in quiet streams (Scott, 1971). Others, particularly those found in Australia, are endemic or are found typically in specialized environments, e.g. *Axelsonia littoralis* is a reef inhabitant (Wallace and Mackerras, 1970).

Collembolans have been classified on the basis of their relative degree of association with, and adaptation to, aquatic habitats. The following three categories have been proposed: (1) *primary aquatic associates* - these are only found in bodies of water; (2) *secondary aquatic associates* - these may occur in or around water bodies but also may be found in areas of high humidity; and (3) *tertiary aquatic associates* - these show the least apparent adaptations to aquatic environments but, for example, may be seen congregating on the surface of temporary rain puddles (Waltz and McCafferty, 1979). Some species, e.g. *Anurida maritima* of the marine intertidal zone, survive temporary inundation by water through their ability to trap a small air bubble against their bodies using their cuticular setae (Joosse, 1966).

The Hemimetabolous Aquatic Insect Orders

2.2   Ephemeroptera

As we saw earlier, the fossil record indicates that mayflies were certainly present in the Permian and that Protephemeroptera (*Lithoneura, Triplosoba*) existed in the Upper Carboniferous. As such, they represent the oldest and most primitive of the existing winged insects; they have, in fact, been likened to "flying Thysanura". Permian mayflies had two pairs of wings of almost the same size but, by the Upper Jurassic (in the Hexagenitidae), the hind wings had become half to two-thirds the length of the forewings. By contrast, the hind wings of modern mayflies vary from six-tenths the length of the forewings to being absent (as, e.g., in the Caenidae, Tricorythidae and many Baetidae). Nymphs of Permian species had gills on abdominal segments one to nine, whereas modern nymphs have them on abdominal segments one to seven, at most (Edmunds, 1972).

Not only do mayflies have a long fossil record but their written record goes back to Aristotle (384 - 322 B.C.). He recorded the emergence of the subimago from the nymphal skin, on the water surface, and remarked on the brief life of the adult. The latter is mentioned by several subsequent chroniclers, including Cicero and Pliny the Elder, and a detailed account of mayfly biology was given by Outgert Cluyt (Augerius Clutius) in 1634 (Fig. 2.2). Not only was this latter work important from the point of view of understanding this particular insect group but, perhaps more importantly, it represented the first extensively documented example of a new way of scientific thinking. Rather than relying on the accounts given by ancient writers, without criticism, Cluyt and his contemporaries favoured their own observations when these were contradicted by earlier opinions. Even so, this method is not without error as, perhaps due to confusion with another insect, Cluyt recorded that "it (the mayfly) originates from a certain worm living in the earth which has feet, but later changes into a small footless animal when it comes out of the earth. It is clothed in a very bright, white skin and when that has burst it starts to fly". This description together with the accompanying figure (Fig. 2.2) clearly indicates that he thought a pupal stage occurred in the life cycle (Franciscen and Mol, 1984).

The Ephemeroptera represents one of the two living orders within the infraclass Palaeopterygota and contains a little over 2,100 species grouped into about 200 genera and 22 families (Table 2.1) (McCafferty and Edmunds, 1979). Nymphs are found in virtually all types of fresh water across the globe, with the

AUGERII CLUTII M.D.
DE

HEMEROBIO

Sive

INSECTO EPHEMERO,

Nec non

DE VERME MAIALI.

Opuſculum II.
Coryphæo Medicinæ celeberrimo

D.NICOLAO PET. TVLPIO,

D. M. Senatori ac fcabino *Amſterodamenſi,*
Amico veteri colendiſſimo & Collegæ dileſtiſſimo inſcriptum.

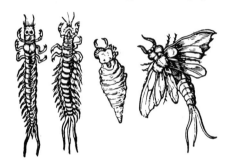

Fig. 2.2   Title page from "De Hemerobio" an early work on Ephemeroptera by Outgert Cluyt (1577-1636) (from Francissen and Mol, 1984). The engraved figures show dorsal and ventral aspects of a nymph together with an adult and a "pupa".

exception of the high Arctic and Antarctica. Mayflies are poorly represented on oceanic islands (apart from some members of the Baetidae and Caenidae) and on isolated mountain tops (particularly above the tree line), and numerous genera and species have failed to cross relatively minor physiographical barriers such as small plains. The latter is thought to be a result of their poor disperal abilities, whereas their absence from polar regions may be related to temperature tolerance. It is well known, for example, that mayfly life cycles show a definite trend with latitude: non-seasonal multivoltine (multiple generations in one year) cycles predominate in the tropics but these are typically replaced by univoltine cycles at higher latitudes. Semivoltinism (two year life cycle) is rare - although *Ephemera danica*, a very common species in Europe, is semivoltine in some British rivers. A few species are able to live in brackish water and one species of baetid in South America is reported to be semiterrestrial. Their greatest diversity is in running water habitats in temperate regions, where they are important members

Table 2.1   The families of Ephemeroptera together with their primary global distributions and descriptions of typical nymphal habitats (after Edmunds, 1972, 1984; McCafferty and Edmunds, 1979; and Campbell, 1988).

Family	Primary distribution	Habitat
Suborder Schistonota		
Superfamily Baetoidea/Heptagenioidea		
Siphlonuridae	largely Holarctic	chiefly running water
Colorburiscidae	S. Hemisphere	stony upland streams
Oniscigastridae	S. Hemisphere	lotic & lentic
Ameletopsidae	S. hemisphere	stony upland streams
Ametropodidae	Holarctic	large rivers
Baetidae	widespread	chiefly running water
Metretopodidae	Holarctic	fast-flowing waters
Oligoneuriidae	esp. Afrotr/Neotrop/ some Holarc. & Orient.	running water
Heptageniidae	largely Holarctic	lotic & lentic-erosional
Superfamily Leptophlebioidea		
Leptophlebiidae	esp. S. Hemisphere/ some Holarc. & Orient.	esp. fast-flowing waters
Superfamily Ephemeroidea		
Behningiidae	Holarctic	burrowers in river sand
Potamanthidae	largely Asian/some N.A.	lotic-depositional
Euthyplociidae	esp. Neotrop./Afrotr.	sand burrowers
Polymitarcyidae	esp. Neotrop./Afrotr./ some in Holarctic	burrowers in stream & river beds
Ephemeridae	Oriental/Afrotr/Holarc.	lotic & lentic burrowers
Palingeniidae	Afrotr/Orient/Palaearc.	burrowers in river beds
Suborder Pannota		
Superfamily Ephemerelloidea		
Ephemerellidae	widespread	chiefly running water
Tricorythidae	esp. Afrotr./Neotrop.	lotic-deposit/lentic littoral
Leptohyphidae	esp. Neotrop./some N.A.	lotic.
Superfamily Caenoidea		
Neoephemeridae	Holarctic & Oriental	lotic-depositional
Caenidae	widespread	lotic-deposit/lentic-litt.
Superfamily Prosopistomatoidea		
Baetiscidae	esp. eastern N.Amer.	lotic-depositional
Prosopistomatidae	Afrotr./ some Oriental	lotic-depositional

of the benthic community (Brittain, 1982).

Mayflies are unique insects in that they have two adult stages. Both are winged, short-lived (1-2 hours to a maximum of 14 days) and do not feed. The subimago (dun) is a stage of sexual maturation in which the wings are semi-opaque and covered in minute hairs (microtrichia) with longer setae on the margins. Beneath the subimaginal cuticle the adult eyes, legs and genitalia are clearly visible. The mature adult (imago) that emerges when the subimago moults has much longer legs and cerci, and wings that are clear and bare. The

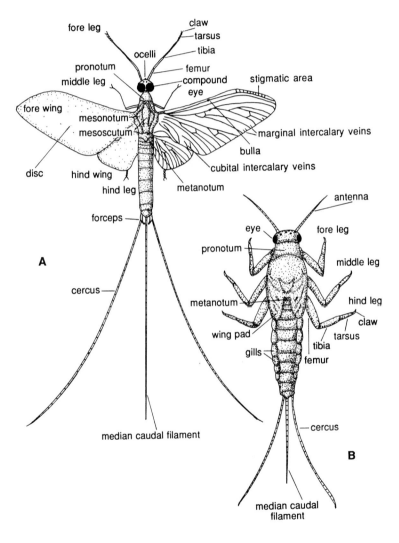

Fig. 2.3   (A) Generalized adult male mayfly showing its basic morphology; (B) dorsal aspect of a generalized nymph (redrawn after Burks, 1953).

females of some species do not shed their subimaginal skin (e.g. the Prosopistomatidae), or only part of it (e.g. some of the Oligoneuriidae). Males typically have very large compound eyes that may be divided into upper and lower portions of differing structure. Also on the head are three ocelli, arranged in a triangle, a pair of short, simple antennae, and mouth parts that are vestigial, but visible.

The thorax is highly specialized in favour of flight rather than walking and, in some species, the legs are reduced, although in others they may be used to hold the female during copulation. The prothorax is small relative to the meso- and metathorax which are robust and fused forming a box around the powerful flight muscles (Fig. 2.3). There are two pairs of thoracic spiracles. The two pairs of wings are roughly triangular in shape and may or may not be patterned. They are held rigidly upright over the thorax when not in use as there is no mechanism to fold them, compactly, as in most other insects. This is a consequence of the primitive articulation of the wingbase with the thorax, a feature that places the Ephemeroptera in the infraclass Palaeopterygota (literally translated as the "old wings"). The wings are characteristically veined (this feature is thus useful in classifying species) and, in flight, act in unison.

The abdomen has ten segments. Each abdominal segment is cylindrical and consists of a dorsal half (the tergum) and a ventral half (sternum) which meet on the sides of the insect. Lateral portions of the 10th segment form two paraprocts each of which bears a single caudal filament, or cercus. A median caudal filament, when present, extends from the single epiproct - the tergal part of segment 10. There are eight pairs of spiracles located laterally, one pair on each of the first eight segments. In males, a pair of thin, jointed appendages, the forceps (or claspers; Fig. 2.4) arise from the corners of the sub-genital plate and these are used to hold the abdomen of the female during mating. Two penes are located on the antero-dorsal border of the sub-genital plate and consist of flattened tubes containing the ends of the seminal ducts. The morphology of the male genitalia is species specific and is therefore a key taxonomic feature. The genitalia are present in the male subimago but are weak and not fully developed. The two oviducts in the female open, separately, between the 7th and 8th sterna. An egg valve (or ovipositor) is formed from the elongation of the 7th sternum which projects posteriorly under the 8th sternum. Underneath, the egg valve is armed with heavily chitinized and complex spines used during mating (Needham *et al.*, 1972).

In contrast to the imagos, mayfly nymphs exhibit considerable diversity in appearance although many stream species are dorso-ventrally flattened. Differences do not always follow taxonomic lines and convergent and parallel evolution is common (Brittain, 1982). As in the adults, the head bears a pair of compound eyes together with three ocelli and a pair of antennae (Fig. 2.3). The mouthparts, however, are well developed (as the nymph is the only feeding stage in the life cycle) although relatively stereotyped, a feature doubtless related to the somewhat conservative diet of detritus and periphyton eaten by most species.

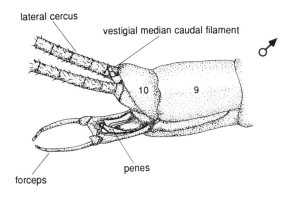

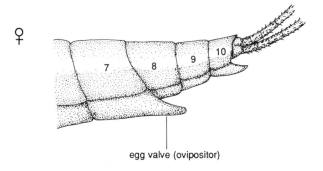

Fig. 2.4 Generalized ends of abdomen of a male and female mayfly; segments are numbered (redrawn after Needham *et al.*, 1972).

The mouthparts (Fig. 2.5) have typical insect components: a labrum (which basically forms an upper lip); a labium (lower lip with sensory palps); a pair of maxillae (sides of the mouth, incorporating sensory palps and a heavily-spined galea, used as a diatom rake); a hypopharynx (a tongue-like structure on the dorsal surface of the labium); and a pair of heavily-chitinized mandibles used for grinding food. Modifications seen among species are generally considered to be a consequence of different food gathering techniques rather than of differences in diet (Soldan, 1979). The mandibles of carnivorous species may be sickle shaped.

Compared with that of the adult, the nymphal thorax is larger but the meso- and metathoracic segments are similarly fused. The dorsal surface is heavily chitinized but has a median suture line running down its length which eventually splits to allow emergence of the subimago. Two pairs of wing buds are visible on the meso- and metathorax which darken prior to emergence. The legs are shorter and more robust than in the adult and are frequently flattened, like the

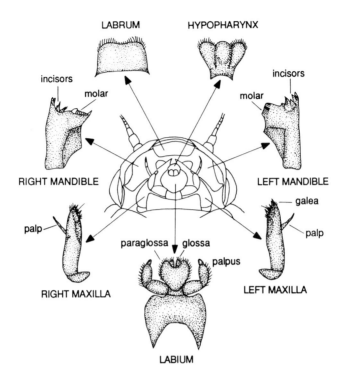

Fig. 2.5   Ventral view of the mouthparts of a nymph of *Ephemerella*
(Ephemerellidae) (redrawn after Edmunds, 1984).

body, in stream species - possibly an adaptation that brings the nymph close to
the substrate surface thus reducing the force of the current. Most segments have
spines and marginal fringes of hairs and each tarsus ends in a single, but
substantial, claw used to cling to rock surfaces.

    The nymphal abdomen is frequently also flattened and tapers towards the
posterior. There are ten distinct segments and the postero-lateral corners of each
commonly bear large spines. Gills are present, in pairs, as large plate-like
structures, or tufts, on up to the first seven segments, depending on species.
Internally, they contain tracheal tubes that connect with those of the main parts
of the body. In a few families (e.g. the Caenidae and Tricorythidae), the gill
plates on the second abdominal segment have become enlarged to form a pair of
opercula that protect the other gills from damage in silty environments. External
genitalia are visible only in mature nymphs but throughout this stage two or
three long, sensory filaments arise from the tenth segment (Needham *et al.*,
1972).

    The primary function of the adult mayfly is to mate and lay eggs. Newly-emerged female imagos typically fly into swarms of males that hover over the water surface, and mating takes place on the wing. These swarms commonly occur at dusk, in temperate regions, but timing is known to be controlled by temperature and light intensity. Once the eggs have been fertilized, the females of most species (particularly those in the families Heptageniidae, Ephemeridae and Leptophlebiidae) lay them in small batches by flying down to the water surface and periodically dipping their abdomens into the water. In other species (e.g. some belonging to the genus *Baetis*, Baetidae) the female crawls down under the water and attaches her eggs to suitable rocks. Still others (e.g. *Ephemerella*, Ephemerellidae) lay all their eggs at once on the water surface where a hydrofugal reaction disperses them widely. The eggs may hatch immediately (in ovoviviparous species) or after varying lengths of development that range from one week to 10 or 11 months in arctic or alpine species (Brittain, 1982). Parthenogenesis is known to occur in about 50 species. Egg development and hatching success rate are very temperature dependent, and many species have quite narrow ranges (e.g. 2 - 10°C) for maximum survival (Elliott and Humpesch, 1980). A summer resting stage (diapause) occurs in the eggs of some species.

Mayfly nymphs go through a large number of moults as they grow, with most species having 15 - 25 instars. Estimates for some species reach as high as 50 (Fink, 1980) and variations occur within single species. In general, mayfly nymphs tend to live mostly in unpolluted lakes, ponds, streams and rivers where, with densities of up to 10,000/m^2, they contribute substantially to secondary production (Clifford, 1982). However, very small amounts of organic pollution can sometimes, initially, increase the numbers and production of certain species while others are exterminated. Species of *Baetis* seem the most tolerant to pollution and these and others are often used as indicators of water quality. Burrowing nymphs such as *Hexagenia bilineata* do particularly well in silted impoundments and the problems associated with their mass emergence from the Mississippi River are notorious - e.g. accumulation of adult bodies on road bridges create slippery surfaces for motorists (Fig. 2.6).

The nymphal diet consists mostly of "scraping" algae from rock surfaces or "collecting" fine-particle detritus from sediments, but some species "filter" suspended particles from the water column, and a very few species are "predators". These feeding techniques have been used to classify mayflies (and other aquatic insects) into functional feeding groups for ecosystem studies (see Chapter 8). The nymphs of some mayflies are, however, carnivorous, for example some genera of Heptageniidae and all members of the Australian family Ameletopsidae.

Transition from the final nymphal instar to the subimago usually takes place at the water surface, making it a particularly dangerous period in the life cycle, as it is exposed to both fish and aerial predators such as dragonflies and birds. Crepuscular (dusk/dawn) emergence peaks may aid in reducing losses to visual predators.

Fig. 2.6   Mass emergence of *Hexagenia bilineata* on a bridge over the
Mississippi River in Minnesota, U.S.A. (from Fremling, 1973).

2.3   Odonata

Dragonfly nymphs are a very conspicuous component of the faunas of many
freshwater habitats, especially permanent lakes, ponds and marshes, but they are
found also in small streams, large rivers, temporary ponds, the water-filled leaf
axils of large tropical plants and even in brackish water. Phylogenetically,
dragonflies are an ancient order having arisen in the Upper Carboniferous (Fig.
1.2) and separated from all other winged insects very early on. During that period
the fossil record indicates the presence of dragonflies with a wingspan exceeding
60 cm which makes them the largest hexapods ever to have existed. Extant
species belong to three suborders: the **Anisozygoptera**, containing only two
species that are restricted to Japan and the Himalayas - although only one family
is now living, fossil evidence of 10 extinct families indicates considerable early
diversity within this suborder; the **Zygoptera**, or damselflies, comprising 17
living families; and the **Anisoptera**, or dragonflies, in which there are eight
living families. Although the terms "damselfly" and "dragonfly" are in common

Table 2.2   Differences between the Anisoptera (dragonflies) and the Zygoptera (damselflies) (after Fox and Fox, 1964).

Anisoptera	Zygoptera
Hindwings broader at base, held horizontally or slightly depressed at rest. Strong fliers.	Wings of equal size, narrow at base, held vertically at rest. Weaker fliers.
Eyes not projecting from sides of head.	Eyes bulbous and prominent.
Most families with reduced or vestigial ovipositors.	Females with well developed ovipositors.
Supra-anal plate present in males.	Supra-anal plate vestigial.
Nymphs robust, with rectal gills.	Nymphs slender, with paddle-like caudal gills.
Eggs usually laid at water surface or on surfaces of aquatic plants.	Eggs inserted into stems of aquatic plants.

usage, members of these latter two suborders are quite similar in structure, appearance and general biology and, as a consequence, many entomologists tend to refer to all odonates as "dragonflies". Differences between the Anisoptera and Zygoptera are listed in Table 2.2. The order is distinguished from other insect orders in that its species have the following combination of features: two pairs of wings, almost equal in size and in which the veins are arranged in a net-like pattern; tarsi that have three segments; short, simple antennae and large compound eyes; and strong chewing mouthparts. Adults may be brightly coloured but nymphs are usually dull or, at best, green or reddish-brown.

Apart from early work by Linnaeus, Fabricius and Rambur, who primarily named species or rearranged their classification, it was not until the middle of the 19th Century that the Classical Period of the study of Odonata began with the studies of Baron Edmond de Selys-Longchamps. This Belgian natural historian showed the importance of wing venation as the basis of classification of the adults (Tillyard, 1917). Odonate wings are primitive in being articulated to the walls of the thorax by humeral and axillary plates that are fused to the proximal ends of the major wing veins (Fig. 2.12A).

A total of around 5,500 species has been described and they are distributed from the tropics, where the greatest numbers and diversity occur, to the tree-line in polar regions. This can be illustrated easily by the following data: only 11 species of odonates occur in New Zealand, for example, and 194 species in the whole of Canada. In contrast, at least 62 species were recorded from a single hill-stream in Malaysia (Bishop, 1973). The Neotropical Region has the greatest diversity of odonates together with the greatest number of endemic species, many of them damselflies. The next richest region is the Oriental where the highly-coloured Calopterygidae (Zygoptera) form a significant part of the fauna, and where many of the major islands have rich and characteristic faunas (Tillyard, 1917). The Afrotropical Region also has a very distinct fauna, rich in both species and genera; endemism is particularly high in Madagascar. The odonate fauna of the Australian Region has several features including a rich complex of archaic forms (Gondwana relicts), particularly amongst the numerous stream-dwelling species; similarity of part of the fauna with the Orient, particularly in tropical Australasia; affinities of some forms with those in the Neotropics; and the Cordulegasteridae are absent (Watson, 1982). The Nearctic is relatively poor in zygopterans (apart from the Coenagrionidae) but certain anisopteran families contain considerable numbers of species, for example the Gomphidae (93 spp.), Libellulidae (93), Corduliidae (50) and Aeshnidae (38) (Westfall, 1984). The Palaearctic Region has the poorest fauna, despite its vast area, and some species are shared with the Nearctic. Japan, however, has a relatively high diversity.

Adult dragonflies are powerful fliers and have been recorded flying distances of several hundred kilometres. As for the mayflies, the adult stage of dragonflies consists of two distinct phases: a prereproductive (maturation) period, that lasts from 2 to 30 days (Zygoptera) or 6 to 45 days (Anisoptera) in temperate regions (depending on species and weather conditions); and a reproductive period that may last from 1 to 8 weeks. However, the physical differences seen in the adult between these two phases are more minor than those seen between the subimago and imago of mayflies. The prereproductive period extends from emergence to sexual maturity and involves changes in the colour of the body and wings, in gonad development, in the size and appearance of certain ectoparasites (e.g. mites), and probably in the number of growth layers in the endocuticle. During this period, adults may disperse widely or not at all, depending on the degree of shelter and habitat continuity. The reproductive period begins when adults start to exhibit sexual behaviour and roughly corresponds to the oviposition and flight period. The timing and duration of the flight period tend to reflect features of the nymphal habitat, especially its permanence. For example, in temperate and high latitude regions it is seasonal, as it is in tropical species whose nymphs live in temporary waterbodies affected by seasonal rains. For those species living in permanent habitats in humid parts of the tropics, the flight period is typically continuous (Corbet, 1980). Unlike mayflies, dragonflies feed throughout their adult life. Prey consists primarily of flying insects which anisopterans tend to take on the wing but which zygopterans tend to capture while the prey is resting. A number of species of dragonfly appear to have the ability to regulate their body temperature, primarily through changes in body posture and degree of exposure to the sun, and this is thought to give them an advantage in that they can begin to hunt for prey early in the morning, perhaps before the latter are fully functional (May, 1976).

The head of the adult is large and concave to the posterior, frequently (in the

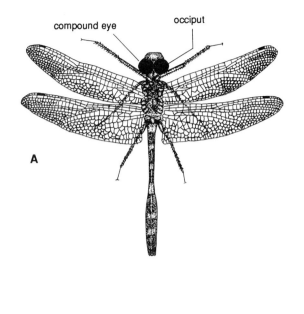

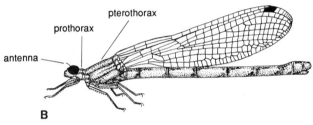

Fig. 2.7 (A) Typical male anisopteran, dorsal view; (B) typical female zygopteran, lateral view (redrawn after Smith and Pritchard, 1971).

Anisoptera) with a modified occipital region. It is connected to the thorax by a thin flexible neck which allows considerable mobility (Fig. 2.7). There is a pair of conspicuous compound eyes, three ocelli and considerably reduced antennae. The mouthparts are robust, of the biting type adapted for predation, and feature a prominent labrum, wide-gape strongly-toothed mandibles, maxillae with spines and unsegmented palps, a hypopharynx, and labial palps modified into large lateral lobes, each of which has a movable hook and a spine near its tip. In the nymph, the eyes are smaller but the antennae are relatively larger. The mouthparts are basically the same as in the adult, except for the labium (mask) which is long and extensible, hinged at its base and used for capturing food (Fig. 2.8). The labial palps are terminal in position and have spines that are used to

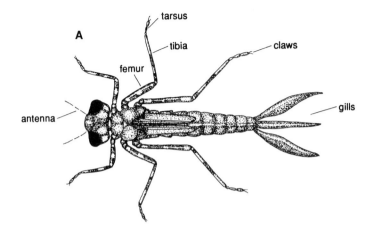

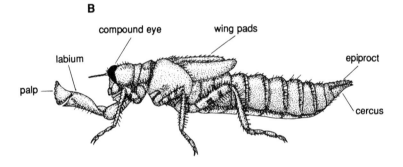

Fig 2.8   (A) Dorsal view of a zygopteran nymph (*Argia* sp.); (B) Lateral view of
             an anisopteran nymph (*Cordulegaster* sp.) with labium extended
             (redrawn after Smith and Pritchard, 1971; and McCafferty, 1981).

grasp prey after the rapid extension of the labium. Extension is brought about by
a local increase in blood pressure resulting from contraction of a diaphragm
located in abdominal segments 4 and 5. It is returned to its folded, ventral resting
position by retractor muscles (O'Farrell, 1970).

    The prothorax of the adult is small, but wide, and is movable with respect
to the large pterothorax which is formed from a fusion of the meso- and
metathorax. In female Zygoptera, the dorsal shield (pronotum) of the prothorax
is subdivided into three lobes and is modified to provide an attachment site for
the anal appendages of the male during mating; in the Anisoptera, the male's
anal appendages attach to the occiput on the female's head. The pterothorax is

strongly humped with the two pairs of wings set well back but the legs placed well forward. The legs themselves are fairly short and weak, being adapted for clinging to perches and catching and holding prey rather than for walking. Each ends in a pair of claws. A pair of spiracles is located on both the meso- and metathoracic segments. The wings are membranous and usually clear although there may be extensive patches of colour in some species. Most of the Zygoptera are capable of raising their wings vertically when at rest, whereas a few, and the majority of the Anisoptera, rest with their wings spread horizontally. Dragonflies cannot fold their wings and thus are classified, together with the mayflies, in the group of "old wing" insects, the Palaeopterygota. In flight, the forewings (mesothoracic) and hindwings (metathoracic) beat almost half a cycle out of phase such that the two pairs brush against each other; this produces a characteristic rustling sound. During flight, the large head may be used as a balancing organ and much of the lift achieved is thought to be produced by unsteady effects associated with the deceleration and rotation of each wing at the bottom of its downstroke (Alexander, 1982). Dragonflies have the unusual ability to fly backwards and, when flying forwards, may reach a maximum speed of 56 km per hour. In the nymph, the prothorax is larger and the pterothorax less humped than in the adult; external wing buds are visible with the hindwings overlying the forewings; and the legs are centrally placed, more robust and adapted for crawling, clinging or burrowing.

The abdomen of the nymph differs among the suborders. In the Zygoptera, the abdomen ends in three large caudal tracheal gills (Fig. 2.8A) which, apart from their obvious function in respiration, also may be used for swimming by lashing the abdomen from side to side. In the Anisoptera (Fig. 2.8B) and Anisozygoptera, external gills are absent but are replaced by an elaborate network of trachea located in the wall of the rectum. Gentle respiratory inhalations and exhalations replace the water in the rectum at regular intervals, whereas rapid exhalation provides a jet-propelled escape response for the nymphs. A spiny epiproct and spiny cerci provide an armed anal pyramid which may be used as an offensive weapon.

The adult abdomen is usually long and cylindrical (although somewhat dorso-ventrally compressed in many anisopterans) and consists of ten flexibly articulated segments. Segments 1 and 2, 8 and 9, or 9 and 10 are frequently shorter than the rest. The ventrally-placed sterna of segments 2 and 3, in males, are modified to form secondary genitalia and, in some anisopterans, these are supplemented by ventrolateral projections (auricles) from the tergum of segment 2 (Fig. 2.9). In both sexes, paired, unsegmented, superior anal appendages on segment 10 may represent modified cerci (although some authorities believe these appendages to be located on a vestigial 11th segment). Paired, inferior anal appendages are present also in male zygopterans whereas a single, median inferior appendage occurs in male Anisoptera. In the male, the gonopore opens ventrally on segment 9. Sperm is transfered from here to a storage vesicle on the anterior end of sternum 3 by bending the abdomen. Sternum 2 is modified to

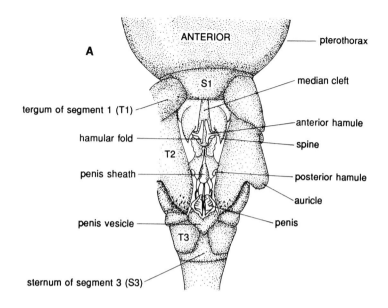

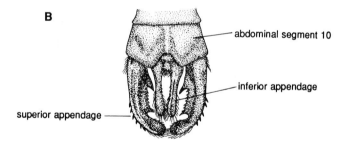

Fig. 2.9   (A) Secondary genitalia in a male anisopteran, ventral view of the 1st,
             2nd and 3rd abdominal segments; (B) anal appendages of a male
             zygopteran (*Lestes* sp.), dorsal view (redrawn after Smith and
             Pritchard, 1971; and O'Farrell, 1970).

form a genital fossa from which the penis extends. On either side of the penis is
a pair of hamules, processes used to hold and guide the female genitalia during
copulation. In the female, the gonopore is located behind sternum 8. Female
Zygoptera, together with some Anisoptera (e.g. the Aeshnidae), have a well
developed ovipositor and a cutting, piercing, or sawing organ (the terebra; Fig.
2.10) which enables them to insert their eggs into the tissues of submerged or
emergent aquatic plants (endophytic oviposition). Most female anisopterans,
however, have an ovipositor that is reduced to a small, vulvar scale, but some

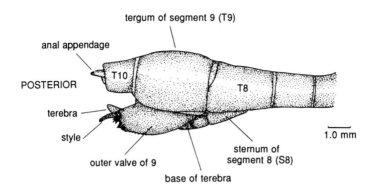

Fig. 2.10   Lateral view of anal appendages of a female zygopteran (redrawn after O'Farrell, 1970).

additionally have sterna 9 and 10 excavated to form a cavity capable of temporarily storing egg masses after they have been extruded from the gonopore. The latter reflects the different egg laying habits (scattering) seen in the majority of anisopterans (O'Farrell, 1970).

Mating in dragonflies is complex. Mature adults tend to congregate at suitable breeding sites where the males may establish territories. Both sexes may mate several times in a day. When a male approaches a receptive female he first grasps her in the "tandem" position (Fig. 2.11); attaching to her occiput, in anisopterans, but to her prothorax, in zygopterans. The pair then forms the "wheel" position in which their abdomens are bent so as to bring the female's genitalia (on the terminal segments of her abdomen) in contact with the the "secondary" genitalia (penis and sperm vesicle located on segments 2 and 3) of the male. In many species, before the male inseminates the female, he performs a series of pumping movements, perhaps lasting up to 20 minutes, with his abdomen. Originally thought to be the actual mechanism of sperm transfer, this action is now known to remove, using the penis, the majority, if not all, of the sperm deposited in the female's spermatheca by another male in a previous mating. Oviposition tends to occur soon after copulation and, in many species of both damselfly and dragonfly, the male remains associated (either in continuous or intermittent tandem) with the female while she does this. Such behaviour may ensure that the eggs laid have been fertilized by the guarding male (Corbet, 1980).

The duration of the egg stage depends on whether it represents the overwintering phase in the life cycle. In some temperate genera such as *Lestes, Aeshna* and *Sympetrum*, delayed development during the winter can mean as many as 80 to 230 days spent as an egg. In other temperate and tropical species there may be no diapause and eggs may hatch after as few as 5 days. The rate of

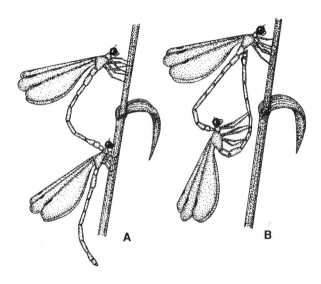

Fig. 2.11   Mating positions in the Zygoptera: (A) tandem position - female grasped by the prothorax; (B) wheel position - female genitalia brought into contact with secondary genitalia of male (on segments 2 and 3 of the abdomen) (redrawn after O'Farrell, 1970).

development and the time of hatching are seasonally regulated according to the phase of development and its response to environmental temperature and, occasionally, to photoperiod. Staggered egg hatching, even from a batch laid by a single female, is common.

For the first few days, newly hatched nymphs live on yolk retained in their bodies. After this has been consumed, they become generalized predators that tend mostly to employ ambush tactics from a perch or burrow. However, if food is scarce locally, nymphs may wander in search of it, especially at night. Food is located primarily by sight, but also by use of mechanoreceptors, and capture is by use of the suddenly extended labium. Prey includes small benthic animals such as chironomid larvae together with zooplankton, while nymphs are small, but includes larger invertebrates and immature vertebrates (fishes and amphibians) as they grow.

In permanent waterbodies at low altitude in tropical and warm-temperate regions, most species complete at least one life cycle per year. In cool-temperate and higher altitude warm-temperate regions, the majority of species are univoltine or semivoltine (one generation in 2 years). At high latitudes, life cycles may take as long as 5 years or more, depending on the species and habitat. Studies of single species in different locations often show clear trends. For example, in Europe at 43 to 44°N, *Ischnura elegans* is primarily trivoltine (three generations per year), at 53 to 54°N it is univoltine, and at 57 to 58°N it is semivoltine (Corbet, 1980). The link between latitude, altitude and voltinism is, obviously, through environmental temperature, as insects are poikilothermic (coldblooded) animals. However, other factors can influence voltinism (see Chapter 5) and one particularly relevant to the Odonata is abundance of food. Nymphal populations subject to diminished food supply will frequently take an extra year to complete their life cycles (Macan, 1964). Nymphs typically pass through 10 to 15 moults before becoming adult but this varies within and among species, again primarily according to temperature and food supply. Emergence from the last nymphal stage to the adult usually takes place after the nymph has climbed above the water surface on an emergent plant or rock. The newly emerged imago rests next to the cast nymphal skin (exuvia) until its wings have been fully extended and hardened. Adult respiration occurs through the two pairs of thoracic spiracles.

There is one known species in which the nymph is terrestrial; *Megalagrion oahuense* is a small damselfly living in the forests of Hawaii.

2.4   Plecoptera

Stoneflies belong to the most primitive order within the infraclass Neopterygota. The latter term, translated literally, means "new wings" and refers to the fact that the two pairs of wings have articulations that allow them to make the same flight movements as the Palaeopterygota (Ephemeroptera and Odonata) but, in addition, to fold their wings back along the abdomen. This is made possible by a series of separate, movable sclerites in the base of the wing (Fig. 2.12B). Typically, the fore and hind wings are weakly coupled in flight but, when the insect lands, the fore wings rest on top of the hind wings, with one fore wing almost completely covering the other. At the base of each wing, the posterior membrane is thickened to form a strong axillary cord and this resists tearing as the wing is folded. When at rest, the folded wings usually project beyond the end of the abdomen. The precise folding mechanism of the wings, together with a primitive pattern of venation, set the Plecoptera apart from the other polyneopteran orders (Table 1.2). In contrast to the Odonata, wing venation in the Plecoptera varies widely, even within individual families. Consequently it is of little use, taxonomically, above the genus level (Zwick, 1981).

Modern plecopterans are thought to have been derived from the Protoperlaria of the Permian (360 - 286 million years B.P.) and the fossil record is quite respectable with more than 30 species described from the different strata of the Permian to the middle Tertiary (Illies, 1965). A fossil, from the Upper Permian, belonging to the still extant Australian family Eustheniidae indicates the primitive nature of many of the modern forms. Adults are terrestrial and live from 3 to 4 weeks. The rest of the life cycle is spent as an aquatic nymph, typically in cool, clear streams, although nymphs are found also in cold, oligotrophic (low nutrient) lakes, but always below 25°C (Harper, 1979). By

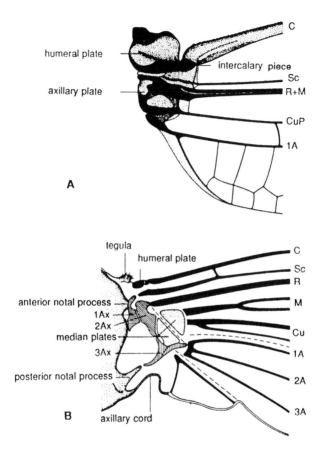

Fig. 2.12   Basal articulations of wings of: (A) a generalized Palaeopterygote
             (e.g. dragonfly); and (B) a generalized Neopterygote (e.g.
             stonefly). 1Ax, first axillary sclerite; 2Ax, second axillary sclerite,
             etc. Major wing veins are also shown: C, costa; Sc, subcosta; R,
             radius; M, media; Cu, cubitus; 1A, first anal vein, etc. (after
             Snodgrass, 1935). Positions of the wing fold regions when the
             insect is at rest are indicated by the dotted lines.

virtue of such temperature requirements, stoneflies are mostly restricted to higher
altitude or circumpolar waterbodies, or spring-fed lowland streams. The nymphs
of a few species are terrestrial but, as might be expected, these are restricted to
cool damp habitats. Adults of *Capnia lacustra* (Capniidae) have been collected at
depths of between 60 and 80 m in Lake Tahoe and are assumed to be the only
known aquatic adults (Jewett, 1971).

    Some 2,000 species, in 15 families, have been described worldwide and
these vary in length from 4 - 5 mm (e.g. many members of the Nemouridae) up
to 5 cm (e.g. Pteronarcidae and Eustheniidae). Most adults are winged, although

a few species are wingless (apterous) or have short wings (brachypterous). None fly well and this has prevented them from crossing even small geographical barriers. Thus, like the mayflies, stoneflies are useful tools in studies of historical biogeography. Low vagility has led to local speciation (Hynes, 1976). The high water quality requirements of the nymphs bars all but a very few species from habitats subject to low oxygen levels, siltation, high temperatures and organic enrichment, and this has led to their effective use as biological indicators of environmental degradation.

Recent phylogenetic reclassification of the Plecoptera proposes two suborders: an exclusively southern suborder, the **Antarctoperlaria**, and a northern suborder, the **Arctoperlaria** (Table 2.3). These groups are thought to be a consequence of the breakup of Pangea into Laurasia and Gondwanaland (Late Triassic) which separated the Plecoptera into these two main lineages, each of which then underwent its own, independent evolution. Members of the Antarctoperlaria are thought to have begun to diversify on the southern continent of Gondwanaland before it broke up by continental drift (Early Jurassic). Multiple sister groups now occur in South America, Australia and New Zealand but absence of this suborder from South Africa and India is probably due to subsequent extinction there. Most families in the Arctoperlaria are Holarctic in distribution (except for a few relict groups) with many similar elements, even species, especially between North America and East Asia. Only the Scopuridae are solely Palearctic. The European stonefly fauna is, however, less similar, sharing only a few genera with Asia and even fewer with North America. This may reflect Pleistocene events such as the temporary separation of Europe from Asia by the Turgai Straits; glacially-induced extinctions in Europe (e.g. the Pteronarcyidae and Peltoperlidae); and faunal exchange between eastern Asia and North America across the Bering Strait. Some Arctoperlaria exist, today, in the

Table 2.3   The families of Plecoptera (after Zwick, 1981).

Suborder **Antarctoperlaria**   (Southern Hemisphere only)
    Eustheniidae
    Diamphipnoidae
    Austroperlidae
    Gripopterygidae
Suborder **Arctoperlaria**   (Northern Hemisphere)

Chloroperlidae	Taeniopterygidae
Perlidae (also S. America)	Capniidae
Perlodidae	Leuctridae
Peltoperlidae	Nemouridae
Pteronarcyidae	Notonemouridae (S. Hemisphere only)
Scopuridae	

Southern Hemisphere but these are believed to be secondary immigrants from the north (Zwick, 1981).

Both adult and immature stoneflies tend to be dull brown or grey in colour but some have contrasting lighter patches (usually yellow) and a few are bright green (Chloroperlidae) or have red, purple or green markings (especially the Eustheniidae of the southern hemisphere) (Hynes, 1976). Diagnostic features of the Plecoptera include relatively primitive wing venation, three tarsal segments and two or three ocelli on the head. Nymphs frequently have external gills at a number of locations on the body according to species, for example on the labium (submentum), the neck, the thoracic segments, the anterior abdominal segments, and near the anus. Gill remnants may be found on some adults and these are useful in classifying species, as are the external genitalia, the armature on the penis and the structure of the chorion of unlaid eggs. Mouthparts, particularly the maxillae and the labium, are useful in separating nymphs.

The adult head has a pair of well developed compound eyes, in addition to the ocelli, and long threadlike antennae consisting of many segments covered with short setae. The mouthparts are of the primitive mandibular (orthopteroid) type although they may be weak in species in which the adult does not feed. In nymphs, the mouthparts are robust, especially in predatory species, and the antennae, compound eyes and ocelli are structurally very similar to those of the adult.

The prothorax of the adult is large with a broad, flat pronotum (Fig. 2.13A). The meso- and metathorax are smaller, about equal in size and are covered dorsally by the meso- and metanotum, respectively; each has a pair of laterally-placed spiracles. The nymphal thorax is very similar except for the presence of long thin, or sausage-shaped external gills, which contain tracheae. Depending on species, they may be located variously on the sternal plates (ventral), pleural plates (lateral), or the coxal segments of the legs (Fig. 2.13B). Sensillae present on the surface of the gills in some nymphs (especially in the Eustheniidae) suggest a chemoreceptive and/or mechanoreceptive function. The legs of both adults and immatures are strongly built, tend to be flattened, always dorso-ventrally, and end in a pair of claws. The wings are unequal in size and membranous.

The abdomen of both the adult and nymphal stonefly is typically soft, cylindrical or slightly flattened dorso-ventrally, and consists of 10 segments. The 10th segment bears a pair of long cerci that are sensory in nature[1], although in some adults they are reduced to one or two segments. When long, they may act to stabilize the body during flight, and when short, they may aid copulation. In the nymph, the abdomen may have segmental or anal gills. In the adult, there is a pair of spiracles located, laterally, on each of segments 1 to 8. Stonefly

---

[1] As is the case for the Ephemeroptera and Odonata, some authorities (e.g. Needham *et al.*, 1972) believe that stonefly cerci are borne not on terminal segment 10, but on the vestigial remains of an 11th segment.

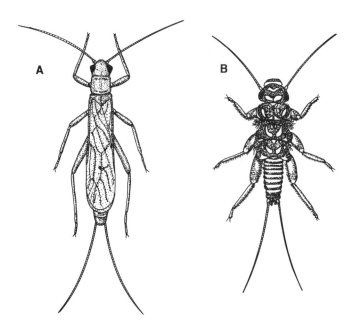

Fig. 2.13   (A) dorsal view of an adult stonefly; (B) dorsal view of a stonefly
              nymph.

genitalia are very diverse. In the female, the gonopore is covered by a modified
sternal plate, the median or sub-genital plate (Fig. 2.14A). Although frequently
it is sternum 8 that is modified (e.g. in the Leuctridae), it may be sternum 7 in
some species (e.g. in the Nemouridae, Capniidae, Perlodidae, Perlidae and
Chloroperlidae), or sternum 9 in others (e.g. Taeniopterygidae). In some species
the sternum may be thickened or slightly elongated but there is no true
ovipositor. Adult males, similarly possess various types of genitalia but these
are mostly highly visible, externally. A pair of paraprocts (Fig. 2.14B) arises
from the postero-lateral margins of segment 10 (where they may be fused with
the bases of the cerci) and these may be long and armed with copulatory hooks.
A single, dorsal epiproct arises from behind the 10th tergum and also serves as
an accessory copulatory organ. It may be complex and project upwards (Fig.
2.14C) or forwards and sometimes occupies a groove or pocket in the tergum.
The position of the male gonopore also varies among species (e.g. between the
7th and 8th sterna, or on the posterior margin of the 9th sternum).   Within the
gonopore is the aedeagus, a partially sclerotized structure, more complex
than the simple penis of the Ephemeroptera, which is everted during sperm
transfer. In addition, in many species (e.g. in the Taeniopterygidae and
Nemouridae), on the anterior margin of the subgenital plate there is a small lobe
used to drum on the substrate during courtship. The positions, shapes and

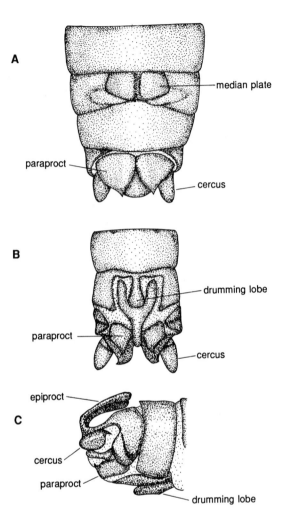

Fig. 2.14   Genitalia of *Protonemura praecox* (Nemouridae): (A) female, ventral
            view; (B) male, ventral view; (C) male, lateral view (redrawn after
            Hynes, 1977).

armature of the various parts of the male genitalia are important taxonomically.
The inability to homologize stonefly genitalia with abdominal structures in
other insect orders suggests that they are secondary developments (Hynes, 1976;
1977).

    Females are attracted to males drumming the ventral lobe on the end of the
abdomen against the substrate. Females also are capable of producing sound and

frequently answer the males. The sound is species specific, as in many other orthopteroids (e.g. crickets, grasshoppers, etc.). Once the male has mounted the female, his abdomen bends to one side and the tip is brought into contact with the sub-genital plate of the female. The hooks on the various accessory copulatory organs hold these two body parts together while the aedeagus is everted into the female's gonopore, generally from beneath the 9th abdominal sternum. Passage of sperm may be aided by a groove in the epiproct or through a pumping action by the female.

Plecopteran families often show distinctive functional as well as morphological traits. The Perlodidae, Perlidae, Eustheniidae and Chloroperlidae are predominantly predators, although smaller instars and smaller species tend to be detritivores or omnivores. Species belonging to the Pteronarcyidae and the Peltoperlidae typically shred large particles of detritus (coarse particulate organic matter, or CPOM) and are often associated with patches of detritus, especially leaf packs, on the stream bed. The Taeniopterygidae and Capniidae tend to emerge as adults in late winter and early spring, in temperate regions, and many avoid the warmer summer water temperatures by diapausing (either as eggs or small nymphs), often buried in the stream bed. Species in both families are detritivores or herbivores (algal grazers). Nymphs of the Leuctridae live on sand and gravel substrates where they feed on fine particles of organic matter (FPOM; consisting of organic debris, precipitated organic matter from the water column, algae, aquatic fungi and bacteria). The Nemouridae live on a wider range of substrates, feed mostly on detritus, and frequently have a long embryonic diapause (Harper, 1979). Several Australian Austroperlidae appear to specialize in eating submerged, rotting wood.

Stonefly nymphs go through a large number of instars, typically 12 to 23. However, such records are based on relatively few studies and, as the number of moults can vary with species, sex and temperature, there is considerable uncertainty. What is more certain is that the earliest instars of most species eat FPOM gathered from the surface of the substratum. Food selection develops in later instars in those species that specialize. In small species, the life cycle is typically completed within one year (univoltine) whereas larger species and species in cold climates (e.g. belonging to the Pteronarcyidae and Perlidae) may take up to two to three years, and occasionally longer. Temperate species that overwinter as nymphs often do not stop growing even in water temperatures close to 0°C. It seems, as we have noted, that it is warm water temperatures rather than cold ones that punctuate stonefly life cycles. The ability to spend the summer in diapause enables some species to live in temporary streams (Williams, 1987).

Restriction of stoneflies to cool, clean habitats in which there is considerable water movement is thought to be connected to the high oxygen requirements of the nymphs, as raising the water temperature or decreasing water circulation distresses them. When subjected to low dissolved oxygen concentration, the nymphs of many species exhibit a characteristic "push-up" behaviour that increases the rate of movement past the gills. As noted earlier, the gills are variously placed among species on the neck, thorax and abdomen. However, some species have no gills and respiration in these is assumed to be across the cuticle surface. In a set of European species, Benedetto (1970) found no correlation between the oxygen requirements of nymphs and the presence or absence of gills. Nevertheless, removal of even a few gill filaments from *Paragnetina media* has been shown to reduce its respiratory ability (Kapoor, 1974). Oxygen consumption may change with season, in concert with life history, and small nymphs tend to consume more oxygen (per mg of body tissue) than large ones (Hynes, 1976).

Field surveys clearly show that the nymphs of many species are associated with particular sections of a stream bed or lake shore. The specific microhabitat occupied depends on a variety of environmental factors such as the nature of the substratum (particle size and configuration), current regime, presence of other organisms, and local variations in water chemistry and temperature (see Chapter 4). Habitat preference often changes as the nymphs develop and with season. Prior to emergence, final instar nymphs tend to migrate towards the bank where they crawl out of the water to shed their skins.

Emergence is controlled by a number of environmental factors, foremost amongst which are light (both intensity and the ratio of the number of hours of light to the number of hours of darkness in 24 h), temperature and humidity. In most species, the emergence of the two sexes overlaps, usually with the males appearing first, and related species in the same habitat tend to emerge serially. Populations of the same species living at higher altitudes or latitudes typically emerge later (Hynes, 1976). Emergence patterns of different species can be assigned to two main categories. In the *synchronous* type, there is a brief emergence period, whereas in the *extended* type emergence is spread more evenly over time   (Harper, 1973). Most species fit into one or the other category. Occasionally, the emergence period of an "extended" species may be bisected by adverse environmental conditions, such as extreme cold, giving the appearance of a bivoltine species.

Shortly after emerging, the adults fly to nearby vegetation. Some species do not feed, drinking only water, but others have been observed to graze on the green encrusting growths on bark, or chew rotting wood or leaves. It seems that, in some species, the adult must feed in order to produce eggs (e.g. some of the Nemouridae) whereas in others (e.g. many of the Perlidae and Perlodidae) all resources have been gathered in the immature stages and the adults emerge with fully formed eggs (Hynes, 1976). Predictably, non-feeding adults are short-lived (just a few days) whereas females of *Capnia bifrons* are reported to live for up to 12 weeks (Khoo, 1964). Some males are short-winged (brachypterous) and therefore cannot fly.

Mating typically takes place on the ground and short-lived species tend to mate as soon as their cuticles have hardened. Winter stoneflies emerge, wingless, while streams are still covered with a thick layer of ice. Often they climb through the small holes left by dead plant stems to mate on the surface of the ice in sub-zero air temperatures. They are quite active and able to survive because their dark bodies preferentially absorb the heat of the sun against the white background of the ice. After mating, the females crawl back down under the ice.

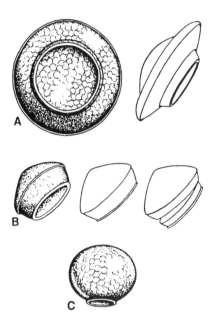

Fig. 2.15   Eggs of the Australian plecopteran family Gripopterygidae: (A) *Riekoperla tuberculata* McLellan; (B) *R. williamsi* McL.; (C) *Leptoperla bifida* McL. (after Hynes, 1974).

In many stonefly species there is a period of maturation, often lasting several days, before the eggs are laid. During oviposition, a mass of eggs is extruded onto the 9th sternum from where they are detached by the current as the female dips her abdomen into the water. Some egg masses break up explosively upon contact with the water and this may serve as a dispersal mechanism. A female may release several such masses within a few days. Alternative methods of laying are rarer, but include submergence of the female to lay her eggs in a jelly-like layer on the stream bed; deposition of single eggs; and insertion of single eggs into crevices at the water margin. There is mounting evidence to indicate that parthenogenesis is not uncommon and that it may serve as a last resort in species which occur in low densities, for example the large carnivorous ones (Hynes, 1976). The eggs themselves are very diverse and often sufficiently ornate to enable specific identification (Fig. 2.15).

2.5   *Orthoptera and Grylloptera*

Orthopterans are not usually thought of as being aquatic, or even semi-aquatic, but a considerable number of species live in association with water (i.e. they are hydrophilous). Recent revision of the group has proposed that it should be split into two: the order Grylloptera, containing the crickets, long-horned grass-hoppers, etc.; and the Orthoptera (*s. str.*, or in the strict sense), containing the short-horned grasshoppers and their allies (Kevan, 1979a; see Table 1.2). Both orders contain hydrophilous species.

Orthopterans generally have two pairs of wings, although the front pair may be large in some species but small and scale-like in others. The hind wings typically are broad and membranous, and can be folded beneath the fore wings which tend to be thicker and parchment-like. The mouthparts are of the generalized chewing type. The most obvious characteristics of the majority of orthopterans are large femoral segments on the metathoracic legs, used for jumping, and the production of sound through a stridulatory organ located at the base of the fore wings.

Within the Grylloptera, several species of Tettigoniidae (katydids) frequent marshes and riparian (bankside) vegetation, and many species of Gryllidae (crickets) are associated with the margins of aquatic habitats. A few species of mole cricket (Gryllotalpidae) are known to burrow in moist riparian sand, using specially flattened front tibia. Within the Orthoptera, many of the Tetrigidae (pygmy locusts) live in moist shoreline areas, and a few species of Tridactylidae (pygmy mole crickets) dig shallow burrows in moist sand. A small number of species of grasshopper (Acrididae) live on emergent aquatic plants and some may even dive beneath the water surface and cling to submerged plants (Cantrall, 1984).

Fig. 2.16   The hydrophilous pygmy mole cricket *Tridactylus minutus* showing modifications to the hind tibiae (redrawn after LaRivers, 1971).

Hydrophilous forms frequently show appropriate morphological or behavioural adaptations. For example, in the Tridactylidae and Acrididae, modifications to the hind tibiae in the form of pairs of long plates, flattened surfaces and rows of long setae, increase the surface area and thus aid swimming

(Fig. 2.16). It has been argued that the basic structure of most saltatorial (jumping) orthopterans preadapts them for movement through water as, for example, even desert grasshoppers swim readily if they accidentally fall into water (La Rivers, 1971).The Australian subfamily Oxyinae (Acrididae) contains species that have an air chamber, formed by doming of the costal area of the fore wing over the first abdominal spiracle, and this provides a supply of air while the insect is submerged (Key, 1970).

The majority of orthopterans feed by chewing green plant material, but some gryllids collect particles of detritus and a few tettigoniids are known to be predators (Cantrall, 1984). Most species are univoltine with the nymphs occupying similar habitats to the adults. Eggs are laid in loose soil, in burrows, or are inserted into plant tissues. Little is known of the general biology of hydrophilous forms.

2.6   Hemiptera

This is a large order of neopterygote insects, that includes both the Heteroptera and the Homoptera (the "true" bugs)[1] . Diversity within the order is high both in terms of structure and habitat. Species range in size from 1 mm to 9 cm and, characteristically, they have mouthparts adapted for piercing and sucking the fluids from plants or animals. The order is thought to be the sister-group of the Thysanoptera (thrips).

The Heteroptera and Homoptera are thought to have been derived from a common ancestral stock sometime in the Upper Carboniferous. Several major, modern subgroups were in existence by the Upper Permian and, in the Mesozoic, extensive diversification occurred concurrent with the emergence of flowering plants (angiosperms). All of the Homoptera are terrestrial and none is carnivorous. Within the Heteroptera there are several wholly aquatic or semi-aquatic families and many of these are predators. Worldwide, there are about 3,200 species of hydrophilic Heteroptera.

With such high diversity in the order, it is difficult to give an account of the structure of a "typical" hemipteran. However, some generalizations can be made (based on Woodward *et al.*, 1970; Savage, 1989; and Carver *et al.*, 1991).

Head - the mouthparts take the form of a tube-like rostrum or beak, although this is much reduced in the Corixidae, and consist of two pairs of thickened but flexible stylets (modified mandibles and maxillae) lying in a groove on the dorsal side of the labium (Fig. 2.17E). Longitudinal grooves on the inner surfaces of the maxillary stylets form two channels when the two stylets are brought together. When feeding, secretions from the salivary glands are pumped down the ventral channel and, usually, liquid food (but particulate in the Corixidae) is

---

[1] Some authors consider the Heteroptera and Homoptera to be separate orders.

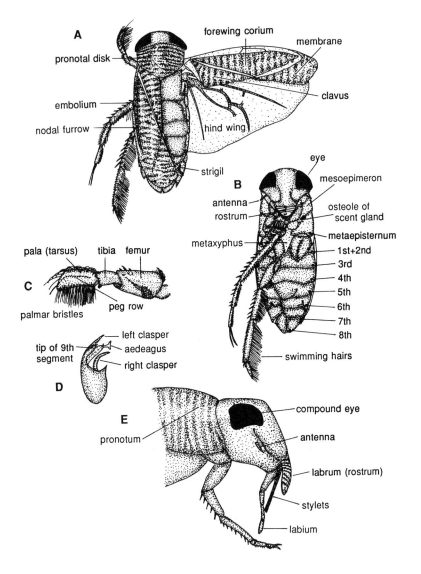

Fig. 2.17   A - D, basic morphology of a heteropteran as illustrated by an adult male corixid; E, generalized mouthparts (redrawn after Usinger, 1971).

sucked up through the dorsal one. The mandibular stylets wrap around the maxillary stylets and, presumably, provide support as well as the means to penetrate prey or plant tissues by use of the spines at their tips. Maxillary and labial palps are absent. Ocelli may be present or absent and there is a pair of compound eyes. The antennae are typically long in surface-dwelling forms but short and inconspicuous in the submerged forms. Corixids are unique amongst the Heteroptera in having buccopharyngeal teeth which, presumably, are used for crushing the particulate food upon which they feed.

Thorax - there are three segments, each bearing a pair of legs. The pronotum tends to be large and, in the Gerridae and some Corixidae, has a longitudinal keel. On the posterior mid-line of the mesonotum, many species have a triangular plate, the scutellum, and on either side of this arises a fore wing, the hemielytron. The latter usually has three parts: near the wing articulation a leathery corium and smaller clavus, separated by a claval suture and, distally, a membranous portion (Fig. 2.17A). On the metanotum is a pair of entirely membranous wings and these fold under the hemielytra. Wing polymorphism (full, reduced, or absent) is common in some heteropterans, particularly in the Gerridae, and this seems to be related to habitat conditions. For example, apterous forms seem to predominate in stable environments whereas unstable ones support a more balanced polymorphism. The Halobatinae (marine water striders), however, are permanently wingless. The metasternum usually has one or more repugnatorial (scent) glands, which produce secretions that repel predators. Both the meso- and metathorax bear a pair of spiracles. Each leg consists of six sections: a short coxa, a small trochanter, a conspicuous femur, tibia and tarsus, and one to two claws. In the Corixidae, the anterior tarsi are modifed to form hair-fringed scoops (pala; Fig. 2.17C) that aid in bringing fine particle detritus to the mouth. The leg segments vary widely among species depending on the mode of life. In gerrids, hydrometrids and mesoveliids, for example, the legs are long and very slender for moving over the water surface.

Abdomen - here, there is considerable variation in structure. Segment 1 (and sometimes also segment 2) is often small and closely associated with the metathorax; the first segment visible is therefore usually segment 2. The subsequent segments are similar, except for segments 8 to 10, which form the genitalia. In the males of some families (e.g. the Corixidae, Gelastocoridae and Ochteridae), the genitalia are asymmetrical and important in taxonomy. Segment 9 forms a genital capsule (pygophore) which, in the male, typically bears a well-formed and often complex aedeagus, and a pair of parameres (claspers) which may be of unequal size (Fig. 2.17D). In females of those heteropteran species that insert their eggs into plant tissues there is a prominent ovipositor, the base of which is enclosed by a small sub-genital plate formed from the 7th sternum. A vestigial 11th segment is visible beneath the 9th tergum, in some species. The number of abdominal spiracles and their structure vary among species. For example, in nymphs of the Nepidae, the spiracles on segments 2 to 8 are situated in hair-lined, ventrolateral grooves that act as air-ways. However, in adult nepids, which breathe through a posterior siphon connected to the spiracles on segment 8, the remaining spiracles are closed or have other functions. Corixids take in air through the first pair of abdominal spiracles and rise to the water surface with a

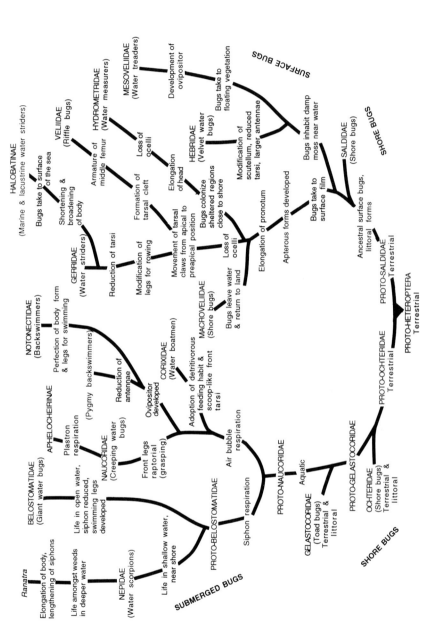

Fig. 2.18   Probable evolutionary lines of the aquatic Heteroptera, showing some of the relevant changes in morphology and habit (based on China, 1955).

characteristic body orientation to achieve this. Notonectids, like nepids, have hair-lined respiratory grooves and take air in through the spiracles on segment 7 but exhale primarily through the thoracic spiracles.

There are 16 families of hydrophilic heteropterans. Figure 2.18 indicates the probable evolutionary lines that resulted in this diversity (note that this does not represent a true phylogenetic tree, nevertheless some groups exhibiting functional similarities are related). Systematically, the suborder Heteroptera can be divided into three infraorders (Table 2.4; Polhemus, 1984), although not all authorities agree on this. The ancestors of the hydrophilic Heteroptera were, undoubtedly, terrestrial. Progression towards a truly aquatic existence was gradual and probably through intermediate forms that lived on freshwater shorelines. Several modern families live in such a riparian fashion, for example the Ochteridae, Gelastocoridae and Saldidae. Broadly speaking, aquatic bugs subsequently evolved along two lines, one leading to families whose members live totally submerged and the other to families whose members live on the water surface. Some aquatic forms have returned to a semi-aquatic, shoreline existence, for example, the Macroveliidae. Such diverse modes of life, with their attendant special adaptations, have undoubtedly led to the high diversity present in the order (Fig. 2.19), although the ancestral mode of feeding (piercing and sucking) has been maintained throughout.

Table 2.4   Families within the suborder Heteroptera that are associated with water (after Miller, 1956; and Polhemus, 1984).

Infraorder **Leptopodomorpha**
    Saldidae (shore bugs) - margins of streams & ponds
Infraorder **Gerromorpha**
    Mesoveliidae (water treaders) - vegetated banks of ponds & lakes
    Macroveliidae (shore bugs) - stream margins
    Gerridae (water striders) - surface of fresh & brackish waters (sea)
    Veliidae (riffle bugs) - surface of ponds & streams, also brackish
    Hydrometridae (water measurers) - surface of calm waters
    Hebridae (velvet water bugs) - marshes & wet riparian mosses
Infraorder **Nepomorpha**
    Nepidae (water scorpions) - ponds, on vegetation
    Belostomatidae (giant water bugs) - ponds, in vegetation
    Corixidae (water boatmen) - fresh & brackish lentic waters
    Notonectidae (backswimmers) - ponds and lakes
    Naucoridae (creeping water bugs) - lentic & lotic, stones & vegetation
    Ochteridae (shore bugs) - stream margins, pond vegetation
    Gelastocoridae (toad bugs) - shorelines, in mud & plant debris
    Pleidae (pygmy backswimmers) - ponds & lakes, in vegetation
    Helotrephidae - ponds & lakes, in vegetation

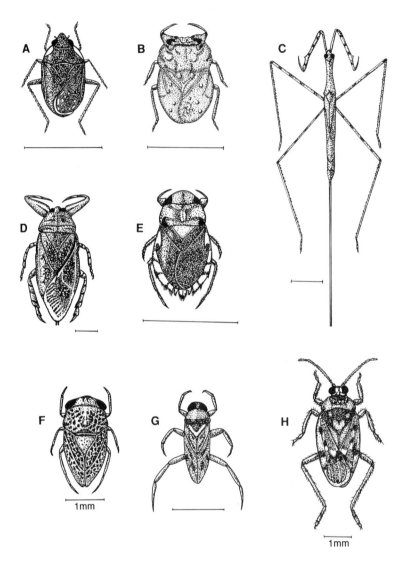

Fig. 2.19   Representative adults of the major aquatic heteropteran families:
             (A) Ochteridae; (B) Gelastocoridae; (C) Nepidae; (D)
             Belostomatidae; (E) Naucoridae; (F) Pleidae; (G) Notonectidae; (H)
             Saldidae (scale line = 1 cm, unless otherwise indicated; redrawn
             after Usinger, 1971; and Woodward *et al.*, 1970).

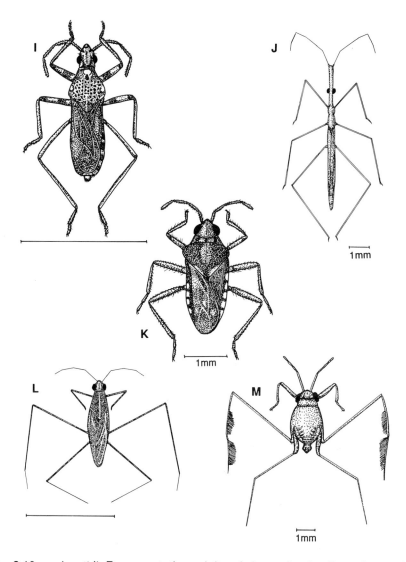

Fig. 2.19   (cont'd) Representative adults of the major families of aquatic Heteroptera: (I) Macroveliidae; (J) Hydrometridae; (K) Veliidae; (L) Gerridae; (M) Halobatinae (scale line = 1 cm, unless otherwise indicated; redrawn after Woodward *et al.*, 1970; and Usinger, 1971).

Typically, the life cycle consists of five nymphal instars, after the egg stage, and the adult. In common with other hemimetabolous insects (no pupal stage), the nymphs are very similar to the adults in terms of their appearance, habitat and behaviour, but are smaller. In temperate regions, life cycles are usually univoltine, or occasionally bivoltine, with reproduction occurring in the summer and adults overwintering (sometimes away from water, as in the Gerridae). Multivoltinism is more common in the tropics. Mating has been poorly studied in most groups, apart from the Gerridae and Corixidae (Savage, 1989). Corixid adults congregate in large clusters near lakeshores in the spring. Attraction of the sexes involves the production of sound (stridulation) in several different modes. In North America, males of *Palmacorixa nana* produce four different signals: a "spontaneous" call (associated with the aggregation behaviour), a courtship call, a mounting call, and a copulatory call. The first two are produced by rubbing a region of small pegs on the prothoracic femora over thickened flanges on the maxillary plate of the head, whereas the latter two are produced by drawing a row of pegs on the inner face of each of the mesothoracic femora over the edges of the hemielytra (fore wings). Females produce only a single "agreement" call in response to the male's courtship call (Aitken, 1982). Stridulatory signals appear to be species specific and must play an important role in separating sympatric species. Sex discrimination in the Gerridae appears to be determined through the frequency of ripples produced on the water surface, chiefly by the males. Three types of ripple signal have been observed in *Limnoporus rufoscutellatus,* in Europe: a high frequency (about 25 Hz) repel signal, an intermediate (c. 10 Hz) threat signal, and a low (c. 3 Hz) courtship signal. Any approaching gerrid that does not respond to a "repel" signal with like kind is judged to be a female and is next confronted with a courtship signal. A receptive female lowers her abdomen and allows the male to mount but a non-receptive female raises her abdomen and may even even emit a ripple signal (Vepsalainen and Nummelin, 1985). A male often remains attached to the same female for the entire reproductive season thus assuring, as do male dragonflies, paternity of the female's eggs. Oviposition generally involves the female submerging to attach her eggs directly to a stable surface such as a stone or plant, although the females of some species lay them at the water's edge. The number laid varies widely among species and individuals (10 to 1000 in the Corixidae), the latter often depending on the amount of food available to the female (Kaitala, 1987). Development rates of both the eggs and nymphs are highly dependent on environmental temperature.

Apart from the corixids which, as we have noted, ingest FPOM, all other hydrophilic heteropterans feed on live or recently dead animals. Gerrids feed on invertebrates, chiefly insects and spiders, that fall onto the water surface. They are attracted from their resting stations by the ripples generated from the struggles of the prey in much the same way as spiders receive vibrations from prey caught in their webs. Veliids and hydrometrids are similarly attracted by ripples produced not only at the water surface but also from just below. The latters' diets thus include animals that fall onto the surface together with zooplankton and the emerging adults of a variety of other aquatic insects, e.g. stoneflies, mayflies, midges, etc. Subsurface predators, such as notonectids and naucorids, detect prey both visually and through water-borne vibrations. Typically, the predator waits at the surface of the water or on some submerged perch (e.g. a plant, or stone) and, once prey is detected, chases after it. Prey is caught with the prothoracic legs, which are specially adapted (armed with teeth and spines) for this purpose and, occasionally, the mesothoracic legs assist in this process, although they are generally not as modified. Once the prey has been subdued, the stylets are inserted through its cuticle and the body fluids are sucked out (Giller, 1986). The longer the predator spends extracting fluid from the same prey, the relatively smaller is the nutritional reward (measured as per unit of effort). It has been shown that, at high densities of prey, notonectids capture more individuals but spend less time extracting the fluids from each individual, than at low densities; this maximizes the overall amount of nutrients obtained. It has been suggested that, like individual honeybees that find a particular flower type in peak nectar production and return time and time again to that same flower type, notonectids also are capable of developing a prey "search image". This involves temporary establishment of a particular foraging sequence which, previously, has proved effective (Cook and Cockrell, 1978; Giller, 1980).

Although heteropterans occur in both running and standing waters, most species are found in lakes and ponds. Within the latter, species distribution is closely correlated with size of the habitat, its water chemistry and the degree of shelter, usually in terms of the presence of aquatic plants. Again, it is the

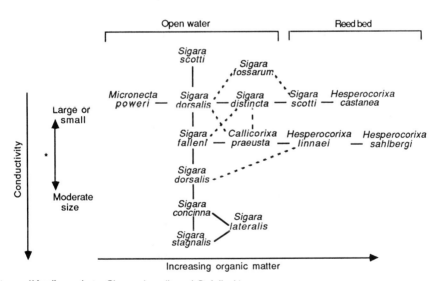

(*Applies only to *Sigara dorsalis* and *S. falleni*.)

Fig. 2.20   Associations of corixids according to habitat size, vegetation, amount of accumulated organic matter and conductivity in English lakes (redrawn after Savage, 1989).

Corixidae and Gerridae that have been studied most in this respect. Early on, Macan (1938) showed a succession in corixid species related to the presence of vegetation and the amount of accumulated organic matter in lakes of the English Lake District. The associations of species of corixid according to habitat size, the type of vegetation, amount of accumulated organic matter and conductivity of the water has been nicely summarized by Savage (1989; Fig. 2.20).

A relatively small number of heteropterans live in salt water: some corixid species occur in inland salt lakes or brackish water tidepools on the seashore, where they may be quite numerous; *Aepophilus*, (Aepophilidae, related to the Saldidae) lives in rock crevices in the marine intertidal zone where it hides under stable, deeply embedded rocks at high tide (Bergroth, 1899); and a few genera of gerrids (e.g. *Rheumatobates, Asclepios, Stenobates* and *Halobates*) live in marine environments. However, only species of *Halobates* are found on the open sea surface. Unlike freshwater gerrids, all species of *Halobates* are permanently wingless and the body is short and broad (Fig. 2.19). The antennae are long and slender, and the limbs are modified in much the same way as freshwater gerrids, that is the front pair specialized for capturing prey, the middle pair providing propulsion, and the hind pair functioning primarily for steering. Little is known of the life history of the open ocean species but it may be similar to that of the coastal species. In the open ocean, the female lays her eggs on floating debris (e.g. pieces of wood and polystyrene, feathers, lumps of tar, seeds) and these, tending to be in short supply, are often covered with the eggs of many females, several layers deep. The emerging nymphs are about 1mm long and it is estimated that between 60 and 70 days are required to reach adulthood. The open ocean species feed on floating insects imported by winds and also on zooplankton; cannibalism of their own nymphs is relatively common. Large prey appear to be shared. In their turn, species of *Halobates* are fed upon by a number of surface-feeding seabirds such as petrels and terns, and perhaps by some fishes (Cheng, 1985).

The Holometabolous Aquatic Insect Orders

2.7   *Megaloptera*

The Megaloptera is a small order (about 300 species) of endopterygote neopterans (Table 1.2) thought to be the most archaic of the Holometabola (Fig. 2.21). Their wings are very similar to those of Palaeozoic insects and the first fossil megalopterans are known from the Lower Permian of Kansas. The adults are terrestrial, whereas the larvae (sometimes known as hellgrammites) are aquatic, inhabiting both lentic and lotic waterbodies. Ancestrally, it is believed that the larvae were also terrestrial, or perhaps inhabitants of moist habitats close to shorelines (Ross, 1955).

All living species belong to one or other of two superfamilies, each

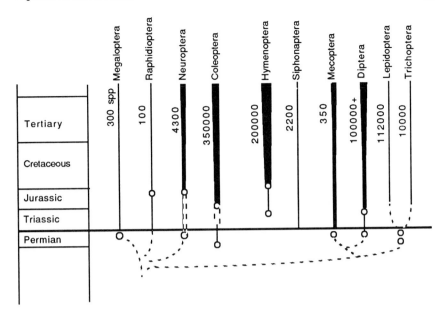

Fig. 2.21   Phylogenetic tree of the Holometabola, showing the roots of the Megaloptera in the Permian (circles represent occurrence and probable lineage of known fossil ancestors; approximate numbers of species known, worldwide, are given; based on Hennig, 1981).

containing a single family. The **Corydalidae** (Superf. Corydaloidea) are widely distributed throughout temperate regions (although they are absent from Europe), with a few species in the tropics. Commonly known as Dobsonflies or fishflies, the adults are large, having a wing span of up to 16 cm (Megaloptera = "large wing"). The larvae live in clear waters and are predaceous. The **Sialidae** (Superf. Sialoidea), or alderflies, are also widely distributed but are confined to temperate latitudes. The larvae tend to live in more turbid waters, or at least those with silty or muddy substrates; they prey upon smaller insects (Kevan, 1979b). Five to 10% of the world megalopteran fauna occurs in Australia.

Adult megalopterans are usually soft-bodied, with long, thin antennae, a pair of prominent compound eyes, and usually three well developed ocelli (Corydalidae only) on the head (Fig. 2.22). The mouthparts are robust, especially the mandibles which, in the males of some species of Corydalidae, are greatly elongated to form tusks. The maxillary palps have 5 segments and those of the labium, two.

The pronotum is large and square-shaped. The meso- and metathorax each bear spiracles and a pair of slightly dissimilar, membranous wings. When at rest, the wings are held in a characteristic, tent-like fashion over the abdomen (Fig. 2.22D). The legs tend to be short with 5-segmented tarsi ending in two simple claws.

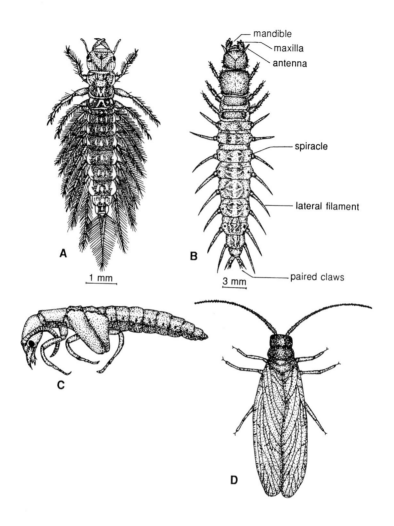

Fig. 2.22   Stages in the life cycle of a megalopteran: (A) Sialidae larva; (B) Corydalidae larva; (C) pupa; (D) adult (redrawn after Riek, 1970b; and Elliott, 1977).

The abdomen is soft, although the cuticle of the first segment may be thickened (sclerotized), and there are spiracles on segments 1 to 8. In the male, tergum 10 is produced to form a pair of anal claspers which generally bear a group of sensory hairs (trichobothria; Fig. 2.23A). Sternum 10 is usually membranous although it may take the form of a sclerotized plate. The aedeagus is hinged to the base of tergum 9 and may be paired or single, and sternum 9 forms a sub-genital plate. In the female, segment 10 forms a pair of anal plates,

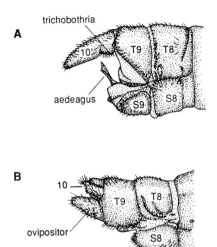

Fig. 2.23   Terminal segments of the abdomen of an adult megalopteran: (A) male; (B) female (T8 = tergum 8; S8 = sternum 8, etc.; after Riek, 1970b).

also with sensory hairs (Fig. 2.23B). Tergum 9 is large and extends downwards, laterally, whereas sternum 9 is small and membranous. In many species of Megaloptera the lower angles of tergum 9 are extended to form the two halves of an ovipositor. In the females of some species, sternum 8 is extended apically to form a sub-genital plate (Riek, 1970b).

Adults appear to be short-lived and are typically active at dusk and dawn (crepuscular) although they may fly during the day in warm weather. Copulation takes place on twigs and branches, often those that overhang a stream. The cylindrical eggs are laid in large, single batches of 1,000 to 3,000 on shaded objects that, again, overhang water. Suitable sites become popular and egg masses from different females frequently abut one another to form mats several square-centimetres in area (Riek, 1970b). Hatching takes between 2 to 4 weeks and the newly-emerged larvae fall off the egg mass into the water.

The larvae are long and somewhat flattened, dorsoventrally, with quadrate (square), well-sclerotized and frequently patterned heads. The mouthparts consist of a labrum, a labium, two stout, robust mandibles, and two elongate maxillae, each with two pairs of palps (Fig. 2.22A,B). The antennae are 4 or 5-segmented (4 only in the Sialidae) and are about the same length as the maxillae. The eyes consist of separate stemmata. In structure these superficially resemble ocelli or the single-faceted compound eyes of more primitive insects but, while they are homologous with the compound eyes of the adult megalopteran, they do not become the adult eyes and are discarded at the final larval moult.

The prothorax consists of a heavily sclerotized pronotum and prosternum; the meso- and metathorax are also thickened but not to the same extent. The legs are of moderate length and the tarsi, which are unsegmented, end in two claws. In the Sialidae, abdominal segments 1 to 7 bear 4- to 5-segmented lateral filaments whereas, in the Corydalidae, 2-segmented lateral filaments are borne on segments 1 to 8. In addition, corydalids have a pair of anal prolegs (each consisting of paired claws and a dorsal filament) compared with the single, long, tapered filament on the terminal segment of sialids (Evans and Neunzig, 1984). In some corydalids, the last pair of abdominal spiracles (on segment 8) may be produced into two tubes so that in *Nigronia* for example, which is found in quiet sometimes stagnant water, they reach to the water surface enabling the larva to breathe air. The lateral filaments contain trachea and serve as respiratory surfaces and may be supplemented, particularly in later instars, by tufts of filamentous gills at their bases.

The larvae are active predators seeking their prey on the stream bed, under rocks and in muddy or silty areas. Not a great deal is known of the biology of the larvae, except that they appear to go through many larval instars and that the life cycle probably extends over two years or more. Prior to pupation, final instar larvae move to the stream bank, coincident with times of flood, and pupate in areas of fine-particle substrate (which retain moisture) under stones, close to the water's edge. The larva constructs a simple pupal cell, in the soil, and may remain "active" for several months before pupating (usually in early summer). The pupal stage is "exarate" and lasts from 2 to 4 weeks. It is unusual in that throughout this time the legs are free and the pupa is able to walk and use its mandibles in defence (Riek, 1970b).

Some species of Megaloptera inhabit temporary streams (Maddux, 1954). In these ephemeral habitats, the female lays her eggs on the surfaces of rocks on the dry stream bed. Upon hatching, the larvae burrow into the bed to await the return of the water. The larvae feed and grow while the stream is flowing and, as it begins to dry up once more, bury themselves under large rocks on the stream bed where they construct their pupal cells.

2.8   Neuroptera

The Neuroptera are very closely allied to the Megaloptera and, indeed, some authorities combine the two groups into one order (Neuroptera). They share many morphological and ontogenetic properties. However, adult neuropterans can be distinguished from adult megalopterans by nature of their wing venation (the former have conspicuous bifurcations of the veins at the wing margins) and the structure of their tarsi (the 3rd or 4th segment of which is bilobed in the Megaloptera, but narrow and straight-sided in the Neuroptera). Larvae may be distinguished on the basis of their mouthparts; neuropterans have apposed

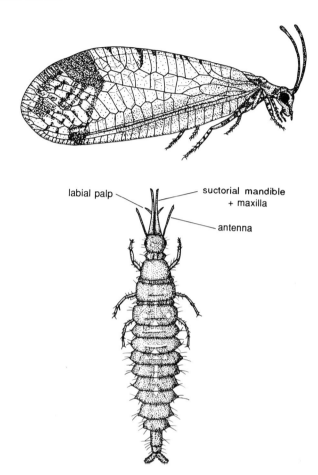

labial palp

suctorial mandible
+ maxilla

antenna

Fig. 2.24   Typical adult and late instar larval neuropteran (redrawn after Riek, 1970b; and Elliott, 1977).

mandibles and maxillae which are hollowed out to form two tubes (suctorial jaws) that project in front of the head (Fig. 2.24), whereas megalopteran larvae have robust mandibles adapted for biting (Elliott, 1977).

Although there are some 4,300 species of Neuroptera, worldwide, very few of them are aquatic. Even then, it is only the larval stage that is aquatic, with the eggs, pupae and adults all being terrestrial. There are two families that have aquatic members, the **Sisyridae**, which is wholly aquatic, and the **Osmylidae**, whose larvae are semi-aquatic.

Larvae of the Sisyridae are parasitic on freshwater sponges of the family Spongillidae and are to be found in both standing and running waters. Their distribution necessarily reflects that of the sponge but is, nevertheless, wide and they occur in Europe, Asia, Africa, North and South America, the West Indies, Australia and the Philippines. Typically, they live attached to the sponge surface or inside the sponge chamber (spongocoel). They feed by inserting their stylet-like mouthparts into sponge cells and sucking out the fluids. As in the Megaloptera, neuropteran eggs are deposited, by the female, on objects overhanging water. The eggs are laid in small batches (2 to 5) and are covered by a blanket of silk threads. They may hatch in as little as eight days, depending on local circumstances, and each larva must free itself by using its "egg-breaker" to cut through the eggshell. Once the larva has dropped onto the water surface, it must force itself through the surface film by flexing its body. A passive drifting phase then follows, punctuated with short bursts of swimming, until contact is made with a sponge. Once it has located a sponge, the larva remains loyal to that particular individual and only moves away if the sponge dies. Respiration appears to be through the general cuticle surface in the first larval instar. Subsequent to this, larvae respire through tracheal appendages, located ventrally on segments 1 to 7. These appendages are vibrated rapidly to create a frequent exchange of water over their surfaces.

The larvae pass through three instars before swimming to the shore and crawling out onto the bank. This is typically done at night and each larva may crawl as much as 20 m away from the water's edge before spinning a pupal shelter in a convenient crevice or under loose bark. This shelter has two layers consisting of an outer, loosely-constructed cover which protects an inner, more finely-woven cocoon. After approximately two weeks, the pupa uses its large mandibles to cut its way out of both the cocoon and cover. The limbs are free as in the megalopteran pupa, but do not become functional until just before the adult emerges. Within about two hours of emergence the adult is fully formed and coloured. This takes place after sunset and is followed by the adult depositing a large faecal pellet containing the accumulated solid wastes from its entire larval existence - as the posterior end of the larval gut is either atrophied or modified to form silk glands (Chandler, 1971). The life cycle is commonly one year, although some species may be multivoltine, and the winter is spent as a larva. Adults live for two weeks or more.

The Osmylidae also have a wide distribution, notably Europe, Asia, Africa, New Zealand, Australia and South America. Larvae typically live in damp moss at the edges of streams or on overhanging banks, or on the damp undersides of stones. Adults are commonly found under bridges and near woodland streams, particularly those with dense riparian vegetation that overhangs the water. The life cycle of one European species, *Osmylus fulvicephalus*, takes one year and has been studied in some detail. The main features (summarized by Elliott, 1977) are as follows. The adults of this species are fairly large (length of the fore wing 1.9 to 2.3 cm) but they are weak flyers. Males live only a few weeks but females can live for up to three months. They are active at dusk when they prey on small insects. The female is attracted to the male in response to his release of a pheromone and once the pair are in contact, an elaborate courtship ensues.

During this, the elongate ventral valves of the female grasp the male's genitalia and lever them for between 10 and 60 minutes, until a single white spermatophore is produced. The female then detaches herself, with the transferred spermatophore projecting from the apex of her abdomen. The spermatophore is large (approximately 4 mm long) and is composed of a stem and four lobes. By bending her head under her body, the female begins to eat the nearest lobe. If she attempts to consume more than this, the male intervenes and strokes her, presumably providing a respite which allows some sperm to enter the female's reproductive system. When the female eventually leaves the male, the remainder of the spermatophore is either consumed or falls away.

Two or three days after mating, the female lays her eggs, in small batches on vegetation (particularly moss) close to the water. Development is highly dependent on environmental temperature (4 to 22 days), and the larva escapes by using its saw-like egg-breaker and burrows into the damp moss. Although osmylids live close to water and have occasionally been observed to enter water to catch prey, they rapidly crawl out if dropped in. Their mode of life must therefore be considered to be semi-aquatic rather than truly aquatic. The larvae pass through three instars during which they feed on small arthropods, typically Collembola and mites during the first instar, and chironomid, tipulid and other dipteran larvae in their second and third instars. Prey is stabbed with the long suctorial jaws and, after injection of a salivary secretion which paralyses the prey, the fluid contents are sucked out. In the autumn, the larvae stop feeding and burrow deep into the moss where they enter a diapause. In this state, they can be totally immersed in water (as happens during winter floods) and survive. In the early spring, feeding recommences and this lasts until late spring when the larva spins a cocoon amongst the moss. Pupation takes place after the larva has been entombed for some 7 to 18 days and at this time the long jaws break off. As in the Sisyridae, the limbs remain free but do not function until just prior to emergence. The pupal stage takes about 10 to 14 days after which the pupa cuts its way out of the cocoon and seeks a firm support. Emergence of the adult occurs in twilight.

2.9 Coleoptera

Of the more than one million described species of insect, at least one-third are beetles, making the Coleoptera the most diverse order of living organisms. They are a monophyletic group of neopterygote insects in which the mesothoracic wings have become modified to form rigid elytra which cannot be folded. When a beetle is in flight, the elytra are held horizontally, at right-angles to the long axis of the body and are thought to act as fixed-wing aerofoils providing additional lift to the body. When the beetle is at rest, they lie along the insect's dorsum (temporarily locked together by a flange and groove or mortice and tenon system) and, at least partially but usually wholly, cover the meso- and metathoracic segments and the abdomen. The metathoracic wings are typically well-developed and membranous, and provide the propulsion during flight. At rest, they are folded and tucked beneath the elytra. Some authorities believe that the acquisition of elytra has been primarily responsible for the dramatic evolutionary success of the Coleoptera. One reason for this could be the increase in protection afforded the body by the elytra in concert with the associated enlargement and thickening of the prothoracic plates (notum, pleuron and sternum) and the abdominal sternites. These structures provide an effective armour, with almost no exposed gaps or membranous regions, against both biotic (e.g. predators) and abiotic (e.g. heat or water loss) elements in the environment. The potential loss of flight capability, normally accompanying weight gain, may be compensated for by the additional lift provided by the elytra.

The oldest fossils, showing unequivocal coleopteran features, are from the Lower Permian of southern Siberia and the Ural Mountains. In addition, Upper Permian species are known from Australia. However, as many families, genera and species of beetle were already in existence in the early Permian, it seems likely that beetles may actually have evolved as early as the Carboniferous (Campbell, 1979). It is thought that the Coleoptera are a sister-group of one of the neuropteroid orders (Megaloptera, Raphidioptera, Neuroptera; Fig. 2.21) (Hennig, 1981).

The 350,000 or so species of beetle are assigned to approximately 151 families in some 23 superfamilies. The latter, in turn, are contained in four suborders as follows:

(1) **Archostemata** - this suborder contains the most primitive beetles still in existence as well as all the early Permian fossil species. Extant species show typical relict, disjunct distributions in all the major faunal regions of the world. None of them is aquatic.

(2) **Myxophaga** - this suborder consists of a few families of small, poorly known species found mostly in the new and old world tropics. Adults and larvae of species in both the families Sphaeriidae and Hydroscaphidae can be found at the edges of streams.

(3) **Adephaga** - this suborder contains eight families. Six of these are aquatic (Haliplidae, Hygrobiidae, Amphizoidae, Noteridae, Dytiscidae and Gyrinidae), one has some species that live at the margins of waterbodies (Carabidae), and the remaining family is terrestrial (Rhysodidae).

(4) **Polyphaga** - this is the largest suborder, containing approximately 137 families grouped in 20 superfamilies. Of the latter, 11 contain families with at least some aquatic or semiaquatic species (summarized in Table 2.5).

Although the zoogeography of terrestrial beetles has been well studied, the same cannot be said of aquatic forms. It is known, however, that each of the families within the Hydrophiloidea (Polyphaga) occurs in all the major zoogeographical regions of the world, and the same is probably true of at least some of the Adephaga (e.g. the Dytiscidae). Within the Hydrophiloidea, many

Table 2.5   Families of Coleoptera that contain aquatic or semiaquatic species (based on Britton, 1970; and Campbell, 1979).

Suborder - **Archostemata** (none)
Suborder - **Myxophaga**
  Superfamily Sphaerioidea
      Sphaeriidae - (minute bog beetles) edges of freshwater bodies, in roots, mud & gravel
      Hydroscaphidae - (skiff beetles) stream margins, often in algae; hot springs
Suborder - **Adephaga**
  Superfamily Caraboidea
      Carabidae - (ground beetles) a few species found at the edges of streams, ponds, swamps; rock crevices on seashores
      Haliplidae - (crawling water beetles) aquatic vegetation at the edges of ponds, lakes & slow streams
      Hygrobiidae - standing, often stagnant, muddy water
      Amphizoidae - (trout stream beetles) fast streams, often on logs
      Noteridae - (burrowing water beetles) shallow margins of standing or slow streams, often in mud or on plants
      Dytiscidae - (predaceous diving beetles) ponds & lakes, esp. near vegetation; slower sections of running waters
      Gyrinidae - (whirligig beetles) ponds & lakes, especially near vegetation; slower sections of streams & rivers
Suborder - **Polyphaga**
  Superfamily Hydrophiloidea
      Hydraenidae - (moss beetles) stream margins, ponds near emergent vegetation; hygropetric (wet rock surface) habitats; marine rockpools & intertidal
      Hydrochidae - (water scavenger beetles) on plants in ponds or slow streams
      Spercheidae - (water scavenger beetles) stagnant ponds on underside of surface film
      Georyssidae - (minute mud-loving beetles) margins of freshwater bodies in sand or mud
      Hydrophilidae - (water scavenger beetles) ponds & lakes, esp. near vegetation; slower sections of streams & rivers
  Superfamily Histeroidea
      Histeridae - (hister beetles) some in ponds, also in damp soil & dung
  Superfamily Staphylinoidea
      Staphylinidae - (rove beetles) some species on the shorelines of fresh and saltwater bodies; marine crevices & intertidal (sand & rocky areas)

Table 2.5 (contd)

Superfamily Scaraboidea
        some groups appear to need very moist environments,
        e.g. Lucanidae, Passalidae & Rutelinae (Scarabeidae)
Superfamily Dascilloidea
    Helodidae (Scirtidae) - (marsh beetles) lentic & slow lotic
        waters, esp. near emergent vegetation; tree holes;
        springs
Superfamily Dryopoidea
    Limnichidae - (marsh-loving beetles) in mud on the margins of
        streams and ponds
    Psephenidae - (water pennies) fast streams, wave-swept shores
        of large lakes
    Ptilodactylidae - (toed-winged beetles) fast & slow water
        regions of streams; stream margins in leaf litter
    Heteroceridae - (mud-loving beetles) tunnels in stiff mud of
        some stream & pond margins
    Elmidae - (riffle beetles) fast & slower sections of streams,
        wave-swept shores of large lakes; some species on
        shoreline
    Dryopidae - (long-toed water beetles) shallow regions of ponds &
        lakes esp. in emergent vegetation; swift streams
Superfamily Cantharoidea
    Melyridae - (flower beetles) semiaquatic on marine beaches &
        intertidal zone
Superfamily Cucujoidea
    Salpingidae - (narrow-waisted bark beetles) marine, on rocks
Superfamily Tenebrionoidea
    Tenebrionidae - (darkling beetles) some species in moist sand on
        beaches at the high tide mark
    Anthicidae - (ant-like flower beetles) some species live in the
        stream-side burrows of staphylinids; salt marshes
Superfamily Chrysomeloidea
    Chrysomelidae - (leaf beetles) ponds & lakes on submerged but
        esp. floating leaves of rooted macrophytes
Superfamily Curculionoidea
    Curculionidae - (weevils) some species of Erirrhininae (e.g.
        Bagous) live on submerged aquatic plants

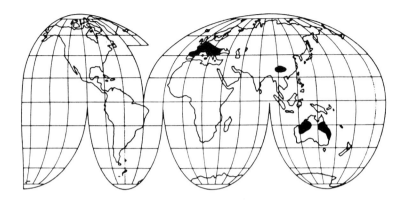

Fig. 2.25   Global distribution of the Hygrobiidae (after Crowson, 1981).

genera show a marked increase in the numbers of species from the temperate zones towards the tropics (Hansen, 1987). The Hygrobiidae (Adephaga), like the Archostemata, show clear disjunct distributions in the Old World but are completely absent (possibly due to local extinction) from the New World (Fig. 2.25). Disjunct distributions, on a much smaller scale, are seen in some flightless aquatic beetles. For example, a population of the flightless form of *Noterus clavicornis* (Noteridae) was discovered in Lindores Loch, in Fife, Scotland in 1950. This was about 200 km northeast from the nearest known colony (also apparently flightless) in Carlingwark Loch. Further south, in England, the species is typically fully winged and capable of flight. The Scottish populations are thought to have been derived from the spread of winged forms from England approximately 5,000 years B.P. when the climate was warmer. Subsequent deterioration of the climate is thought to have been followed by retraction of the populations into a very few, older, climatically-buffered waterbodies and the eventual loss of the capacity to fly (Jackson, 1950).

The typical adult beetle head is a rigid capsule divided into reasonably well-defined regions. The eyes vary in size among species but are generally large and compound. In some instances, for example in the Gyrinidae, each eye is divided into an upper and a lower part, designed for differential focusing on objects above and below the water surface, respectively. Ocelli are not common. Between the compound eyes is a large plate (the frons) which forms the upper/anterior surface of the head. The anterior portion of the frons is separated from the clypeus by a suture (Fig. 2.26A). The antennae typically consist of 11 segments, of which the basal two segments (scape and pedicel) are quite distinct, and vary greatly in length amongst species, although tending to be short in aquatic forms. Antennal morphology also varies and can be assigned to different types, for example *filiform* (long and cylindrical), *lamellate* (with each segment drawn out into a

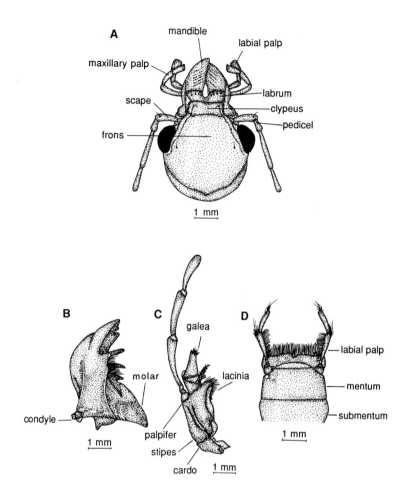

Fig. 2.26   (A) Generalized dorsal view of head of an adult beetle; (B) ventral
            view of mandible, (C) ventral view of maxilla, (D) ventral view of
            labium of an adult hydrophilid beetle (redrawn after Britton, 1970).

flat plate), *moniliform* (like a string of beads), *geniculate* (bent into a right-
angle), *clavate* (with a clubbed end), and *serrate/pectinate* (with saw-like edges).
The sides of the head below and behind the eyes are the genae which are separated
on the mid-ventral line by the gula. The mouthparts are of the generalized
chewing type and are robust (Fig. 2.26B,C,D). The mandibles articulate at their
dorsolateral and ventrolateral angles which provides opposed movement on a
transverse axis. The internal, basal portion of each mandible may be thickened
and enlarged (particularly in the Polyphaga) to form a molar. This is particularly

well developed in herbivorous species and is used for grinding plant material. In carnivorous species, the tips of the mandibles are sharp and curved inwards as an adaptation for seizing prey. In the majority of beetles, the maxillae have become adapted to assist in chewing food. Sensory palps are present on both the maxillae and labium.

As previously noted, the prothorax of the adult beetle is well developed and thickened and, together with the abutting head, often appears as a "forebody" marked off from the rest of the thorax and abdomen (Fig. 2.27). The meso- and metathorax are rigidly fused (to form a "pterothorax") and bear the elytra and

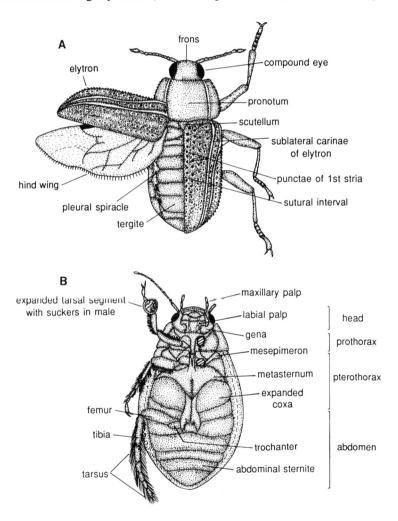

Fig. 2.27   Basic morphology of an adult aquatic beetle as illustrated by: (A) dorsal view of an elmid; (B) ventral view of a dytiscid (redrawn partly after Brigham *et al.*, 1982).

wings. Mesothoracic spiracles are tucked away in the cavity between the pro- and mesothorax and take the form of large, oval structures on the intersegmental membrane attached to the posterior of the prothorax, a position that is particularly important to aquatic species. The metathoracic spiracles are similarly large and hidden, one beneath each of the two mesepimera. The mesothorax tends to be smaller than the metathorax.

The legs of adult beetles are usually adapted for walking or running but in some families are modified for other, more specialized, functions, for example burrowing, jumping and swimming. The basal coxae are normally capable of at least some degree of rotation with respect to the sternal plates but, in the Adephaga (which includes some of the most powerful and agile swimmers, e.g. the Dytiscidae and Gyrinidae) the metathoracic coxae are immovable. The trochanter articulates with the coxa and is proximal to the femur. The tibiae tend to be broader towards their distal ends where they bear spines or spurs. The tarsi are typically 5-segmented, may be differentiated between the sexes (including tarsal suckers in the male) and end in a pair of claws. In aquatic species which cling tightly to the substratum, the claws and last tarsal segment tend to be enlarged as, for example, in the Elmidae.

The general dorsal surface (known as the "disc") of each elytron is commonly marked by striae, corresponding to longitudinal rows of sclerotized pillars linking its upper and lower surfaces. Together with the overall thickness of the cuticle, these give the elytra its characteristic rigidity. The areas between the striae are known as interstriae and the odd-numbered of these (counting from the inner to the outer edge of the elytron) correspond to the primitive wing veins. Additionally, these odd-numbered interstriae are frequently marked on the elytral surface by bands of setae or distinctive sculpturing, important taxonomic aids. The cavity between the upper and lower surfaces of each elytron is lined by epidermis and contains blood channels, tracheae and nerves.

The membranous metathoracic wings are generally longer than the elytra and therefore must be folded to fit under the elytra when the insect is at rest. Although most wings are folded both longitudinally and transversely, the actual mechanics varies amongst families. A typical opening and closing sequence consists of: the wing being rotated forward, on its base, into the flight position by the action of the direct flight muscles. This same action spreads the wing, opening the longitudinal folds, which results, by virtue of the stiffness and springiness of the wing membrane, in the automatic opening of the transverse folds. As a consequence, the costal (leading) edge, which is acutely folded at the hinge, at rest, straightens as the wing rotates forwards. At the same time, the anal fold opens and the whole wing flattens as the membrane tightens. In folding, relaxation of the direct flight muscles allows the wing to fold longitudinally and this, automatically, leads to the transverse folding, centred on the costal hinge. The final stage of closure is aided by movements of the abdomen and elytra.

The positions of the actual fold lines are determined by the lines of stiffness produced by adjacent veins. Thus longitudinal folds generally lie alongside primary veins, whereas the angles of the V-shaped transverse folds are determined by the ends and weak hinge-points of veins (Britton, 1970). As in the Hemiptera, wing polymorphism is evident in some beetle species and may range from being shortened and not folded (brachypterous), to being highly reduced (micropterous), to totally absent (apterous); corresponding changes are seen in the elytra, too.

The adult beetle abdomen consists of 10 segments, although 11 segments have been seen in the embryonic development of many genera including some aquatic ones (e.g. *Hydrophilus*). In *Dytiscus*, a 12th segment (representing the telson) is also present early on. Subsequent fusion produces 10 discernible segments, some of which are composite. In many families, however, only eight are readily visible, externally, as segment 9 (modified as the genital segment, in both sexes) is withdrawn inside the body and segment 10 is much reduced (Fig. 2.28A,C). In contrast, in the Adephaga and some Polyphaga (including the aquatic Hydraenidae and the semiaquatic Staphylinidae), segment 9 is fully developed and surrounds both the rectum and the aedeagus in the male. In the female it forms the armature at the tip of the genital tube. In most families, there are 8 pairs of functional spiracles on the abdomen and these are situated on the membranous side (pleura) or dorsal plates (terga) and, consequently, are covered by the elytra when the insect is a rest. The sterna are massive and, in most winged species, extend upwards to touch the ventrolateral edges of the elytra thus effectively encasing the abdomen in a strong shell.

In the adult female, abdominal segment 9 is primarily membranous and supports a pair of appendages. Each appendage consists of a valvifer (sclerotized part of sternum 9), an adjacent, segmented coxite, and an apical style bearing setae Fig. 2.28D). Between these appendages is located the vulva. In species that have an ovipositor, it may be formed either from enlarged styles or from extensions of segments 8 and 9. Internally, lies the spermatheca a sac-like structure that receives and stores sperm from the male. It is generally small and membranous, particularly in the Hydrophiloidea, but may be more sclerotized in other groups, for example the Hydraenidae.

In the male, the copulatory organ is the aedeagus which represents an extension of the posterior end of the ejaculatory duct which opens just behind invaginated sternum 9 (Fig. 2.28B). Although there are many variations in structure, typically the aedeagus consists of: (1) a membranous basal section that allows protrusion and withdrawal of (2) a rigid middle section consisting of a tubular tegmen connected by a membrane to a tubular penis which, in turn, contains (3) an internal sac representing a specialized section of the duct. During mating, these reproductive organs are everted in two stages. First, the aedeagus is protruded from the body through the genital opening. Second, the internal sac is everted from the penis through the ostium. Spines erected on the surface of the sac upon eversion may hold the sac within the female's genitalia or may rupture the spermatophore, releasing sperm. In many species, the ejaculatory duct opens

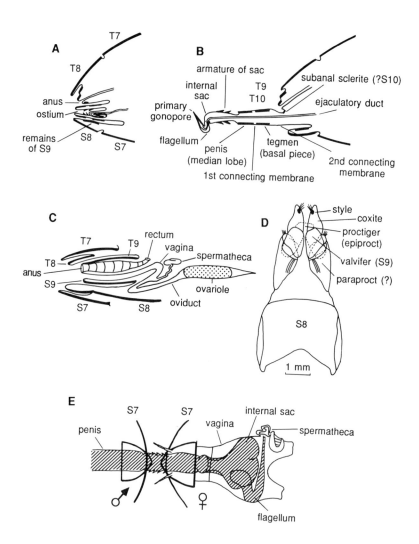

Fig. 2.28   Reproductive system of adult beetles: (A) diagrammatic sagittal
             section through male with aedeagus retracted; (B) male with
             aedeagus evaginated; (C) sagittal section through female; (D)
             ventral view of female genitalia; (E) position of organs during
             copulation (note close proximity of the male's flagellum to the
             female's spermatheca; redrawn after Evans, 1961; and Britton,
             1970).

at the tip of a flagellum which protrudes from the distal end of the sac. In these
cases, at copulation, the flagellum is inserted directly into the opening of the
female's spermatheca (Fig. 2.28E) and this, it is thought, greatly conserves

sperm. As in many other groups of insect, the structure of the male genitalia is of considerable taxonomic importance (Britton, 1970).

Adults of the aquatic Adephaga respire by taking a supply of air with them when they submerge. This is stored under the elytra and is renewed when the beetle breaks through the surface film with the tips of its elytra and abdomen. Adult aquatic Polyphaga also carry air between the dorsum of the abdomen and the elytra but, in addition, most have their ventral surface covered with short hairs. This traps a sheet of air as the beetle dives and, because it is connected to the elytral reservoirs, provides an additional store (see Chapter 6).

Unlike the Hemiptera, the larvae of Coleoptera are morphologically and behaviourally different from the adults, and their diversity is high (Fig. 2.29). Most aquatic larvae have a distinct head, three more or less similar thoracic segments, and an abdomen consisting of 10 segments. All but a few (e.g. some hydrophilids and curculionids) have three pairs of legs on the thorax, each ending in either one or two claws (one in the Polyphaga).

The head is sclerotized and has mandibles, maxillae, a labium, a labrum (sometimes fused with the clypeus to form a nasale), antennae that are 2- or 3-segmented, and eyes typically in the form of stemmata (Fig. 2.29D). The mouthparts are variable, depending on diet, and may resemble the differences seen among adults that feed on different items, for example grinding mandibles with molar lobes in herbivores such as haliplids and elmids, but sharp, curved mandibles in predators like gyrinids; in some predators the mandibles may even be suctorial (e.g. dytiscids and some hydrophilids). The labium consists of two movable parts, the mentum and the submentum, the latter bearing a pair of 2-segmented palps. A distinctive feature of many aquatic larvae, particularly the predaceous ones, is the ventral closure of the head capsule behind the labium. In herbivorous species, a shorter head, without ventral closure, often predominates and this is commonly associated with the presence of the more primitive orthopteroid-type chewing mouthparts.

On the thorax, there is usually just a single pair of spiracles located near the junction of the pro- and mesothoracic segments. There is one less segment in the legs of the Polyphaga and Myxophaga compared with the Adephaga (and Archostemata). Most often, the limbs of the Adephaga terminate in a pair of claws, whereas those of the Polyphaga end in a single claw.

On the abdomen, there is a variety of segmentally arranged, gill-like appendages, and these can be single or in groups, simple or branched, segmented or unsegmented. In groups such as the elmids, these "gills" are tucked into a pocket formed beneath the last tergum. On the apical or preapical segments of the abdomen, some species have urogomphi, thought to be homologous with the cerci of other endopterygotes. Semiterrestrial larvae usually have 9 pairs of functional spiracles (on the mesothorax and the first 8 abdominal segments) but this number may be reduced in truly aquatic species. In some hydrophilids, fleshy ventral prolegs are present on the tip of the abdomen.

The pupal stage of aquatic beetles is often passed on land, in earthen cells.

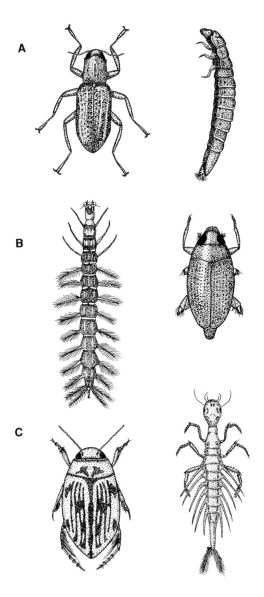

Fig. 2.29   Representative adults and larvae from selected aquatic coleopteran families: (A) Elmidae; (B) Gyrinidae; (C) Dytiscidae (redrawn after various sources, especially McCafferty, 1981; and White *et al.*, 1984).

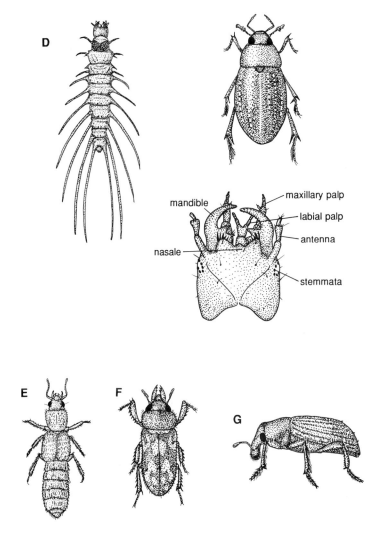

Fig. 2.29   (cont'd) (D) Hydrophilidae, with expanded dorsal view of larval head; (E) Staphylinidae; (F) Heteroceridae; (G) Curculionidae.

Later larval instars tend to be less dependent on water than earlier ones and may crawl out of the water as pupation approaches. Some species (e.g. those of the Noteridae) are known to pupate under water but the pupa is generally surrounded by an air-filled cocoon. Pupae of waterpennies (Psephenidae) also remain fully submerged and breathe through abdominal spiracles which acquire plastrons and function as "physical gills" (Hinton, 1955). Aquatic beetle pupae are typically

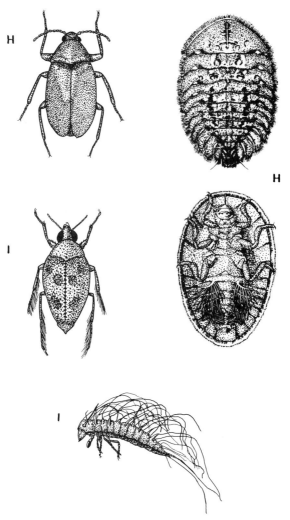

Fig. 2.29   (cont'd) (H) Psephenidae (with dorsal and ventral views of larva; (I) Haliplidae (with lateral view of larva).

exarate and have various bumps and setae on their surface; the latter are believed to keep the pupa away from the walls of the pupal cell. In several families, the pupa may be enclosed, at least partially, in the last larval skin which may afford it some protection by way of long, sharp setae or toxic secretions. Some abdominal segments, of which there are typically nine, may have functional muscles which allow the pupa to readjust its position slightly. As in the larvae, coleopteran pupae usually have one pair of thoracic spiracles on the anterior mesothorax. There are others on the abdomen, on segments 1 to 6 in the

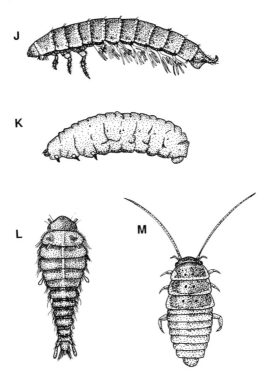

Fig. 2.29   (cont'd) larvae of: (J) Ptilodactylidae; (K) Chrysomelidae; (L) Hydroscaphidae; (M) Helodidae.

Adephaga and on segments 1 to 7 or 8 (or, more rarely, 1 to 6, 1 to 5, 1 to 4, or even 1 to 3) in the Polyphaga; presumably most are functional. A peculiar defence mechanism seen in the pupae of many species takes the form of the margins of certain abdominal terga becoming heavily sclerotized and toothed. These "gin trap" structures bite together upon stimulation by a foreign intrusion (such as, perhaps, parasitic mites) and are present in at least one aquatic family, the Psephenidae.

Aquatic beetles show the holometabolous development typical of endopterygotes. The eggs are usually rather soft and have a smooth shell (chorion), and many have a tendency to take up some water from the environment. In some aquatic species, the chorion is thin and is augmented by a tough, sclerotized cuticle which the embryo secretes around itself. Most aquatic species deposit their eggs in one of three places: (1) in air-filled cocoons (particularly the Hydrophiloidea); (2) out of the water; or (3) inside the air-containing stems of aquatic macrophytes, for which the female is equipped with a sharp ovipositor (e.g. Dytiscidae and Dryopidae). In *Hydrophilus piceus*

(Hydrophilidae), however, the female constructs a cocoon made from silk secreted by the Malpighian tubules of the hind-gut. This cocoon, containing the eggs, is attached to a floating leaf or twig. Other hydrophilids, such as *Spercheus* and *Helochares*, carry the cocoon beneath their abdomens as they move around. The Hydraenidae lay their eggs singly, on stones or algae, in or out of the water.

The pattern of embryonic development varies considerably among species and this seems to be related to the amount of yolk with which they are provisioned. *Dytiscus*, for example, lays large eggs, rich in yolk, and its general pattern of development resembles that of typical orthopteroids. When the larva hatches, it must break out of the shell and this is accomplished using special "oviruptors" (egg-bursters), situated on the head, thorax or abdomen, or the mandibles. The number of instars through which the larva passes ranges, in beetles in general, from 1 to 29 (but is, more usually, 4 to 15) and seems to depend on two factors: first, the size ratio of the egg to the adult, and second, the degree of sclerotization and consolidation of the larval cuticle. Between instars the larva must shed its skin, and this moulting process always involves a period of inactivity during which the larva ceases to feed. The transition from the last larval instar to the pupa is usually prolonged and it is in the latter stage that many external and skeletal differences become apparent (e.g. male and female genitalia, outline of the wings, modification of the limbs, etc.). Expulsion of a meconium, matter which has accumulated during the pupal stage (e.g. the larval gut plus various waste products), by the newly emerged adult is known to occur in at least a few species of terrestrial beetle (Crowson, 1981).

Location of the sexes prior to mating is aided, in many beetles, by the production of a chemical attractant (sex pheromone) by the female. This scent is detected by the antennae of the male which are frequently large and complex. In other species, both sexes produce a pheromone which has an "assembling" effect rather than a sex attracting one. The extent to which aquatic beetles employ such techniques is largely unknown, however. Location is usually followed by a simple courtship behaviour which primarily seems to establish species identity. Some species, particularly in the Hydrophilidae, have the ability to stridulate and the sound produced is known to attract the sexes during the reproductive period. These sounds appear to be species specific and this, again, provides a mechanism that prevents interbreeding of closely related species.

In temperate regions, beetles from most major groups commonly exhibit univoltine life cycles, and this is thought to represent the ancestral condition. This permits close adaptation of particular stages in the life cycle to seasonal and climatic changes; however, multivoltinism is, as might be expected, more common in the more stable tropics. Where temperate aquatic species take more than one year to develop, adverse conditions are usually passed in a particular growth stage, for example as the larva in the cold water of winter, or as diapausing eggs or winged adults in the case of summer droughts. A striking feature of the life history of many beetle species, in contrast to that of many other insects, is the longevity of the adults. Compared with adult lives in the order of days or weeks (e.g. mayflies and dipterans), many beetle adults live for one or more years and breed at least twice. Adult elmids have been kept alive, in captivity for up to 9 years (Brown, 1973). Long-lived adults tend to be found in the same type of habitat as the immature stages where they eat similar, protein-rich foods, whereas short-lived adults often live in quite different habitats from their larvae. Many aquatic species are of the "pioneering" type and consequently show a predominance of the r-type selection traits (Pianka, 1970) suitable for the rapid colonization and exploitation of new habitats (see Chapter 5). This contrasts with the more conservative K-selection seen in most terrestrial beetles.

A particularly interesting suite of aquatic and semiaquatic habitats inhabited by beetles occurs at the edge of the sea. In general, insects have not made major inroads into salt water (see Chapter 4), but a considerable number of beetles are able to tolerate such environmental conditions by either physiological tolerance or behavioural adaptation.

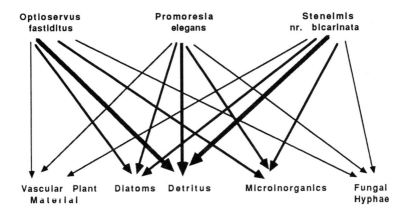

Fig. 2.30   Diets of three species of elmid beetle in a Canadian stream (relative importance of the various food items in the diet are represented by the thickness of the lines; after Tavares and Williams, 1990).

Although the diets of predaceous species have been studied in some detail, particularly with respect to the pressures that aquatic beetles can exert on mosquito larvae in pools (James, 1969), information on herbivores is rare. In a study of three sympatric species of Elmidae in a Canadian stream, detritus was identified as the most important food source for all size classes of larvae, in all species, throughout the year. Biofilm-coated, microinorganic particles ranked second, with diatoms, fungal hyphae and vascular plant material being less important (Fig. 2.30). In none of the species was there any evidence of change of diet among instars. Food niche overlap indices indicated almost complete dietary overlap among species. Severe competition for the same food resource appeared to be avoided by the sheer abundance of detritus combined with limitation of the beetle populations by physical factors in the stream (Tavares and Williams, 1990).

2.10   Diptera

The Diptera, or true flies, are a large order of endopterygote Neoptera. The adults have a single pair of membranous wings, borne on the mesothorax, the metathoracic wings having been reduced to a pair of club-like halteres that aid manoeuvring in flight. The pro- and metathorax are substantially reduced, and the adult mouthparts are suctorial and frequently modified for piercing, especially in the female. Larval body form is diverse, ranging from the primitive, cylindrical body with complete head capsule, seen in many nematocerans, to the maggot-like body with no externally discernible head, seen in many of the brachycerans. The larvae have no true limbs, although there may be fleshy prolegs or segmentally arranged protrusions used to assist the oftentimes peristaltic method of locomotion. The pupal stage may be tightly enclosed in the last larval skin, which forms a sclerotized, protective puparium or, in the more primitive families, it may have its limbs free and be more active (exarate condition). In aquatic species, only the larval and pupal stages live in the water, the adults, with very few exceptions, being terrestrial. Aquatic dipterans represent some of the best known insect forms, including mosquitoes, black flies, midges, crane flies and horse flies, many of which are the most troublesome of all insect pests, particularly in terms of human health and economics. Despite this, many groups of aquatic Diptera play pivotal roles in the processing of food energy in aquatic environments and in supporting populations of fishes and waterfowl.

It is estimated that the order contains about 200,000 species, worldwide, although only just over half of these have been described (Hardy, 1977). Clearly, the Diptera have been much more successful than their presumed sister-group, the Mecoptera (scorpionflies) which has less than 400 described species. Fossil records date back to the Mesozoic when, it is thought, the Diptera were better represented than any other insect order - although most are preserved in amber which tends to select smaller flying insects.

Some 32 families of Diptera contain species whose larvae are either aquatic or semiaquatic. These are contained in two suborders, the more primitive **Nematocera** (literally translated as "thread-like horn", and referring to the nature of the adult antenna) and the **Brachycera** ("short horn" = short antenna). It is beyond the scope of this book to attempt to give a comprehensive account of all these families. Instead, detailed coverage will be confined to four groups that are important from ecological and/or human welfare perspectives, namely the Tipulidae, Culicidae, Simuliidae and Chironomidae. Habitat descriptions and relationships of the remaining families are summarized in Table 2.6.

Table 2.6   Families of Diptera that contain aquatic or semiaquatic species, together with descriptions of larval habitats (based on McAlpine, 1979).

Suborder - **Nematocera**
  Infraorder Tipulomorpha
   Superfamily Tipuloidea
    Tipulidae - (crane flies) fast & slow-flowing streams & rivers; springs; hygropetric habitats; ponds & lakes, esp. in shallow water; fresh & brackish water marshes; wet moss; tree holes & other phytotelmata; marine intertidal zone; saturated soil; temporary pools; floodplains
  Infraorder Psychodomorpha
   Superfamily Blephariceroidea
    Blephariceridae - (net-winged midges) fast-flowing, cool streams, generally at high altitude
    Deuterophlebiidae - (mountain midges) fast-flowing, cool streams, generally at high altitude
    Nymphomyiidae - among moss in cool, fast-flowing streams
   Superfamily Tanyderoidea
    Tanyderidae - (primitive crane flies) shallow water at margins of stream & rivers
    Ptychopteridae
     Ptychopterinae - drainage ditches, esp. those contaminated with manure; beaver ponds
     Bittacomorphinae - (phantom crane flies) very shallow woodland pools
   Superfamily Psychodoidea
    Psychodidae - (moth flies) fast streams, esp. margins; littoral zone of lakes, esp. in detritus; marine beaches
  Infraorder Culicomorpha
   Superfamily Culicoidea
    Dixidae - (dixid midges) sheltered regions of streams, also in detritus; pond & lake at margins or under surface film
    Chaoboridae - (phantom midges) lakes in profundal, littoral or open water zones; boggy pools; temporary fresh waters; small springs, esp. limnocrenes
    Culicidae - (mosquitoes) ponds & lakes; pools and slow sections of streams & rivers; bogs; marshes; woodland pools; temporary waters; salt marshes; marine rockpools; phytotelmata; water tanks; small container habitats e.g. tin cans, tyres, bottles, coconut shells

---

Table 2.6 (contd)

Superfamily Chironomoidea
  Thaumaleidae - (solitary midges) hygropetric habitats; small,
        cold streams, esp. margins; wet moss
  Simuliidae - (black flies) fast-water regions of streams &
        rivers; wave-swept, littoral zone of large lakes
  Ceratopogonidae - (biting midges, no-see-ums) littoral & open
        water zones of lakes, also at margins esp. in algal
        mats; streams & rivers, esp. margins & in detritus;
        tree holes; temporary pools; moist soil; salt marshes;
        marine beaches; some species associated with pollution
  Chironomidae - (midges) most types of waterbody, including
        intertidal rockpools & coral reefs; moist soil;
        phytotelmata; dung
Suborder - **Brachycera**
Infraorder Tabanomorpha
  Superfamily Stratiomyoidea
    Stratiomyidae - (soldier flies) streams & rivers, esp. margins;
        littoral zones of ponds & lakes, esp. around submerged
        & emergent macrophytes; hot springs
  Superfamily Tabanoidea
    Rhagionidae - (snipe flies) moist soil & moss in woodlands
    Pelecorhynchidae - small, foothill streams, on sand substrates
    Tabanidae - (horse & deer flies) wetlands; littoral zones &
        margins of ponds & lakes; damp soil; slow-water
        regions of streams & rivers; tree holes; marine
        beaches; estuaries
    Athericidae - streams & rivers
Infraorder Asilomorpha
  Superfamily Empidoidea
    Empididae - (dance flies) streams & rivers, esp. in detritus;
        littoral zones of ponds & lakes; wet soil
    Dolichopodidae - margins of ponds, lakes, rivers & streams;
        leaf-miners in aquatic macrophytes; marine intertidal;
        estuaries
Infraorder Cyclorrhapha
  (Section Aschiza)
  Superfamily Phoroidea
    Phoridae - (humpback flies) burrowers; predators on Psychodidae

---

Table 2.6 (contd)

Superfamily Syrphoidea
    Syrphidae - (rattail maggots, flower flies) pond & lake margins,
        esp. in fine particle detritus & on water plants; shallow
        marsh & bog pools; tree holes; polluted water
(Section Schizophora: Subsection Acalyptratae)
Superfamily Sciomyzoidea
    Coelopidae - (seaweed flies) breed in decaying seaweed around
        the high water mark
    Dryomyzidae - moist places, including the seashore
    Sciomyzidae - (marsh flies) marshes: margins of streams, ponds
        & lakes, esp. near emergent macrophytes; temporary
        ponds; salt marshes; larvae predators or parasites of
        aquatic & terrestrial snails
    Sepsidae - (black scavenger flies) decaying organic matter;
        terminal drying phase of temporary pools; dung
Superfamily Heleomyzoidea
    Sphaeroceridae - (small dung flies) decaying organic matter;
        terminal drying phase of temporary ponds; dung; septic
        tanks; wet waste disposal sites
Superfamily Drosophiloidea
    Ephydridae - (shore flies, brine flies) littoral zones & margins of
        lotic & lentic habitats, often near or within stems of
        aquatic macrophytes; temporary waters; saline lakes
        & pools; salt marshes; marine intertidal; pools of crude
        petroleum; hot springs
    Canaceidae - (beach flies) on algae in intertidal zone
(Section Schizophora: Subsection Calyptratae)
Superfamily Muscoidea
    Scathophagidae - (dung flies) burrowers in tissues of aquatic
        macrophytes in lakes & ponds; in rotting seaweed;
        damp soil; sewage treatment beds
    Anthomyiidae - (root maggot flies) littoral zones of lakes &
        ponds, often those that have been enriched by organic
        wastes; tree holes; a few species in streams
    Sarcophagidae - (flesh flies) semiaquatic; burrowers in
        macrophytes

Tipulidae

This is the largest of the dipteran families, with approximately 14,000 species described to date. Commonly known as crane flies or daddy-long-legs, tipulids are worldwide in distribution, although their greatest diversity is in the humid tropics. Adults are slender and easily recognized by a V-shaped suture on the

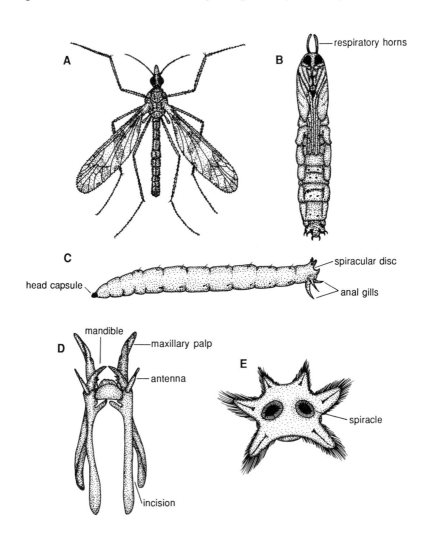

Fig. 2.31   Stages in the life cycle of a tipulid: (A) adult, dorsal view; (B) pupa, ventral view; (C) larva, lateral view; (D) dorsal view of larval head capsule; (E) posterior spiracular disc of larva (redrawn after various sources, especially Byers, 1984).

mesonotum, the presence of two anal veins reaching the wing margin, and lack of ocelli (Fig. 2.31A). They also have brittle legs that tend to break off between the femur and trochanter if the animal is stressed. On occasion, they may be confused with members of the Tanyderidae (primitive crane flies) and the terrestrial Trichoceridae (winter crane flies). Size varies greatly, from a wing length of 2 mm (*Dasymolophilus*) to one of 4 cm (*Holorusia*). Adults tend to be short-lived and are typically found in shaded, humid areas of woodland where they feed on honeydew and nectar. The larvae are much longer-lived, up to a year or more, and are primarily aquatic or semi-aquatic, although some species live in decaying wood, soil and fungi. The larvae are variously phytophagous, saprophagous or carnivorous (McAlpine, 1979). The larvae of some soil-dwelling species of *Tipula,* known as "leather jackets", may cause severe damage to moist pastureland by eating the roots of grass. Tipulids are important, both as larvae and adults, in providing food for other species as, besides being eaten by other invertebrates, fishes and amphibians, at least 91 species of bird are known to eat them in New York State, alone (Alexander, 1919). In many freshwater habitats, especially ponds, streams and floodplains, tipulid larvae play an important role in "shredding" riparian leaf litter, thus making it available to other species that can feed only by "gathering" smaller organic particles.

Tipulids represent an ancient group of Diptera, having many features that resemble those of the Mecoptera and, together with the Trichoceridae, are considered to be the sister group of all other dipterans. Tipulids probably evolved from ancestors similar to, or included in, the Architipulidae of the Upper Jurassic. Two of the three presently extant subfamilies (Tipulinae and Limoniinae) are thought to have differentiated in the mid- to late Palaeocene, whereas the Cylindrotominae seems to represent a relict group that was more prominent in the Tertiary (Pritchard, 1983).

Crane fly larvae are elongate and cylindrical (Fig. 2.31C). The anterior portion of the head is heavily sclerotized and the mouthparts are well formed. Posteriorly, the head appears incomplete and extends into the thorax as sclerotized plates or rods which are separated by deep incisions (Fig. 2.31D). By muscular contraction, the entire head capsule can be withdrawn into the thorax, for protection. Three thoracic and eight abdominal segments are visible externally. At the posterior end of the body, the 9th abdominal segment is represented by a spiracular disc (Fig. 2.31E) surrounded by a number of lobes, and the 10th segment by the anal gills. In some species, the ventral portions of some of the abdominal segments are thickened and bear rows of teeth or dense hairs. These "creeping welts" provide temporary anchorage of successive parts of the body to the substratum as the larva crawls in peristaltic fashion. Morphological features important in the taxonomy of tipulid larvae include the mouthparts, the nature of the skin and creeping welts, and the arrangement of the lobes on the spiracular disc.

The aquatic habitats of the larvae range from all types of fresh waters to the marine intertidal zone and the water-filled leaf axils of plants (phytotelmata). A certain minimum level of moisture in the habitat seems to be crucial to larval distribution (even those that live in apparently dry soil tend to aggregate under moist cowpats), but the availability of suitable food is also important, especially in the case of detritivores. The larvae of several aquatic species remain totally submerged at all times, presumably respiring cutaneously, but other aquatic tipulids make frequent trips to the surface to take in air through the, typically, single pair of spiracles on the posterior disc. These have no closing mechanism but water loss from the body of semiaquatic species is reduced by microtrichia on the walls of the spiracular atrium (the "felt chamber"). The exact role of the anal gills is not proven, although they are well supplied with tracheae and therefore a respiratory function seems likely. However, their function could well be osmoregulatory.

Larvae of the Limoniinae and Tipulinae pass through four instars before pupating, whereas those of the Cylindrotominae reportedly have more. Much of the growth seems to take place in the first three instars and it is related to environmental temperature. In *Tipula oleracea*, for example, development is fastest above 20°C but in the Alaskan species *T. carinifrons*, it is fastest at 5°C. Growth is also affected by the amount and quality of the food available and, in some species, by photoperiod. In *Tipula subnodicornis*, for example, growth of the final instar is retarded at short daylengths, thereby causing larvae to overwinter in this stage until spring temperatures rise above the threshold for growth (Pritchard, 1983).

Tipulids have pharate pupae (i.e., a "prepupal" stage formed upon cessation of feeding by the last larval instar). The larval to pupal moult coincides with the detachment of the adult epidermis from the old pupal cuticle (apolysis) and thus the exact timing of these various stages is frequently difficult to follow. In some aquatic species, the last larval instar moves onto land to "pupate" in burrows or moss and subsequent respiration takes place through the persisting larval spiracles. In those species in which the pupa remains submerged, generally those living in swift streams, cutaneous respiration through two cuticles is thought to occur, but this may be feasible only in such well-oxygenated environments.

A pharate adult phase also occurs. This stage is frequently quite active, particularly just prior to the final moult to the adult proper (sometimes referred to as the non-pharate adult). Movement is aided by spines and processes on the abdominal segments of the pupal cuticle. Breathing is achieved through two thoracic horns, each of which is connected by a trachea to the mesothoracic spiracles of the pharate adult. In forms such as *Prionocera* that live in habitats liable to flooding, the thoracic horns may be very long. Other interesting respiratory adaptations are seen in species in which the pharate adult remains submerged, and these include, for example, burrowing into hollow plant stems, development of spiracular gills that operate as a plastron, and tapping gas spaces in the stems of aquatic plants through insertion of the tips of the breathing horns.

The non-pharate adult is a wholly terrestrial animal. The preference of these adults for shaded, humid habitats may partly reflect proximity to the larval habitats but also seems to be related to the fact that many adults become desiccated quite quickly. *Tipula* adults, for example, are known to lose 10% of their body weight in 2 to 6 hours at 20°C at a relative humidity of 60% (Freeman, 1968). Drinking plant fluids can partially compensate such losses but selection of moist habitats, combined with nocturnal activity, also helps. Interestingly, the longest lived adults (up to 2 months) are those like *Chionea* which inhabit the cool, moist cavities under snowfields. Female tipulids typically emerge containing mature eggs and mate almost immediately. In fact, males have been observed to grasp females as the latter are extracting themselves from their pupal skins. In some species, especially in the Limoniinae, the males form mating swarms, consisting sometimes of several hundred individuals, into which females are attracted. Copulation frequently takes place in flight or after a pair have settled and, particularly in the case of the latter, may last up to several hours (Pritchard, 1983).

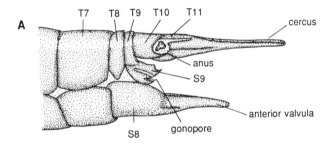

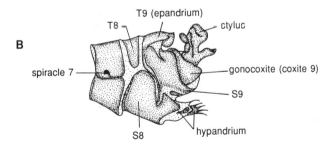

Fig. 2.32   Lateral views of genitalia of adult tipulids: (A) female; (B) male (redrawn after Matsuda, 1976).

Egg laying occurs soon after mating. The female's ovipositor is formed from the cerci or the latero-sternal portions of the 9th abdominal segment (Fig. 2.32A) and its exact form varies according to the oviposition habits of individual species. Some species insert their eggs directly into wet soil or algal mats, whereas others lay them on the water surface or drop them from the air.

Temperate tipulid species are typically univoltine, although many are bivoltine. At higher latitudes, the life cycle is often spread over two years (semivoltine) and, near the northern-most limit of crane fly distribution, in Alaska, species like *T. carinifrons* may take as long as five years (merovoltine). In temperate bivoltine species, such as those of the damp-seeking *Dolichopeza*, the adults of the autumn generation are smaller than those emerging in the spring, presumably because the former are not able to feed as well during the relative dryness of the summer (Pritchard, 1983). In species living in intermittent habitats, the life cycle is commonly synchronized with changes in the environment. For example, the peak emergence of floodplain tipulids in Michigan occurs in mid-May when favourable water, temperature and food conditions of floodplain pools are available for the soon-to-be-produced larvae. Despite the abundance of tipulid species in the tropics, there have been no studies on their life cycles. Possibly, there, as in subtropical Florida, the majority of species would show no obvious seasonality (Rogers, 1933).

Culicidae

The infraorder **Culicomorpha** contains the majority of the so-called "biting flies". It contains what are thought to be two sister-groups, the **Culicoidea** and the **Chironomoidea**, which have followed separate evolutionary paths in terms of their mode of life and larval morphology. The superfamily Culicoidea encompasses the Dixidae (dixid midges), the Chaoboridae (phantom midges), and the Culicidae (mosquitoes) (Table 2.6). At one time, the chaoborids and dixids were treated as subfamilies within the Culicidae, a view still held by some researchers. The three groups are certainly related but adult chaoborids and dixids differ from mosquitoes in many respects, foremost amongst which is the prominent, elongate proboscis peculiar to mosquitoes.

Although the Culicidae is a relatively small family (only some 3,065 species worldwide), it is the best known of all the Diptera and, in many parts of the world, mosquitoes are the most familiar type of insect to Man. The reasons for this revolve around the blood-sucking habits of the females and the associated spread of many debilitating and often fatal disease organisms, particularly those responsible for malaria, dengue and yellow fever (see Chapter 11). More than any other blood-feeding group, mosquitoes frequently breed near towns and cities which the adults invade in vast numbers. Apart from the diseases spread throughout settlements in warm climates, in temperate and high latitude regions these suburban invasions create considerable human discomfort from their bites (McAlpine, 1979). The family Culicidae is divided into three subfamilies, the **Anophelinae**, the **Culicinae** and the **Toxorhynchitinae**.

Mosquitoes occur almost everywhere, from the Arctic to the most remote desert oases. Although populations of certain northern species may be very large, diversity in temperate regions is generally not high, for example only 74 species occur in Canada (compared with 520 species of Tipulidae and 546 species of

Trichoptera). Mosquitoes are clearly predominantly a tropical group with the richest development of genera and species in the Oriental and Neotropical regions. Although the faunas of these latter two regions are comparable in size, 893 and 874 species respectively, that of the neotropics shows a greater degree of isolation and, consequently, a higher number of endemic species.

The Anophelinae contains three genera. Whereas *Bironella* and *Chagasia* contain less than 10 species each and are highly endemic, there are almost 400 species of *Anopheles*, worldwide. The latter are important as the sole vectors of malaria.

There are more than 2,500 species of Culicinae including the important genera *Aedes* (> 900 species) and *Culex* (almost 750 species). Many culicine genera are widespread and they are the primary vectors of several human diseases, particularly those caused by arboviruses and filarial worms. The Culicinae have been partitioned into four ecologically-different groups (Mattingly, 1969): (1) the aedine genera *Aedes* and *Psorophora* which, possessing drought-resistant eggs, breed in temporary pools and container habitats; (2) the quasi-sabethines which have both aedine and sabethine features and breed in natural containers like tree holes, leaf axils and bamboo stems, and include the genera *Haemagogus, Eretmapodites* and *Armigeres*; (3) a group, associated with dense aquatic vegetation, containing *Mansonia, Coquillettidia* and *Ficalbia*; and (4) a group of miscellaneous genera including *Culex, Culiseta* and *Orthopodomyia*.

The Toxorhynchitinae contains one genus, *Toxorhynchites*, which contains about 69 species chiefly confined to the tropics and subtropics, particularly in the Neotropical, Oriental and Afrotropical regions. None of the adults sucks blood and the larvae, which live in small bodies of water, are predaceous on the larvae of other mosquito species.

The adults of most species have a long proboscis formed from the fleshy labium wrapped around the fine stylets which, in blood-sucking species, are used to pierce the host's skin. Males do not suck blood but may drink nectar and honeydew, particularly in the longer lived species; females may also feed on these plant juices in order to obtain sugar for flight energy. Both sexes tend to have their wings, legs and other body parts more or less covered with scales (flattened setae) and males can be distinguished from females by their large bushy antennae and by differences in the size and shape of their maxillary palps. In addition, the stylets of the male are typically less well-developed. Mosquito adults are slender in form and range in length from 4 to 10 mm. The head lacks ocelli but has a pair of large compound eyes between which are the long antennae (Fig. 2.33A).

The thorax is slightly humped and covered dorsally and laterally with coloured scales, the pattern and distribution of which are important taxonomically. The three thoracic segments are fused but the mesothorax predominates and, dorsally, forms a large plate, the scutum. The mesothoracic wings are always fully developed and bear characteristic venation highlighted by longer scales. The legs are long and slender and each consists of a short coxa and

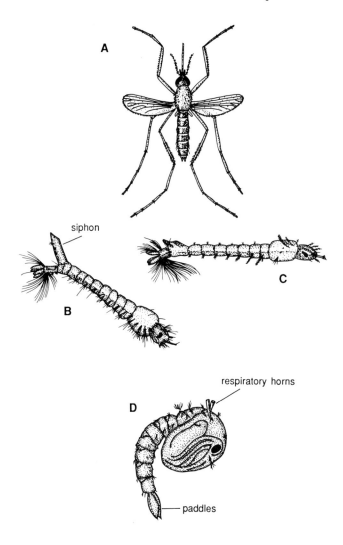

Fig. 2.33   Stages in the life cycle of a mosquito: (A) adult female, dorsal view;
(B) culicine larva, lateral view; (C) anopheline larva, lateral view; (D)
pupa, lateral view .

trochanter, a long femur and tibia, and a very long tarsus consisting of five
segments (tarsomeres). The distal (5th) tarsomere ends in two claws and, in some
genera, there may be also two pads of short setae (pulvilli) which aid adhesion to
smooth surfaces.

The abdomen consists of ten segments, the first eight of which are obvious.
Segments 9 and 10 are modified for sexual functions. In the female, segment 10
is produced to form a pair of cerci, whereas, in the male, the 9th and 10th

segments form the external genitalia or hypopygium. The latter consists of a central phallosome surrounding the aedeagus (penis), both lying ventral on the 10th segment and flanked by a pair of basal gonocoxites, each of which articulates with a gonostylus which ends in a claw. The sternum of segment 10 forms a paraproct which supports the anal lobe (Cranston *et al.*, 1987). The relative proportions and shapes of the various components of the hypopygium are important taxonomic aids.

Mosquito larvae are all aquatic and take in atmospheric oxygen at the water surface through spiracles borne on the dorsal side of abdominal segment 8. These spiracles may be plain (as in the Anophelinae) or located at the end of a sclerotized tube, known as the siphon (as in the Culicinae; Fig. 2.33B). Larvae of the genus *Mansonia* extract air from the tissues of submerged macrophytes and are equipped with a siphon adapted for piercing through plant stems. Other distinguishing features of larval mosquitoes are a well-developed, non-retractable head, a bulbous thorax, posterior gill-like anal papillae, and absence of legs. The head and body are covered with numerous tufts of long setae and the nature and distribution of these (chaetotaxy) are used in species identification, as too are features of the ventro-lateral rows of teeth (pecten) on the siphon.

Mosquitoes are typical holometabolous insects, passing through four life cycle stages: egg; four larval instars; pupa; and adult. The adults tend to hide in dimly-lit refuges with high humidity, such as the secluded corners around buildings or amongst low vegetation, until they are ready to feed. Feeding typically commences at dusk but may occur throughout the day in shaded areas. In the case of the females of most species, this involves a blood meal, usually necessary for the development of the eggs. There is considerable host specificity and not all species attack humans. Many specifically suck the blood of birds, or amphibians, or reptiles, or even fishes (e.g. the amphibious mudskippers) as well as a host of mammals (Colless and McAlpine, 1970). Mating generally occurs on the wing. In many species, a female that has successfully laid one batch of eggs, may take further blood meals and may mature one or more subsequent batches of eggs. The females of most species show at least some degree of preference in oviposition sites which include both permanent and temporary water bodies, and salt water (Table 2.6), but especially those rich in organic matter.

The eggs are generally small (less than 1 mm), dark and elongate-oval in shape in the Culicinae but boat-shaped with lateral floats in the Anophelinae. Some are laid together in floating rafts whereas others are deposited singly on the water surface or on solid objects just above the water line. Eggs laid on the surface of water typically hatch within a few days but others may be laid in depressions on the ground and these overwinter and hatch the following spring when the depressions fill with rain. Such is the resilience of the eggs that they can often withstand years of drought before hatching *en masse* when suitable conditions return. In fact, it is now thought that the eggs of many temporary water mosquitoes require a certain amount of drying, and sometimes exposure to cold, before they can hatch.

First instar larvae break out of their eggshells using a structure on the dorsal surface of their heads. Larvae feed on algae and detritus in both the water column and on the bottom, and they gather these using an elaborate pair of brushes on either side of the labrum. Each brush consists of a dense mass of long, flexible, closely-packed setae that is repeatedly and rapidly extended and reflexed under the control of a pair of large muscles in the head. This action creates minute currents in the water that bring further particles within reach. The food is combed off each labral brush by a row of long stiff setae located on the outer edge of the adjacent mandible which flexes (and thus opposes) every time the labral brush extends. The particles are, in turn, combed from the mandibles by another brush (the epipharyngeal apparatus) situated in the centre of the ventral surface of the labrum. This transfer is aided by the maxillae which work below. Ultimately, the food particles are scraped from the maxillae and epipharynx and pushed into the pharynx by yet another stiff brush on the inner surface of the mandibles, as the latter close again (Wood, 1979). A few species of mosquito are predators.

The 4th larval instar moults to become an active, but non-feeding, pupa. This stage is of short duration (usually 3 or 4 days) and is easily recognizable by its distinctive "comma" shape (Fig. 2.33D) in which the bulk of the body is formed from the combined head and thorax (cephalothorax). On the dorsal side of the cephalothorax is a pair of short respiratory horns and these break through the water surface each time the pupa rises to breathe. Having an active pupal stage makes the mosquitoes unusual amongst dipterans. The abdomen consists of nine segments, the last one being small but supporting a pair of paddles. It is the latter that provide the motive force enabling the pupa to dive rapidly if danger threatens. Both the cephalothorax and abdomen have patches of setae useful in identification. Emergence of the adult occurs through a dorsal longitudinal split in the cephalothorax. Once free of the pupal skin, the adult rests for about an hour either on the water surface or on riparian vegetation. During this time the wings unfold and harden.

Many temperate and higher latitude species have only one generation per year, but some others have as many as time, temperature, food resources and rainfall permit. In the tropics and subtropics, most species are multivoltine with life cycles as short as two weeks. Temperate and circumpolar species tend to overwinter as eggs, although a few of the more southerly ones may hibernate as adult females. Species rarely overwinter as larvae, but one Canadian species does so frozen solid in ice. Where there is just surface ice, the larvae become inactive and remain in debris on the bottom where they respire cutaneously. Diapause is seen in some tropical species, too, particularly those whose larval habitats are seasonal in occurrence, for example floodplain pools or monsoonal rainpools. Here, it is the egg stage that persists.

Simuliidae

The Simuliidae, or black flies, comprise a cosmopolitan family of biting flies of great importance in many parts of the world as bloodsuckers and vectors of certain parasitic organisms (e.g. filarial worms; see Chapter 11). The larvae and pupae are aquatic but are confined to running waters where they attach themselves to firm, usually smooth, substrates. Different species often exhibit preferences for certain current regimes and/or substrate types, and the outlets of ponds and lakes are particularly favoured and productive habitats.

The oldest known fossil black fly is a pupa found in rocks of Middle Jurassic age (dated approximately 165 to 170 million years B.P.) from the U.S.S.R, near the Mongolian border (Crosskey, 1990).

Contemporary zoogeographical studies show that black flies occur on all the major land masses (apart from Antarctica) as well as on many archipelagos and isolated islands where there is sufficient running water to sustain the immature stages. Representatives of the estimated 1,554 known species are known from as far north as Bear Island in the Arctic Ocean (74°30'N) to as far south as the subantarctic islands of Crozet and Campbell (46°27'S and 52°30'S, respectively). However, at high latitudes, diversity is low as is the number of blood-feeding species.

Speciation in simuliids is not always detectable solely on the basis of morphology. Modern black fly taxonomy uses cytological studies, which evaluate chromosome structure in the larval salivary gland, and enzyme electrophoresis of tissues to establish differences known previously only from ecological or behavioural observations of various adult "populations". Simuliids exhibit many such "species complexes" whose components are biologically distinct but morphologically inseparable (Crosskey, 1981a).

The distribution of species across the globe is summarized in Table 2.7. This indicates that the Palaearctic Region is the richest in terms of the number of species, but this may be somewhat biased in that it is perhaps the best studied region, taxonomically. The Palaearctic together with the Nearctic (collectively termed the Holarctic) contain a high proportion of the more primitive prosimuliine forms as well as a large number of species in the genus *Simulium*. In addition, the genera *Twinnia* and *Gymnopais* occur in these two regions and nowhere else. Some species have genuine Holarctic distributions, for example ranging from Canada to Siberia. *Simulium vittatum* (which may be a species complex) ranges from Mexico to Greenland, Iceland and the Faeroe Islands. The Oriental Region is characterized by very low diversity and complete absence of prosimuliine forms. Australasia, too, supports relatively few species but the genus *Austrosimulium* is endemic. A few primitive prosimuliine species occur in Australia and Tasmania. Black flies are present on many of the Pacific islands but have not been recorded from Hawaii, Tonga or Samoa.

Simuliids are found throughout most of the Afrotropical Region, especially south of the Sahara. The fauna is diverse and contains many endemic subgenera,

Table 2.7   Approximate numbers of species and genera/subgenera* of black
              flies in the world fauna, together with the numbers of species in the
              genus *Simulium* (after Crosskey, 1990).

Zoogeographical region	Total no. of species	Species of *Simulium*	Number of genera/subgenera
Palaearctic	571	410	29
Neotropical	355	269	23
Afrotropical	194	184	15
Nearctic	163	85	28
Oriental	178	178	9
Australasian	120	86	9
World fauna**	1554	1212	113

* The term genera/subgenera is necessary because taxonomists tend to differ in the rank
that they assign to named aggregates of species.
** These totals are less than the sums of the constituent regions because a few species and
genera/subgenera occur in more than one region.

for example *Simulium (Lewisellum)* and *S. (Edwardsellum)*, some of which are
major disease vectors. There is a small, relict group of prosimuliine forms,
chiefly in the south. An unusual feature of the Afrotropical fauna is the common
occurrence of phoretic species whose larvae and pupae live on the surfaces of
freshwater decapod crustaceans and mayfly nymphs.

The Neotropical Region supports a very rich black fly fauna, with over 350
species described to date. These include 86 species of Prosimuliini of which the
majority are endemic (e.g. the large genus *Gigantodax* which ranges from
Colombia to Tierra del Fuego). A number of species of *Simulium* occur in the
West Indies but there are none on the Galapagos or Falkland islands. Few
neotropical species extend into North America, north of Mexico (Crosskey,
1981b).

Adult black flies are readily recognized by their stout bodies, humped thorax
and broad, short wings which have prominent veins, concentrated near the
anterior wing margin, and a large anal lobe (Fig. 2.34A). They are generally dark
brown or black but some may be rust-coloured, grey or even yellow; many
species have paler markings on their legs. The head supports a pair of large
compound eyes, but no ocelli, and the antennae, although short, have between 9
and 11 segments which lack long setae. In blood-sucking species, the
mouthparts of the females consist of a short proboscis together with mandibles
and the laciniae of the maxillae; both of the latter are serrated and are used for
cutting into the host's skin to release internal fluids. In the females of nectar-
feeding species and all males, the mouthparts are suitable only for collecting
exposed fluids. The legs tend to be short and stout and each ends in a pair of

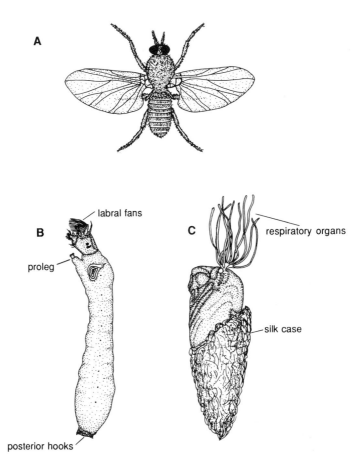

Fig. 2.34   Stages in the life cycle of a black fly: (A) adult female, dorsal view; (B) larva, lateral view; (C) pupa, lateral view (redrawn partly after Peterson, 1981 and 1984).

simple claws. The first tergum of the abdomen is modified to form a collar-like scale. Adults range in length between 1.0 and 5.5 mm.

Black fly larvae are slender and somewhat club-shaped and attach themselves to the stream or river bed by means of several incomplete circles of minute hooks located on their posterior ends. These hooks are inserted amongst the threads of a small silken pad which the larva spins on the rock surface. Such is the strength of this bond that the larvae are able to hold station in the fastest of currents where they may be the only animals present, often occurring in densities of many thousands/ m^2. Displaced larvae pay out silk lifelines that enable them to return rapidly to the substrate and this technique is frequently employed as a means of relocation. Larvae can also move shorter distances by "mountaineering" their way across rock surfaces using silk and a looping behaviour similar to that of some caterpillars. This looping is aided by a second array of hooks borne on the apex of a fleshy proleg on the 1st segment of the thorax (Fig. 2.34B). The larval head capsule is heavily sclerotized, may be patterned, and bears a pair of short antennae and usually two pairs of eyespots. The most obvious feature, in most species, is a pair of large, dorsal labral fans that are used for capturing food particles from the passing water (see Chapter 6). The abdomen has eight indistinct segments and usually an anal sclerite together with anal papillae (thought to have an osmoregulatory function) adjacent to the posterior ring of hooks.

In later instar larvae, developing legs and wings are visible under the cuticle of the thorax. This is because, in the Simuliidae, metamorphosis to the pupal stage occurs within the body of the larva. The pupa is covered by a silk cocoon attached firmly to the substrate (Fig. 2.34C). This cocoon is spun by the pharate pupa while it is still enclosed in the larval cuticle, the latter being shed upon completion of the cocoon (Colbo and Wotton, 1981). From the antero-lateral corners of the pupal thorax arise two respiratory organs that typically consist of many thin plastron-covered filaments which function as gills. The orientation of the cocoon, open end downstream, is such that these filaments stream out into the passing water.

Most temperate and subarctic black fly species are univoltine but those in the tropics and subtropics, which may breed continually, may have four or more generations in a year. Larvae may moult from six to nine times (typically seven) before pupating, with the rate of growth depending largely on water temperature and the amount and quality of food that each larva can catch. Temperature also determines the duration of the pupal stage, although it is commonly 1 week or less. At emergence, the adult pulls itself through a T-shaped slit in the dorsum of the pupal skin and rises to the surface in a bubble of air. Immediately capable of flight, the adult seeks some convenient, riparian vegetation upon which to alight and allow its cuticle to harden. Synchronized, mass emergences frequently occur, particularly in temperate regions, and these may persist for much of the spring and early summer as successive species mature. The sheer numbers of blood-seeking adults make such outbreaks especially dangerous to livestock and humans. For example, in Algonquin Park, Ontario, Canada, maximum attack rates of *Simulium venustum* on humans, in June, have been recorded at 78 flies/6.5 cm^2 of skin/min (landing rate) and 17 flies/6.5 cm^2/min (biting rate).

Host location by the females follows mating and is known to consist of a series of behavioural steps (Table 2.8). The females of some species have specific habitat preferences which can usually be linked with a specific host upon which they feed. For example, *Simulium anatinum* remains close to bodies of water, seldom flying higher than 2 m, and feeds on waterfowl. Other species fly higher and disperse farther afield, perhaps to open pasture or woodland, and may be attracted to either large mammals or small mammals and smaller birds. The

Table 2.8   The behavioural steps by which female blackflies locate a vertebrate host and obtain a blood meal (after Davies, 1978).

- initial releasing stimulus, possibly as a result of mating
- activation flight through intrinsic factors (e.g. hunger) or extrinsic factors (e.g. changes in wind speed, atmospheric pressure or light)
- random flying
- far-distance orientation (by olfaction, especially to the body odour [amino acids &/or oils] or CO_2 emissions of the host; female flies upwind as long as flight is reinforced by the scent)
- middle-distance orientation (olfaction & vision); bright colours contrasting with background and/or moving objects are preferred
- near orientation (mainly by vision - dark colours preferred)
- landing (mainly vision, again dark colours preferred)
- crawling and probing (olfactory & gustation)
- piercing and feeding (gustation)
- withdrawal and leaving the host
- resting and digesting the blood meal

---

shaded side of a host is liable to receive significantly more bites than the side exposed to the sun, and biting by those females that land increases dramatically during periods of rapidly changing (especially falling) atmospheric pressure (Davies, 1978).

Eggs are commonly laid around sunset but also during the day if it is overcast. Oviposition technique varies among species and ranges from release of one or two eggs each time a flying female's abdomen taps the water surface to strings of eggs attached to submerged or floating substrates. The eggs are small (0.2 to 0.46 mm long) and generally oval, and darken with age. Given favourable environmental conditions, they may hatch in from 4 to 30 days, depending on water temperature, but may undergo a lengthy diapause if adverse conditions prevail.

Chironomidae

The Chironomidae, commonly known as non-biting midges, is a large, cosmopolitan family of nematocerans whose adults are small and delicate and superficially resemble mosquitoes. Chironomids have been treated somewhat cursorily in most ecological studies of fresh water, despite the fact that they are undoubtedly the most ubiquitous and usually the most abundant insect group in all types of fresh water. This is perhaps largely due to the earlier difficulties in identifying the various and diverse forms, a situation which is now changing with the recent publication of some excellent keys (e.g. Wiederholm, 1983).

Like the black flies, larval chironomids often reach population densities of several thousand/ m², but such levels can be attained in a much wider range of habitat types than is the case for black flies.

The distribution of chironomids extends to both the northern and southern limits of land, and they are the dominant group in the Arctic. As well as occurring in all the "usual" types of freshwater habitat (streams, rivers, lakes and ponds), many are terrestrial or semi-terrestrial. Others live in pitcherplants, leaf axils or tree holes, and some are marine, living in tidepools or even on tropical coral-heads to a depth of 30m (Bretschko, 1982). Two species are known from Antarctica and these represent the southernmost, free-living, holometabolous insects.

The earliest known chironomids are fossils of the subfamily Podonominae from the Lower Cretaceous of Lebanon, some 130 million years B.P. Currently, the family is partitioned into ten subfamilies and at least 16 tribes; the general habitats of most of these are summarized in Table 2.9. Most species belong to three main subfamilies: the Tanypodinae, Chironominae and Orthocladiinae. The first two tend to be warm adapted and thus their diversities increase towards the lower latitudes, whereas the latter, together with the Diamesinae, Podonominae and Prodiamesinae, tend to be more cold adapted and thus are predominant towards the poles. This does not preclude certain species from occurring outside what might be thought of as the normal range for their subfamily. There are, for example, species in the Orthocladiinae and Podonominae that live in the tropics, although they are mostly confined to cool mountain streams, and a number of species within the Chironominae and Tanypodinae that live in very cold climates and even in cold springs.

Adult chironomids are minute (e.g. wing length 0.8 mm in *Orthosmittia reyei*) to medium-sized (wing length 7.5 mm in *Chironomus alternans*) insects. The antennae are often longer than the head and frequently possess long setae, especially in males (Fig. 2.35A). The mouthparts are generally reduced, as few adults live for more than a few days. The legs are long, especially those on the prothorax, and the wings are slender and lack scales. Apterous forms occur in certain habitats. Most adults are black or brown but many are green, or more rarely red or yellow. Adults often emerge, simultaneously, in huge numbers, and proceed to form vast mating clouds. They are especially attracted to lights. Females typically produce only a single batch of eggs and these may be broadcast at the water surface or, more commonly, deposited in a gelatinous mass, often on emergent vegetation.

The larvae are long (2 to 30 mm, depending on species) and slender and often assume a slightly curved posture, particularly when preserved. The body bears two pairs of fleshy prolegs, one pair on the ventral side of the prothorax and the other at the end of the abdomen (Fig. 2.35C). There is a pair of tufted papillae on the dorsal side of the anal segment adjacent to which may be two pairs of tubules and, on the preanal segment, a pair of fleshy tubercles (procerci). The head capsule is complete and heavily sclerotized. Sense organs include the

Table 2.9   The major subdivisions of the Chironomidae together with the
             typical habitats in which they are found (based on Coffman and
             Ferrington, 1984).

Subfamily	Tribe	Habitat
Tanypodinae	Coelotanypodini	littoral zone of ponds & lakes (lentic)
	Macropelopiini	streams & rivers (lotic); some lentic littoral & profundal
	Natarsiini	fast-flowing waters
	Pentaneurini	fast-flowing waters; lentic littoral; a few hygropetric
	Tanypodini	lentic littoral
Podonominae	Boreochlini	fast-flowing waters; lentic littoral; esp. cold waters
	Podonomini	fast-flowing, cold waters
Telmatogetoninae		saltmarshes & tidepools, estuaries
Buchonomiinae		unknown, but probably in rivers in Oriental & Palaearctic regions
Diamesinae	Boreoheptagyini	cold, fast streams
	Diamesini	fast-flowing, cold waters; springs
	Protanypini	profundal zone of lakes
Prodiamesinae		fast-flowing waters, often in detritus
Orthocladiinae	Clunionini	marine, rocky shores
	Corynoneurini	lotic fast & slow water; lentic littoral
	Metriocnemini	wide range of lentic & lotic habitats, including springs, pitcherplants, dung, interstitial, marine intertidal & semi-terrestrial
	Orthocladiini	wide range of lentic & lotic habitats, including marine intertidal,
Chironominae	Chironomini	lentic, littoral/profundal; slow lotic; especially on sandy substrates & associated with aquatic macrophytes
	Tanytarsini	lotic fast & slow water; lentic littoral; occasionally in brackish water
Chilenomyiinae		unknown; restricted to Chile
Aphroteniinae		lentic & lotic in S. Hemisphere; esp. in sandy areas overlain with FPOM; also swift mountain streams

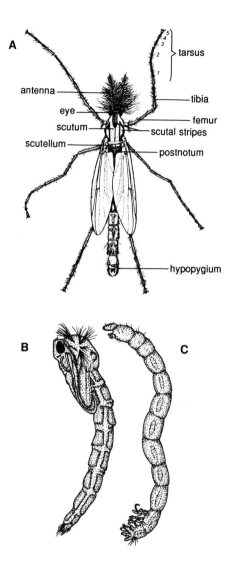

Fig. 2.35   Stages in the life cycle of a chironomid: (A) adult male, dorsal view;
              (B) pupa, lateral view; (C) larva, lateral view (redrawn partly after
              Pinder, 1978).

eyespots, antennae (which are retractile in the Tanypodinae), and various setae,
scales and lamellae.

    The larvae of most species are small particle feeders, typically ingesting
unicellular algae, FPOM, and small animals which they graze off rock surfaces,
gather from minute backwaters, or filter from the passing water. Larvae of other

species may be predators (e.g. some of the Tanypodinae and larger Chironomini), plant feeders, or even parasites. Apart from the latter, however, few chironomids appear to be obligate feeders on a single food type and most tend to be opportunists. Despite these convergences in diet, the larval mouthparts are quite complex and extremely variable among chironomid genera. These structures are situated on the antero-ventral aspect of the head and include mandibles, premandibles, maxillae, a prementohypopharyngeal complex, the mentum, and the pecten epi-pharyngis. The primordial larval habitat is thought to have been one rich in diatoms and this may be why these constitute a conspicuous component in present day larval guts (Brundin, 1966). However, some studies have shown a distinct avoidance of abundant algae in favour of detritus (Ward and Williams, 1986); it is clear that, as yet, we understand little of the feeding intricacies of this important group of dipterans.

Chironomid larvae often show a distinct preference for certain types of substrate and this creates distinct microdistribution patterns. For example, the Diamesinae and Orthocladiinae, in general, prefer hard rock surfaces and gravel, whereas finer sediments of sand and silt tend to be dominated by the Tanypodinae and the Chironominae. These relationships are not always immediately apparent as they may be masked by the interactive effects of other influential environmental factors together with the physiological state of the animal. The larvae of many Chironominae, and some Tanypodinae, possess haemoglobin in their blood and this enables them to exist in waters that have very little oxygen. Respiration is typically cutaneous.

Larvae pass through four instars before pupating. The pupae of many species live inside cylindrical tubes whereas those of others are free-swimming. The external morphology of the pupa is clearly visible since the adult develops inside a modified larval skin (Fig. 2.35B). There are three basic body parts: the head, composed primarily of eye, antennae and mouthpart sheaths; the thorax, bearing the leg, wing and halter sheaths, and the taxonomically important respiratory horns; and the abdomen, which consists of eight similar segments plus one or more others modified into anal lobes and genital sheaths. The anal lobes often have fringes of setae that enhance swimming ability (Coffman and Ferrington, 1984).

In temperate regions, many chironomid species are uni- or bivoltine, but up to four generations in a year are not uncommon. Multivoltine species, particularly in the Orthocladiinae, frequently exhibit continuous recruitment during much of the year. Species living in the cold, profundal zones of deep lakes may take more than one year to complete their life cycles, and circumpolar species require at least two years and, occasionally, as many as seven (Butler, 1982; Pinder, 1986). In such long-lived species, the larvae pass the winter dormant and tightly folded in special cocoons. In highly transient, tropical habitats such as rainpools, life cycles may be as short as a few weeks.

The immature stages of chironomids are important elements in the food webs of aquatic systems. Larvae assist in the decomposition and recycling of nutrients and, together with adults, are fundamental to the diets of many other aquatic invertebrates, fishes and birds. A largely unappreciated fact is the contribution that emerging adults make to the diets of riparian species of spider. Species or species groups of chironomid larvae are often characteristic of various types of standing or running water, and may be associated with different degrees of depth, levels of dissolved oxygen and/or nutrients, and different temperature regimes. Detailed knowledge of these environmental requirements has been used extensively in Europe to classify lakes and rivers. The predictable responses of populations of certain species to different levels of a variety of pollutants has resulted in the use of larval chironomids as biological indicators of water quality (Oliver, 1979). Additionally, chironomid larvae are essential components in the efficient biological processing that takes place in the oxidation ponds of sewage treatment plants.

2.11   Lepidoptera

The butterflies and moths make up one of the largest orders of insects, with well in excess of 100,000 species in the world fauna. Although traditionally thought of as a terrestrial group, there are never-the-less some aquatic and semi-aquatic species, perhaps as many as 600 in North America, for example. It is typically the immature stages that are aquatic, with the adult female only coming into contact with water when she emerges and oviposits. An exception to this is the nearctic *Acentria niveus*, whose female is brachypterous (short-winged) and fully aquatic. There are, however, a greater number of semi-aquatic species whose larvae seldom actually become submerged but are miners and borers in the tissues of emergent macrophytes. Hydrophilic lepidopterans are found in at least five families (Table 2.10) but most species belong to the Pyralidae (tribe Argyractini).

The adults are typically moth-like, with overlapping, flat scales on the body, legs and both pairs of wings. In some species, the metathoracic legs are equipped with an oar-like fringe of hairs used to increase the efficiency of swimming (e.g. in the Nymphulinae). The mouthparts include a tube-like siphon formed from the modified galeae of the maxillae. The antennae are usually well-developed but are shorter than the body.

The larvae are caterpillar-like with a long, cyclindrical body and a distinct head. The head has short antennae and two lateral rings of stemmata, or simple eyes. The mouthparts are adapted for chewing, with opposable mandibles, and have a silk-producing structure, the spinneret. There are three pairs of articulated legs on the thorax and five pairs of fleshy prolegs on the abdomen (segments 3, 4, 5, 6, and 10), each of which usually ends in a series of minute hooks (crochets). Filamentous gills are present on many of the thoracic and abdominal segments in some species.

The pupa is of the typical lepidopteran, obtect, type (i.e. having appendages

Table 2.10   The major families of aquatic and semi-aquatic Lepidoptera together with the typical habitats in which they are found (after Lange, 1984).

Family	Habitat
Pyralidae	lakes & ponds, on aquatic macrophytes (submerged & floating); emergent vegetation; rapid streams & rivers, on cobbles or bedrock; bog pools - includes both free-living forms & leaf miners/stem borers
Nepticulidae	lakes & ponds, on aquatic macrophytes especially in emergent zone - primarily leaf miners & stem borers
Cosmopterigidae	lakes & ponds, on aquatic macrophytes especially in emergent zone; often in mosses, lichens & algae at margins - primarily stem borers & leaf miners, with some free-living forms
Noctuidae	lakes & ponds, on floating and emergent macrophytes - generally leaf miners/stem borers, with some free-living forms
Tortricidae	lakes & ponds, on floating & emergent macrophytes - leaf miners & stem borers

attached to the body) and may or may not develop inside a cocoon. Gills and respiratory horns are absent but, in truly aquatic forms, the external spiracular openings on abdominal segments 3 and 4 are much enlarged. A hook-like cremaster on the terminal abdominal segment is used to anchor the pupa (Lange, 1984).

Larvae pass through between five and seven instars, depending on species. Silk is spun for a variety of purposes including the building of protective retreats and flat sheets (many square centimetres in area) under which larvae graze algae in competition-free space. Others may construct flat or tubular cases from plant material in much the same style as the larvae of their sister order, the Trichoptera.

Transition from the pupa to the adult takes less than a month. The adult swims to the surface after it has freed itself from the pupal skin and cocoon, by using an exit slit made by the larva before pupation. Adults live from about 24 hours to two months, according to species, and the longer-lived ones feed on plant juices. Males are thought to be attracted to females by pheromones. The eggs of stream species are generally deposited on submerged rock surfaces, whereas those of pond species are attached to the undersides of floating leaves. They hatch within two weeks. Most hydrophilic lepidopterans are uni- or bivoltine, and, typically, it is the larval stage that overwinters (McCafferty, 1981).

2.12   *Trichoptera*

The Trichoptera, or caddisflies, are an order of holometabolous neopterans whose terrestrial adults resemble small moths, and whose aquatic larvae spin silk. The larvae and pupae live in most types of waterbody, including both cold and warm (up to 34°C) springs, temporary waters and (rarely) the seashore, and there are even a few terrestrial species. The 10,000 or so described species are distributed over much of the globe (except Antarctica) and, while some families are worldwide in occurrence, others are restricted to either the northern or southern hemisphere (Table 2.11).

Systematically, the Trichoptera are considered, by most authorities, to be the sister-group of the Lepidoptera, this being based primarily on similarities between the adults. It is thought that these two orders must have arisen from a common ancestor in which the adult was much like a trichopteran but the larva much like a lepidopteran. In the line leading to the caddisflies, the larvae became highly modified for an aquatic existence with, initially, few changes to the adult. In the line leading to the Lepidoptera, the adults lost more ancestral characters but the larvae remained relatively unchanged, except perhaps for the evolution of prolegs (Ross, 1967). Precise interpretation of the origins of the Trichoptera are not, as yet, possible because of the relative sparseness and nature of their early fossil record. However, biogeographical evidence suggests that typical trichopterans had evolved by the Jurassic. The evidence for earlier caddisflies seems tenuous, except for ancestral forms of austral families such as the Plectrotarsidae, Tasimiidae, Philorheithridae and Helicophidae, known to have occurred in the Upper Permian and Triassic of Australia (Riek, 1970c).

Existing caddisfly families have been placed, by some authors (e.g. Ross, 1967), into two large and fairly homogeneous groups (the superfamilies Hydropsychoidea and Limnephiloidea), with a smaller, more primitive group (Rhyacophiloidea) remaining. The Hydropsychoidea contains families whose larvae spin silk to form nets for food capture and to form non-portable retreats. The Limnephiloidea contains families whose larvae use their silk, in combination with pieces of plant material and/or mineral particles, to construct cylindrical or flattened cases which they carry about with them. The Rhyacophiloidea consists of the free-living families Rhyacophilidae and Hydrobiosidae, the saddle-case-making Glossosomatidae, and the purse-case-making Hydroptilidae; primitive members of these families show sufficient similarities to suggest a common relationship. Other authors (Weaver and Morse, 1986) have created further superfamilies to contain subsets of families

Table 2.11   Distribution of the families of Trichoptera, together with typical larval habitats (after Wiggins, 1982).

Family	Distribution and Habitat
Rhyacophiloidea	
Rhyacophilidae	All major zoogeographical zones* except Australian, Neotropical & Afrotropical; in cool running waters
Hydrobiosidae	Mainly confined to Australian & Neotropical regions; running waters
Glossosomatidae	Cosmopolitan; running waters
Hydroptilidae	Cosmopolitan; running & standing waters
Hydropsychoidea	(Net-spinning caddisflies)
Philopotamidae	Cosmopolitan; running waters
Stenopsychidae	Oriental, Australian, Afrotrop. & Asian Palaearctic regions; fast-flowing rivers
Hydropsychidae	Cosmopolitan; running waters & wave-swept shores of lakes
Polycentropodidae	Cosmopolitan; running & standing waters
Dipseudopsidae	Afrotropical & Oriental regions (1 genus in the Nearctic); lakes & slow-flowing waters
Ecnomidae	All regions except the Nearctic; lakes, ponds & slow-flowing waters
Psychomyiidae	All regions except Australian & Neotropical; cool running waters, some in lakes
Xiphocentronidae	Afrotropical, Oriental, Neotropical & extreme south of Nearctic (Mexico, Texas); small streams
Limnephiloidea	(Tube-case-building caddisflies)
Phryganeidae	Confined to Nearctic, Palaearctic & Oriental regions; mainly lakes & marshes, slow-flowing streams, temporary pools
Phryganopsychidae	Himalayas & China to Japan, Korea & adjacent Siberia
Brachycentridae	Confined to Nearctic, Palaearctic & Oriental regions; running waters (cool streams to large rivers, depending on genus)
Limnocentropodidae	Oriental Region & Japan; rapid streams
Chathamiidae	Restricted to Australian Region; embryogenesis takes place in coelom of starfishes, larvae become free living in coastal waters (marine)
Tasimiidae	Australian & Neotropical regions only; clear mountain streams

* caddisflies have not been found in Antarctica

Table 2.11 (contd)

Limnephilidae	Mostly in cooler parts of Nearctic & Palaearctic, some in adjacent Oriental, some in temperate Neotropics (Dicosmoecinae), a few in Australian & Afrotropic regions; most types of running & standing waters, including temporary & brackish waters
Goeridae	All regions except Australian & Neotropical; running waters, especially spring seeps
Thremmatidae	Confined to southern Europe; cold mountain streams
Uenoidae	Western North America, Japan & Himalayas; rapid streams
Lepidostomatidae	All regions except Australian, but in Neotropical only in montane Central America, not South America; mainly slow, cool running waters; littoral of lakes
Oeconesidae	Confined to Australian Region; forested streams, in plant debris
Kokiriidae	Confined to Australian & Neotropical regions; sandy substrates in streams & lakes
Plectrotarsidae	Confined to Australia; larvae unknown
Beraeidae	Eastern Nearctic & European Palaearctic; cool streams, springs & organic muck in spring seeps
Sericostomatidae	All regions except Australian; flowing & standing waters
Conoesucidae	Confined to Australian Region; streams
Antipodoeciidae	Confined to Australia, larvae unknown
Calocidae	Confined to Australia & New Zealand; small forested streams
Helicophidae	Confined to Australia & N.Z.; clear, fast streams
Molannidae	In Holarctic & Oriental regions; sandy substrates in standing or slow-flowing waters
Odontoceridae	All regions, except Afrotropic; running waters
Atriplectididae	Known from Australia & Seychelles; bottom sediments in lakes & slow rivers
Philorheithridae	Confined to Australian & Neotropical regions; cool, rocky streams
Helicopsychidae	Cosmopolitan, but with greater diversity in tropics; cool & warm running waters; littoral zone of lakes
Calamoceratidae	All regions (but sparse), but mainly subtropical; slow streams, coastal lakes, swamps; phytotelmata
Leptoceridae	All regions (abundant); mainly standing waters; slower sections of rivers

thought to show closer relationships, for example the Philopotamoidea containing the Philopotamidae and Stenopsychidae, and the Phryganoidea containing the Phryganeidae and Phryganopsychidae.

Although caddisfly larvae are found in a wide range of aquatic habitats, the greatest diversity occurs in cool running waters. Furthermore, in families represented in both lotic and lentic habitats, the genera exhibiting more ancestral characters tend to be found in cool streams whereas those showing more derived characters tend to occur in warm, lentic waters. These two findings point to cool, running waters as the most likely primordial caddisfly habitat, the one in which the ancestors of the Trichoptera first became aquatic and the one in which differentiation into the basic groups (superfamilies) took place (Ross, 1956; Wiggins, 1977). Lately, this view has been challenged on the basis of a perceived closer relationship between the Rhyacophiloidea and the Hydropsychoidea than between the Rhyacophiloidea and the Limnephiloidea. This latter phylogeny has prompted Weaver and Morse (1986) to propose that the ancestral caddisfly larva was more probably a tube-dwelling detritivore inhabiting pockets of detritus near the shores of lentic or slow-flowing habitats. This ancestor is then thought to have evolved into a tube-case-making detritivore and scraper-feeder leading, eventually, to the Limnephiloidea, and into a retreat-making collector-gatherer leading to the Rhyacophiloidea and Hydropsychoidea. Yet another proposal, placing the free-living Trichoptera as a sister-group of the combined retreat-making/net spinning and tube-case-making caddisflies, based on the osmoregulatory physiology of the eggs, larvae and pupae (Wiggins and Wichard, 1989; Wichard, 1991), indicates the heated debate currently surrounding the evolution of this particular group of aquatic insects.

Adult caddisflies differ from moths in a number of aspects foremost amongst which are patterns of wing venation and structure of the mouthparts. Adult caddisflies are small (1.5 mm body length) to moderate-sized (4.0 cm), tend to be drab in colour, and are mostly active at night, especially around lights. During the day they hide in riparian vegetation. The compound eyes are well developed and there may be up to three ocelli. The mouthparts are weak and are capable only of ingesting liquids. The maxillary palps are typically 5-segmented and may be highly modified (lengthened, reduced, or multiarticulated) in the males of some species. Similar modifications may be seen in the 3-segmented labial palps. Eversible scent glands (pilifers) are sometimes present between the bases of the antennae (Riek, 1970c).

All three thoracic segments are distinct. The legs are long and slender and the tibiae have varying numbers of apical and preapical spurs which are a useful aid in the taxonomy of this group. The wings are covered with setae, from which the name of the order is derived (from the Greek "trichos" = hair, and "pteron" = wing), although there may be regions of flattened setae, or scales, along some of the veins. Venation is simple and is characterized by a forked anterior cubitus vein (CuA). When at rest, the wings are held in a tent-like fashion over the body (Fig. 2.36A) and, when flying, the fore and hind wings of strong flying species

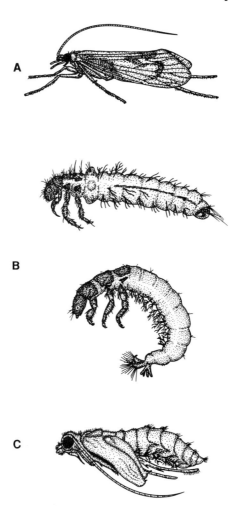

Fig. 2.36   Stages in the life cycle of a caddisfly: (A) an adult limnephilid; (B) case-building larva (top), free-living larva (bottom); (C) pupa; all in lateral view.

tend to be physically coupled together. The latter is achieved through the temporary linking of large setae (macrotrichia) on the anterior border of the hind wing with either similar setae or a longitudinal ridge on the posterior of the fore wing. Strong fliers also tend to have narrow fore wings but expanded hind wings, and long antennae.

Segmentation of the abdomen is also distinct. The male has nine clearly defined segments, segment 10 being incorporated into the external genitalia (terminalia). The female usually has 10 defined segments, often with a pair of small cerci on segment 10 or the terminal segments drawn out to form an ovipositor. The female's genital opening lies between segments 8 and 9 (Riek, 1970c).

Caddisfly larvae are quite similar to lepidopteran caterpillars but have only a single pair of abdominal prolegs which are located on the terminal segment and are each equipped with an apical anal claw (Fig. 2.36B). On the well-sclerotized head capsule, are a pair of eyes made up of no more than seven closely-grouped stemmata, and two very small antennae. The mandibles typically have distinct tooth-like points but, in species that feed by scraping, there is a broad flat edge for this purpose. Silk is produced by a gland that opens onto the tip of the labium. The bulk of the head capsule is composed of three plates, separated by Y-shaped suture lines: a single, dorsal, somewhat flattened frontoclypeus, and two lateral, deeply-curved parietals which meet posteriorly along the coronal suture; ventrally the parietals are typically separated by an apotome.

The larval thorax is well developed, with at least the pronotum covered dorsally by a pair of sclerotized plates. In some families, there is a membranous horn on the prosternum. The mesonotum may have sclerotized plates or small sclerites or may be naked, and the metanotum is typically naked in most species. The legs are well-developed and, in many species, the middle and hind pairs are longer than the forelegs. The legs are not only used for locomotion but, in case-builders, are used to hold, measure and position pieces of plant and mineral material during case construction and, in net-spinners, aid in the positioning of the threads of silk as they emerge from the mouth. In some larvae, notably those in the Brachycentridae, the meso- and metathoracic legs are very long and have fringes of fine setae used to filter food particles out of the water. In others, especially the Leptoceridae, the metathoracic legs are long and equipped with setal fringes that enable the larvae, in their cases, to swim.

In portable-case-building larvae, the first segment of the abdomen usually has three expandable, membranous humps, one on the dorsum and two laterally. These are thought to position the larva centrally in its case so as to enable oxygen-bearing water to flow effectively, and equally, past the filamentous abdominal gills. There are nine abdominal segments all of which are membranous except for segment 9 which, in some families, has a dorso-median sclerite. The anal prolegs are typically large and mobile in free-living species, in which they aid in clinging to substrate surfaces, but are small and immobile in case-building species in which their primary function seems to be to grip the silken lining of the case. The tracheal gills are extensions of the body wall that contain fine tracheoles in the epithelium. Not all species possess gills and those that do not presumably respire across the general cuticle surface. Gills may be single or branched and are typically arranged in dorsal, ventral and lateral pairs on each side of a segment (Wiggins, 1982, 1984).

The trichopteran pupa is exarate and, in case-building species, develops within the larval case after it has been secured to the substrate and sealed with silk. In free-living species, the final instar larva builds a special pupal case, generally made from silk and mineral particles, again this is firmly attached to the substrate. In most species, the pupa is equipped with heavily sclerotized mandibles which enable it to cut an opening in the case so that it may escape and swim to the water surface. This ascent is aided by dense fringes of swimming hairs on the mesothoracic tarsi (Fig. 2.36C). In those species lacking or having reduced mandibles (e.g. some of the Phryganeidae), the larval case is not sealed. Pupae have hooked, dorsal sclerites which engage the silk lining of the case thus enabling the pupa to move within and also exit from the case (Wiggins, 1982).

Most caddisflies mate at dusk or in the early evening and copulation begins on the wing but continues after the pair has settled. The eggs are small (0.1 to 0.5 mm diameter) and more or less spherical, and are laid as strings or in a mass. The females of many free-living species crawl beneath the water to attach their eggs directly onto the substrate. Others may deposit their eggs on objects above the water or on the water surface. The adults of some spring-emerging species living in temporary waters exhibit delayed sexual maturity which postpones egg laying until the autumn when dried up basins begin to refill. In the latter case, there is an additional safeguard in that the larvae can actually hatch within the gelatinous egg mass and thus, protected from desiccation, survive until the water returns, even for as long as several months. The eggs of the New Zealand caddisfly *Philaniscus plebeius* are deposited in the coelomic cavity of the starfish *Patiriella regularis* and complete their development there until the young larvae emerge to become free-living in the intertidal zone (Winterbourn and Anderson, 1980).

Most caddisflies have five larval instars and the diversity of larval form and case structure is quite remarkable. The latter varies from the precise, regularly-structured, cylindrical, leaf cases of the Phryganeidae to the more irregular, "log-cabin" type cases of many of the Limnephilidae; the sleek, horn-shaped, uniform sand grain cases of the Uenoidae; the fine-grained "purses" of the Hydroptilidae; the flattened sand grain cases, with dorsal hood, of the Molannidae; and the geometric complexity of the spiral, sand grain cases of the Helicopsychidae (Fig. 2.37). Trichopteran diversity has been interpreted as an expression of the ecological opportunities made possible by the ability to spin silk (Mackay and Wiggins, 1979). This has enabled the larvae to produce a variety of fixed and portable shelters, together with some specialized feeding techniques, especially filter-feeding. These, in turn, have allowed the exploitation of a wide range of aquatic resources. Construction of a portable case may aid larval respiration (see Chapter 6) and, consequently, may have been a major factor in allowing caddisflies to spread from their ancestral, cool, running water habitats into standing waters (Wiggins, 1977; Williams *et al.*, 1987).

Caddisfly larvae consume a wide range of foods and different groups show both specialist and generalist diets. Generalists, such as many of the limnephilids, consume detritus, diatoms and other algae which they collect or scrape from substrate surfaces. Other limnephilids, such as *Pycnopsyche*, tend to be more selective and feed on dead riparian leaves which they reduce (shred) to

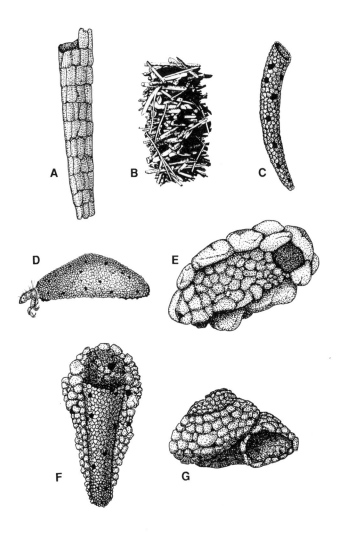

Fig. 2.37   Examples of the diversity in caddisfly larval cases: (A)
Phryganeidae; (B) Limnephilidae; (C) Uenoidae; (D) Hydroptilidae;
(E) Tasimiidae; (F) Molannidae; (G) Helicopsychidae (not to scale).

smaller particles in the process. Rhyacophilids are typically predators,
sometimes capable of engulfing quite large prey. Of the net-spinning families,
the Hydropsychidae filter a wide range of materials (both plant and animal) and
particle sizes from the passing water. Within the hydropsychids, *Arctopsyche* for
example, collects larger particles, whereas *Macrostemum* tends to collect
smaller particles. Most of the Philopotamidae specialize in collecting very fine
particles (typically less than 2 μm in diameter) and their net meshes are

correspondingly very fine. Because of the broad trophic habits of the larvae, caddisflies play a major role in energy transfer at several levels in aquatic ecosystems. In particular, the larvae, pupae and adults form a significant portion of the diets of many fishes, and the immature stages are fed on by many species of waterfowl.

Adult caddisflies have been regarded as pests where they emerge in large numbers close to lights, and the larvae of some leptocerids are reported to damage the young shoots of rice plants in paddy fields. The larvae of a few species are known to eat fish eggs. On the beneficial side, many hydropsychids prey on black fly larvae.

Most temperate species of caddisfly are univoltine, with the bulk of the year passed in the larval stages. As in most other aquatic insects, the rate of development is controlled primarily by the quantity and quality of available food in combination with water temperature and other important environmental factors such as photoperiod. Therefore, as in other groups, there is a tendency for multivoltinism to become more common in the tropics whereas life cycles lasting for two or more years are common at higher latitudes.

A final point worth mentioning about the Trichoptera, in general, and about those that live in small streams in particular, concerns habitat specificity. Many species that are restricted to small streams reflect the ecological characteristics of the surrounding terrestrial community. In such streams, conditions for the larvae are affected by shade in summer or winter, the amount and periodicity of leaf-fall, and the distribution of local precipitation, all three of which are integrated with the type of climax community occupying the general area. As a result, there is, in general, a marked correlation between these terrestrial biomes and the ecological affinities of their respective small-stream caddisfly faunas (Ross, 1963). Such correlations are of enormous importance to palaeoecology, and caddisfly remains are proving to be a powerful tool in the interpretation and reconstruction of past environments (Williams, N.E., 1988; see also Chapter 11).

2.13   Hymenoptera

The Hymenoptera is a large order of insects containing the bees, wasps, ants, sawflies and other similar forms, the vast majority of which are terrestrial. Existing primitive families are known from Triassic fossils and it is believed that the group arose sometime in the Permian. Since then, hymenopterans have diversified dramatically to the estimated 200,000 or more species that make up the present-day world fauna. A few of these have become aquatic.

The order is divided into two suborders, the **Symphyta** (sawflies and horntails), and the **Apocrita** (bees, wasps and ants). There appear to be no aquatic symphytans, but several families within the Apocrita contain species that are, in some way, associated with water. The latter are all wasps and they are all

Table 2.12   Major families of aquatic wasp, together with the hosts that they typically parasitize (based on Hagen, 1984).

Family	Hosts
Apocrita	
Section Parasitica	
Braconidae	Diptera (Ephydridae), Lepidoptera (Noctuidae)
Ichneumonidae	Coleoptera, Lepidoptera, Trichoptera, Diptera
Mymaridae	Hemiptera (esp. Gerridae), Odonata, Coleoptera
Trichogrammatidae	Odonata, Diptera, Hemiptera, Coleoptera, Megalopt.
Eulophidae	Odonata (esp. Lestidae), Coleoptera
Pteromalidae	Diptera (Ephydridae), Coleoptera, Neuroptera
Diapriidae	Diptera (Ephydridae, Sciomyzidae), Coleoptera
Scelionidae	Hemiptera (esp. Gerridae), Lepidoptera, Diptera
Agriotypidae	Trichoptera (ectoparasites)
Eucoilidae	Diptera (Ephydridae)
Section Aculeata	
Pompilidae	Araneae (pisaurid spiders)

parasitic on a variety of aquatic hosts (Table 2.12). Many females actually enter water to lay their eggs directly in the immature stages (egg, larva or pupa) of their host. Others deposit their eggs in the adults of their host species soon after the latter have emerged from the water; these types of wasp are, therefore, strictly speaking, not aquatic.

With few exceptions, aquatic wasps are small but, otherwise, they show little in the way of morphological adaptations to water. The female, as in terrestrial parasitic species, generally has a long ovipositor, and the larvae, as internal parasites, are already adapted to living in a liquid environment. In *Caraphractus cinctus*, a tiny mymarid wasp whose life cycle is perhaps the best known, the adults spend quite some time under water, swimming with their wings and respiring cutaneously. They mate in water, on the underside of the surface film, and the female seeks out and lays her eggs in the eggs of dytiscid beetles. The larvae feed on the host's egg yolk and several generations may occur throughout a summer. In the autumn, larvae enter a diapause in the final instar which carries them over to the next spring when they pupate and emerge (Jackson, 1961a,b). The life cycle of *Agriotypus armatus* (Agriotypidae), in Britain, has been described by Elliott (1982). It is one of three species in the family, the other two occurring in Japan and India. All species are oligophagous ectoparasites that prey on only two families of trichopterans, the Odontoceridae and the Goeridae.

3 KEYS TO THE AQUATIC INSECT ORDERS

It is not possible within the confines of this book to provide detailed keys for the identification of the multitude of species and genera of aquatic insects. Indeed, for those groups in which the taxonomy is little studied or presently in a state of flux, meaningful keys are not available. It would be a formidable and lengthy task even to provide identification guides to the level of family, particularly as many aquatic insect families show high degrees of regional endemism and diversity which may be of less interest to the general reader. For lower level and regional identification of aquatic insects, the reader must therefore seek out more specialized works. In Britain, for example, there is a series of guides to the identification of aquatic insects published by the Freshwater Biological Association; in North America, the book "An Introduction to the Aquatic Insects of North America" edited by Merritt and Cummins (1984) is an excellent guide to the level of genus; in Australia, the C.S.I.R.O. publication "The Insects of Australia" (1970) provides keys to the identification of aquatic insects in Australia and New Zealand, to the family level, and provides references for further study; in the Palaearctic, there is a series of publications on the "Fauna of the U.S.S.R." produced by the Zoological Institute of the Academy of Sciences of the U.S.S.R. published for the Smithsonian Institution and the National Science Foundation, Washington, D.C. by the Israel Programme for Scientific Translation. Most of the world's major museums publish series of taxonomic works of a more regional nature. In addition, there are taxonomic works published in both international and regional entomological journals, a comprehensive list of which is given in Gilbert and Hamilton (1983).

The following keys to the aquatic stages of insects (based on Croft, 1986 and Daly, 1984) have the traditional dichotomous format. From each couplet choose the description that best fits the specimen. This choice will result in either the name of the insect order or the number of another couplet to follow, further down the page. Identifications should be checked against the more detailed descriptions and diagrams in Chapter 2.

- Active insect with legs freely movable; not in a cocoon or sealed, capsule-like case .....*Nymphs, Larvae & Aquatic Adults*..............**Key A**
- Typically an inactive insect in a "mummy-like" state; appendages may be fused to the body or free; sometimes in a cocoon or sealed in a case or puparium ....*Pupae* ....**Key B**

Key A: *Nymphs, Larvae and Aquatic Adults*

1. Insect living inside an open-ended case or tube constructed from
    silk, sand, small stones, or vegetation 2
    Insect without a case or tube, free-living 6

2. Case covers the abdomen & is usually carried around with the
    insect, but sometimes attached to a rock or plant surface 3
    Insect in a long, silk tube (may be covered with silt or debris)
    attached to, or buried in, substrate; or, if made from leaf pieces,
    may be loosely attached to a plant 4

3. Insect with jointed legs ....(Fig. 2.36B) larvae of **Trichoptera**
    Insect without jointed legs ....(Fig. 2.35C) larvae of **Diptera**

4. Insect with a pair of hooks on the end of the abdomen ....
    ....(Fig. 2.36B) larvae of **Trichoptera**
    Such hooks absent 5

5. Insect with 3 pairs of jointed legs & several pairs of soft,
    abdominal prolegs ....larvae of aquatic **Lepidoptera**
    Insect without jointed legs, but with both anterior & posterior
    prolegs ....(Fig. 2.35C) larvae of **Diptera**

6. Insect with or without jointed legs; may or may not have prolegs 7
    Insects with 3 pairs of jointed legs 8

7. Living attached to a plant & piercing the air space with a posterior
    tube, or living inside the air space ....larvae of **Coleoptera**
    Insect without jointed legs; most are free-living but a few live in
    plant air spaces; others have a posterior tube but this is used only
    to reach the water surface ....(Figs 2.31C; 2.34B) larvae of **Diptera**

8. Insect with 2 pairs of fully developed wings (adult). Anterior pair
    frequently form hardened covers for the hind wings 9
    Insect without fully developed wings, although there may be
    wing buds but these do not cover more than half the abdomen 11

9. Fore wings membranous, hind wings forming a fringe of hairs.
    Tiny wasps which are typically found either on the water surface
    or swimming under water (not common) ....aquatic **Hymenoptera**
    Fore wings modified to form covers for the hind wings &
    frequently match the rest of the exoskeleton 10

10. Wing covers (hemielytra) overlap at the hind end; mouthparts
    may or may not be modified to form a piercing "beak"; never
    with segmented palps ....(Fig. 2.19) adult **Hemiptera**
    Wing covers (tegmina) overlap more extensively; mouthparts
    of the chewing-biting type; hind legs very long with enlarged
    femora, suited for jumping; palps segmented; tarsi with 1 to 4
    segments ....(Fig. 2.16) (rare) adult aquatic **Orthoptera**
    Wing covers (elytra) meet along the midline of the back all the
    way down the abdomen; mouthparts never form a piercing beak;
    palps segmented; tarsi typically 5-segmented (except in
    Chrysomeloidea & Curculionoidea in which segment 4 is
    very small & fused to segment 5) ....(Fig. 2.27) adult **Coleoptera**

11. Insect with 1 to 5 "tails" (may appear as hair-like filaments,
    hooks or points) projecting from the posterior end of abdomen 12
    Insect without any such tails, hooks or points (except in some
    species of Collembola in which a bifurcated jumping organ
    arises from near the tip of the abdomen) 22

12. Abdomen terminating in one "tail" 13
    Abdomen ending in from 2 to 5 "tails" 15

13. Insect with gills along sides of abdomen ....
    ....(Fig. 2.22A,B) larvae of **Megaloptera**
    No abdominal gills, although there may be backward
    protruding spines on all segments 14

14. Fore legs modified for grasping prey (raptorial); mouthparts
    form a beak; "tail" used as snorkel for obtaining air at water
    surface ....(Fig. 2.19) **Hemiptera**
    None of the above features; edges of body segments have spines
    that overlay the next segment ....
    ....(Fig. 2.29I) larvae of **Haliplidae (Coleoptera)**

15. Insect with 2 or 4 "tails" 16
    Insect with 3 or 5 "tails" 19

16. Tails short (often limb-like), associated with tiny hooks which
    may be partly hidden by tufts of setae 17
    Tails longer, but never terminating in hooks 18

17. Two hooks, only, each on a fleshy proleg ....
    ....(Fig. 2.36B) larvae of **Trichoptera**

Four hooks (between bases of "tails") ....
..............................(Fig. 2.29B) larvae of **Gyrinidae** (Coleoptera)

18. Tails in the form of segmented filaments which, when intact,
are at least one-third the length of the insect ....
....(Fig. 2.13B) nymphs of **Plecoptera**
Tails never segmented & typically less than one-third the length
of insect; head usually rounded or pointed ....larvae of **Coleoptera**

19. Tails in the form of 3 or 5 sharp points; head angular in
dorsal view; mouthparts form an extensible mask ....
...................................(Fig. 2.8B) nymphs of **Anisoptera** (Odonata)
Three tails, which may be hair-like filaments or flattened, leaf-like
structures (gills); angular head & mask present or not 20

20. Tails flattened & leaf-like; head angular in dorsal view; mouth-
parts form an extensible mask ....
...................................(Fig. 2.8A) nymphs of **Zygoptera** (Odonata)
Tails filamentous; head frequently rounded; no mask 21

21. Gills extend from lateral margins of abdominal segments; may lie
on dorsum of abdomen in some species, or be partly hidden by
covers (opercula) in others ....(Fig. 2.3B) nymphs of **Ephemeroptera**
Abdominal gills may or may not be present; if present, then they
are ventral & not seen from above ....larvae of **Coleoptera**

22. Insects with at least one pair of legs elongated & used
for rowing ....(Fig. 2.19G,L) **Hemiptera**
No such modification 23

23. Mouthparts modified to form a single, piercing beak ....
....(Fig. 2.17E) **Hemiptera**
No such modification 24

24. Mouthparts modified to form long slender tubes; insect typically
found in a sponge or moss ....(Fig. 2.24) larvae of **Neuroptera**
No such modification 25

25. Insect typically less than 1 mm long & has a short tube-like organ
(collophore) on the ventral side of the first abdominal segment;
usually found on the water surface; bifurcated jumping organ
folded under the abdomen in some species ....(Fig. 2.1) **Collembola**
Insect has no collophore or jumping organs ....
....(Fig. 2.29) larvae of **Coleoptera**

Key B: *Pupae*

1.   Pupa exarate (appendages free, distinct & not fused to body) 2
      Pupa obtect (appendages fused to body) or coarctate (appendages
      concealed in a hardened capsule) 7

2.   Abdomen broadly joined to thorax 3
      Abdomen joined to thorax at a constriction ....**Hymenoptera**

3.   One pair of wing pads visible ....(Fig. 2.34C) **Diptera**
      Two pairs of wing pads visible 4

4.   Pads of fore wings thickened; antennae usually with 11
      segments or less ....**Coleoptera**
      Pads of fore wings normal; antennae with 12 or more segments 5

5.   Mandibles stout & not crossing each other; pupa typically
      found near water's edge but not normally submerged 6
      Mandibles curved, pointing forward and usually crossing; pupa
      in case & typically submerged (occasionally semi-terrestrial) ....
      ....(Fig. 2.36C) **Trichoptera**

6.   Smaller (typically < 10 mm long), in double-walled, mesh-
      like cocoon in sheltered places ....**Neuroptera**
      Larger (typically > 10 mm), without a cocoon but in a chamber
      in soil or rotten wood ....(Fig. 2.22C) **Megaloptera**

7.   Appendages visible on pupal surface; 2 pairs of wing pads,
      although hind pair may be concealed by the fore pair; no
      obvious gills or breathing tubes ....**Lepidoptera**
      Appendages may be visible or completely concealed in a
      barrel-shaped puparium; if visible, then usually projecting
      respiratory organs or paired, dorsal breathing tubes on
      prothorax; occasionally gills at the tip of the abdomen; one
      pair of wing pads ....(Fig. 2.33D) **Diptera**

4 HABITATS AND COMMUNITIES

The habitats of aquatic insects may be classified according to a number of different criteria depending on the aims of the particular exercise. The system chosen here is based primarily on the major physical and chemical features of the environment. There are three broad categories, each differing in their degree of stability and/or continuity: (1) permanent habitats; (2) temporary habitats; and (3) man-made habitats.

4.1 Permanent habitats

These aquatic insect habitats contain water on a permanent basis, although perhaps not when measured on a geological time scale. On a global scale, most are cosmopolitan in occurrence although some are more regional because of local differences in one or a combination of factors, foremost amongst which are geology, climate, precipitation profile, vegetation, soil, drainage and water chemistry.

4.1.1 Running waters

Rivers and streams exhibit features resulting from their role as channels for the transport of excess water, derived from precipitation, that terrestrial environments cannot hold. Consequently, most river valleys are formed by erosion. In general, running (lotic) waters have the following properties: unidirectional movement, downhill; variable levels of discharge and associated parameters such as current velocity, depth, width and turbidity; continual turbulence and mixing of water layers, except at low altitudes; and relative instability of bottom sediments.

The distinction between rivers and streams is often vague. Macan (1974) has suggested that streams can be forded by a person wearing hip-waders, but rivers are deeper. However, this does not deal satisfactorily with the many channels that may be 10 m or more wide but only 0.5 m or so deep. Instinctively, most limnologists would categorize the latter as rivers. In a classification of streams in Southern Ontario, Canada, Ricker (1934) defined rivers as having a discharge of greater than 0.28 m^3/s on the 1st of June, with a width greater than 3 m. Geologists tend to use the terms interchangeably, but often reserve the term river for the main stream, or larger branches of a drainage system (Morisawa, 1968).

Classifying running waters has always been problematic. Various schemes have involved local geology, the source of the water, size, current speed, gradient, discharge, substrate type, temperature, dissolved oxygen and carbon dioxide concentrations, fauna (both vertebrate and invertebrate) and flora, and productivity - either singly or in combination. However, as Hynes (1960) points out, most of these classifications apply only locally and run into problems on a global, or even a continental, scale. Running waters differ individually and will only accept broad, general classification.

One system with seemingly universal application is based on physical parameters. This is the stream order system of Horton (1945). In it, the smallest, unbranched tributaries of a drainage system are designated lst order streams. A stream arising from the confluence of two first order streams is designated a 2nd order stream; the result of the meeting of two second order streams, a 3rd order stream, and so on. This system correlates closely with factors used in other classification schemes and, from a biological point of view, has the advantage of categorizing streams in an objective way.

Another workable system is based on successive faunal zones that appear to be distinct ecological entities (Illies and Botosaneanu, 1963). Although originally proposed for streams in Germany, the classification seems to have wider application, at least in temperate regions. It involves four zones along the entire length of a watercourse:

Eucrenon - the spring source zone;

Hypocrenon - the short part of a stream flowing directly from the spring (sometimes known as the "springbrook");

Rhithron - the zone extending from the hypocrenon to the point where the annual range of monthly mean temperature does not exceed 20°C; where the current is fast and turbulent; where the substrate is composed of large particles, mainly fixed boulders, stable cobble, and gravels with some sand, silt and mud in sheltered patches; and where dissolved oxygen is near saturation;

Potamon - the zone below the rhithron extending to the sea or a large lake, where the annual range of monthly mean temperature rises above 20°C (above 25°C in tropical latitudes); where the flow is slow (except during seasonal peaks such as spring floods in temperate and polar regions, and during the monsoons in the tropics) and nearly laminar over substrates of predominantly sand with some mud and silt; where there may be partial light extinction resulting from suspended solids; and where oxygen deficiencies may develop.

A particular problem in studying the headwater regions of watercourses is deciding the boundary of a spring, although the bounds of certain spring-fed systems (e.g. the basin of a limnocrene, see below) are easier to define. But in other spring-fed streams conditions change gradually downstream from a point or multiple source. With increasing distance, conditions become less springlike and more streamlike by dilution and warming, and the rate at which these changes take place depends on factors such as flow rate, channel geometry and season. Entomologists sampling springs and spring-fed streams have tended to decide habitat boundaries in different ways and, consequently, the *hypocrenon* (spring-brook) is frequently merged into either spring studies or stream studies. Perhaps, its usefulness as a distinct biological zone has been lost. Further, Illies and

Botosaneanu (1963) originally proposed that both the rhithron and the potamon could be subdivided into an epirhithron, a metarhithron, a hyporhithron, an epi-potamon, etc. It is becoming increasingly evident, however, that a greater level of resolution does not have universal application. For present purposes, and to keep matters simple, only three of the primary divisions are retained: the eucrenon, which we equate to a *spring* source; the rhithron, which we equate to a *stream*; and the potamon, which we equate to a *river*.

Coldwater springs

Springs are the sites of emergence of groundwater flowing along an impervious rock stratum and, accordingly, can be found anywhere such conditions arise. Most permanent freshwater springs are highly stable environments. Here, many physical and chemical features, particularly water temperature, discharge and water chemistry, fluctuate less than in any of the other lotic zones. Associated with stable discharge is relative stability of inorganic bed materials. The diversity of the insect fauna of running waters generally increases downstream. For example, compared with the 189 species of aquatic Coleoptera known from lotic habitats in Canada, only 61 species (including those of lentic as well as lotic affinities) have been reported from springs (Roughley and Larson, 1991).

It is now becoming clear that, with respect to conditions for insect life, the term "spring" refers to several different habitat types. Conditions in any particular spring depend, to a large extent, on the properties of its groundwater source (e.g. size, chemical nature and residence time) together with a matrix of other influences such as the slope, latitude, the rate of discharge of the spring, and the nature of riparian vegetation (Danks and Williams, 1991). Flow rate, for example, varies from rapidly rushing springs to seeps, and flow rate together with the geometry of the source area determines the three well-known spring types: (1) a source opening to a rapidly flowing stream of steep gradient, which does not permit settling of fine particles - *rheocrene*; (2) a source first entering a basin, which frequently becomes lined with mud and fine debris and overgrown with aquatic plants, and from which water tends to flow out more slowly into a channel - *limnocrene*; and (3) one percolating into a marshy holding area - *helocrene* (Bornhauser, 1913).

Considering the insect fauna of coldwater springs as a whole, the Diptera, Coleoptera and Trichoptera are best represented, but also evident are characteristic representatives of other aquatic orders, particularly the Plecoptera, Odonata and Ephemeroptera. It seems that the representation of these orders, in springs, may simply reflect their overall diversity in the aquatic fauna. However, many of the insect genera found in springs are different from those downstream (the former tending to be stenothermal, i.e. they have a narrow temperature range for survival) indicating that there is a distinctive spring fauna. Within the Trichoptera, for example, the number of species present in springs increases with habitat diversity. Up to nine species have been found in small (< 1 m wide), but microhabitat-rich, rheocrenes in Canada, compared with just one or two species in microhabitat-poor systems (Williams, N.E., 1991). Several caddisfly species, in Canada, seem to be predominantly associated with springs even though they have continent-wide ranges (e.g. *Hesperophylax designatus* and *Chyranda centralis*, Limnephilidae), and some entire genera seem to be restricted to springs and springbrooks (e.g. *Anagapetus*, Glossosomatidae). Other genera are frequently found in springs but are not restricted to these habitats (e.g., in North America, *Lepidostoma*, Lepidostomatidae; *Rhyacophila*, Rhyacophilidae; *Parapsyche*, Hydropsychidae; and *Neophylax*, Limnephilidae) (Williams and Williams, 1987). Limnephilids (*Apatania, Potamophylax* and *Chaetopteryx*), glossosomatids (*Agapetus* and *Glossosoma*), and rhyacophilids (*Rhyacophila*) also tend to feature prominently in the trichopteran faunas of springs in the Palaearctic together with polycentropodids (e.g. *Plectrocnemia*) and sericostomatids (*Sericostoma*) (Lepneva, 1970; Lindegaard, *et al.*, 1975; Verdonschot and Schot, 1986). Rhyacophilids and sericostomatids are common, too, in springs in New Zealand where they occur alongside some polycentropodids, leptocerids, hydroptilids, helicophids (endemic) and the glacial relict *Rakiura vernale* (Helicopsychidae) (Michaelis, 1977). Temperate glacial relicts such as *R. vernale*, and *Apatania muliebris* and the stonefly *Capnia bifrons* (in Europe), are not uncommon in coldwater springs which they colonized during the advance of the polar ice caps in the Pleistocene, but in which they subsequently became marooned after the ice retreated.

Spring-inhabiting beetles have not been well studied, but a review of the species found in Canada and Alaska showed some 63 species distributed among six families: Dytiscidae (38 spp.), Hydrophilidae (9), Hydraenidae (8), Chrysomelidae (Donaciinae) (6), Haliplidae (1), and Dryopidae (1). Of the 38 species of dytiscid found in springs, only nine (24%) are restricted to springs. The diversity of these families only loosely parallels their diversity in the total fauna. Moreover, several relatively diverse families (Gyrinidae, Scirtidae and Curculionidae) appear to be absent from springs, as are some predominantly lotic families (Amphizoidae, Elmidae and Psephenidae) (Roughley and Larson, 1991).

The Diptera, with larvae generally adapted to aqueous habitats and with many families, are well represented in coldwater springs. Chironomids are prominent, especially those belonging to the subfamily Orthocladiinae, which tend to be cold-adapted and dominate habitats with coarse substrates, like rheocrenes. The subfamilies Podonominae, Diamesinae and Prodiamesinae have similar requirements, in general terms, but contain far fewer genera. The Chironominae and Tanypodinae have relatively few genera associated with springs as they are more common in warm, lentic habitats with organically-rich sediments (Colbo, 1991). One genus, however, *Lithotanytarsus* (Tribe Tanytarsini), is attracted to limestone springs. In these springs, the emerging groundwater is high in calcium bicarbonate and, as the carbon dioxide in the water is removed by both equilibration with the air and photosynthesis around the spring boil, calcium carbonate is precipitated on the spring bed.

Lithotanytarsus uses this material to construct calcareous tubes. Algae growing on these precipitates may also attract other dipterans, for example psychodids (Geijskes, 1935). A group of non-insect arthropods, the water mites, is especially diverse and abundant in springs. Since the larvae of most genera found in springs parasitize chironomids, this reinforces the perception of the dominance of this group of dipterans in these habitats (Smith, 1991).

Other dipterans found in coldwater springs belong to the Tipulidae (crane flies), Empididae (dance flies), Dolichopodidae (particularly in small seeps), Ptychopteridae (phantom crane flies) and the Chaoboridae (phantom midges), the latter especially in limnocrenes.

Among the remaining taxa, any given spring may support typically a maximum of one or two species of stonefly, mayfly or dragonfly. Coldwater odonates belong, especially, to the families Petaluridae, Gomphidae, Corduliidae (Anisoptera) and the Coenagrionidae (Zygoptera). The Plecoptera, in general, are adapted to cool lotic habitats, but there are also specific, spring-dwelling forms, particularly belonging to the families Leuctridae and Nemouridae in the Northern Hemisphere and to the Eustheniidae and Notonemouridae in the Southern Hemisphere.

The composition of spring communities may be related to the longevity of the habitat. In the mid-Appalachian Mountains (U.S.A.), for example, frequently -disturbed springs are dominated by insects, which have more rapid colonization abilities than non-insects. Given sufficient spring longevity in this region, flatworms (Tricladida), pericarid crustaceans and other slow colonizers arrive, windborne, by surface and subterranean stream movements, or by phoresy (Gooch and Glazier, 1991). Riparian vegetation, current regime, bed substrate particle size, the type of food available (diatoms vs allochthonous input), pH and the chemical composition of the emerging groundwater are other factors that seem to have strong influences upon the composition of spring communities (Williams, N.E., 1991). However, as yet and as a consequence of the small number of ecological studies done on springs, most of the evidence supporting such relationships is correlative.

Streams

Streams can be roughly equated to the *rhithron* in the classification scheme of Illies and Botosaneanu (1963). The wide range of substrate particle sizes and configurations on the stream bed, together with the associated, complex patterns of microcurrents and variety of available food, result in a mixture of habitats that is dominated by a high diversity of aquatic insects. The fauna contains representatives of most aquatic insect orders and includes several groups that only occur in running water (the dipterans Blephariceridae, Simuliidae and Deuterophlebiidae) or that reach their maximum development and diversity there. Survival of many ancient insect stocks in these habitats is thought to be a consequence of the great permanence of streams compared with lakes and ponds, most of which fill in over time. In contrast, most watercourses have been in continuous existence for millions of years. Once formed, the stream in a valley seldom disappears, thus allowing greater continuity of the fauna (Hynes, 1970).

Running water insects are largely cool-water, stenotherms as streams are typically located at high altitude or latitude, or, elsewhere, are fed by cool groundwaters. Since the habitat is a dynamic one, the constant flow of water has resulted in forms that are closely associated with the bed (benthic) rather than the water column (planktonic), as the latter mode of life in moving water calls for greater expenditure of energy. Apart from zoogeographically isolated regions (e.g. oceanic islands) where certain elements may be absent because of colonization limitations, the faunas of streams show exceptional similarity the world over. This can be illustrated by comparing the insects identified in faunal surveys of streams in each of the major zoogeographical regions (Table 4.1). The taxa shown do not represent all those included in the original surveys, but rather the ones that consistently occurred among sites. Not only is there remarkable uniformity at the family level but, in many cases, cosmopolitan genera are evident, for example the mayfly *Baetis*, the stonefly *Nemoura*, the dipterans *Limonia*, *Atherix* and many of the chironomids, and the free-living caddisfly *Rhyacophila*. There are instances where, for historical reasons, certain groups are absent from a particular area (e.g. the perlid and perlodid stoneflies from Australasia) but they are typically replaced by locally endemic forms (in this case the Southern Hemisphere stonefly families Gripopterygidae, Eustheniidae and Austroperlidae). The niches occupied in the Northern Hemisphere by the characteristically dorso-ventrally flattened mayfly family Heptageniidae are, in New Zealand for example, filled by species of the Leptophlebiidae which have become flattened and have the typical heptageniid brushing-type mouthparts (McLellan, 1975).

Comparison of lotic communities from the two hemispheres has been the subject of much heated debate. Because of the unpredictable and arid nature of the Australian climate and the supposed aseasonal nature of the leaf litter input, it was believed, initially, that the diversity of insects in southern streams was not as high as in the north (Williams, W.D., 1976). More recently, however, streams in southeastern Australia have been shown to have comparably high diversity. Certain taxa, especially the mayflies and stoneflies, in this region are, indeed, not as richly represented as in temperate Northern Hemisphere streams. However, this is more than compensated for by the very high diversity of caddisflies, beetles and dipterans. Previous estimates of diversity in the south may well have been low because of weaknesses in sampling protocol, differences in habitat heterogeneity and, perhaps most important, lack of taxonomic precision, particularly in groups like the chironomids. It has been argued that longitudinal zonation in Australian streams may be less consistent than elsewhere, as some Australian systems show marked spatial and temporal variation in community structure, with no consistent trends, within the same stream order (Lake *et al.*, 1986). Nevertheless, several surveys of Australian

Table 4.1 Typical components of the insect fauna of a stream as illustrated by a comparison of the families collected in streams from around the world; prominent genera are indicated where appropriate (• = present).

Taxa	Wales	U.S.S.R.	Canada	U.S.A.	S. Africa	S. America	New Zealand	Australia	Malaysia	China
Collembola	•	•	•	•	•	•	•	•	•	•
Ephemeroptera										
Baetidae	•	•	•	•	•	•	•	•	•	•
(esp. *Baetis*)										
Siphlonuridae	•	•	•	•		•	•	•	•	?
Heptageniidae	•	•	•	•	•	•			•	•
(esp. *Epeorus*)										
Leptophlebiidae	•	•	•	•	•	•	•	•	•	•
Ephemerellidae	•	•	•	•	•	?	•	•	•	•
(esp. *Ephemerella*)										
Caenidae (*Caenis*)	•	•	•	•	•	•			•	•
Odonata	•	•	•	•	•	•	•	•		?
(esp. Gomphidae, Agrionidae & Cordulegastridae)										
Plecoptera										
Nemouridae	•	•	•	•	•	•	•	•	•	•
(esp. *Nemoura*)										
Capniidae	•	•	•	•	•	•	•	•	•	•
Perlidae	•	•	•	•	•	•		•		•
Perlodidae	•	•	•	•						
Gripopterygidae/Eustheniidae/Austroperlidae						•	•	•		
Hemiptera										
(esp. Gerridae)	•	•	•	•	•	?		•	•	?
Megaloptera	•	•	•	•	•	•	•	•	•	?
Coleoptera										
Hydraenidae	•	•	•	•			•	•	•	
Psephenidae		?	•	•		•	•	•	•	•
Elmidae	•	•	•	•	•	•	•	•	•	•
Diptera: *Nematocera*										
Deuterophlebiidae			•	•	•					
Blephariceridae			•	•	•	•	•	•	•	*
Tipulidae	•	•	•	•	•	•	•	•		*
(esp. *Dicranota, Antocha* & *Limonia*)										
Psychodidae	•	•	•	•	•		•	•	•	*
Ceratopogonidae	•	•	•	•	•	•	•	•	•	*
(esp. *Bezzia*)										

Table 4.1 (contd)

Taxa	Wales	U.S.S.R.	Canada	U.S.A.	S. Africa	S. America	New Zealand	Australia	Malaysia	China
Simuliidae (esp. *Simulium*, *Prosimulium* & *Austrosimulium*)	•	•	•	•	•	•	•	•	•	*
Chironomidae: Tanypodinae										*
Procladius	•		•	•			•	•	•	
Ablabesmyia	•	•	•	•			•	•		
Diamesinae							•	•		*
Diamesa	•	•	•	•			•		?	
Orthocladiinae							•			*
Eukiefferiella	•	•	•	•	•	•		?		
Orthocladius	•	•	•	•	•	•			?	
Cricotopus	•	•	•	•	•	•		•	•	
Psectrocladius	•	•	•	•	•			•		
Thienemaniella	•	•	•	•	•			•	?	
Corynoneura	•	•	•	•	•	•			•	
Chironominae										*
Chironomini										
Microtendipes	•	•	•					•	•	
Polypedilum	•	•	•	•	•	•	•	•	•	
Tanytarsini										
Micropsectra	•	•	•				•	•	•	
Tanytarsus	•		•	•	•	•	•	•	•	
Dixidae	•	•	•	•	•	•	•	•	•	*
Diptera: *Brachycera*										
Tabanidae	•	•	•	•	•	•	•	•	•	•
Rhagionidae (esp. *Atherix*)	•	•	•	•	•	•		•	•	*
Empididae	•	•	•	•			•	•	•	*
Lepidoptera (Pyralidae)	•	•	•	•		•		•	•	
Trichoptera										
Rhyacophilidae (esp. *Rhyacophila*)	•					•		•	•	•
Glossosomatidae (esp. *Agapetus*)	•	•			•	•	•	•	•	•
Hydroptilidae	•	•	•	•	•	•	•	•	•	•
Philopotamidae (esp. *Chimarra* & *Dolophilodes*)	•	•	•	•	•		•	•	•	•

Table 4.1 (contd)

Taxa	Wales	U.S.S.R.	Canada	U.S.A.	S. Africa	S. America	New Zealand	Australia	Malaysia	China
Psychomyiidae (esp. *Lype, Psychomyia* & *Tinodes*)	•	•	•	•	•	•	•	•	•	•
Polycentropodidae (esp. *Polycentropus*)	•	•	•	•	•	•	•	•	•	•
Hydropsychidae (esp. *Hydropsyche* & *Cheumatopsyche*)	•	•	•	•	•	•	•	•	•	•
Brachycentridae	•	•	•	•		•				?
Limnephilidae	•	•	•	•	•	•	•	?		•
Sericostomatidae	•	•	•	•	•	•	•	•	?	?
Helicopsychidae (esp. *Helicopsyche*)			•	•	•	•	•	•	•	
Leptoceridae (esp. *Oecetis*)	•	•	•	•	•		•	•	•	•

Sources: *Wales* - Hynes, 1961; Brooker & Morris, 1980; *U.S.S.R. (Tien Shan)* - Brodsky, 1980; *Canada* - Bishop & Hynes, 1969; Tavares & Williams, 1990; *U.S.A.* - Minckley, 1963; Ward, 1986; *South Africa* - Harrison & Barnard, 1972; King, 1981; *South America* - Edwards, 1931; Illies, 1969; Turcotte & Harper, 1982; *New Zealand* - Winterbourn, 1978; Towns, 1979; *Australia* - C.S.I.R.O., 1970; Chessman, 1986; Outridge, 1987; ; *Malaysia* - Bishop, 1973; *China* - Dudgeon, 1982.
* running water Diptera from this region are not well known
? indicates uncertainty or ambiguity in the records

running waters have been able to discern the rhithron clearly from the potamon (e.g. Gooley, 1977; Malipatil and Blyth, 1982) although further subdivision of both zones was not possible. Bishop's (1973) study of a hill-stream in West Malaysia showed that whereas many insect families known to have a predilection for the rhithron occurred in the upper reaches of the Sungai Gombak, division of the fauna into two distinct biocoenoses was not clear. However, the stream could be fitted into the Illies and Botosaneanu system after some modification of the criteria defining sections. This had previously been proposed for rivers in South Africa (Harrison, 1965). A consensus among many authors now advocates that, as is done here, on a global scale the rhithron/potamon distinction be treated as a useful but first order definition only.

Rivers

Rivers are "potamonic" in nature. The fauna is eurythermic (i.e. it has a wide temperature tolerance) or warm stenothermic and contains forms which reach their maximum development in lentic habitats. In some instances (e.g. pools in large rivers), planktonic forms are present. The marked uniformity noted amongst the faunas of streams throughout the world is less evident amongst the faunas of the softer substrates in rivers. Certain elements, including some genera, are cosmopolitan but the "less-rigorous" habitats created by slower, less turbulent currents enable less-specialized species to occur and local species to dominate (Hynes, 1970). It is, consequently, more difficult to generalize as to the typical insect fauna of rivers compared with that of streams.

There are, nevertheless, some typical components of the insect fauna of rivers (Table 4.2). Not all of these are likely to be found alongside each other in the same part of a river, however. Substrate type varies, as in streams, but the extent of different patches is usually much greater. Sand, mud, bedrock or submerged macrophytes frequently cover large areas and may support somewhat different faunas.

Table 4.2   Insect groups commonly found in large rivers.

Collembola
Ephemeroptera: Siphlonuridae; Potamanthidae; Polymitarcyidae; Caenidae; Ephemeridae
Odonata: many families
Hemiptera: Corixidae; Notonectidae; Gerridae
Coleoptera: Dytiscidae; Haliplidae; Hydrophilidae; Gyrinidae
Diptera: Chironomidae (esp. Chironomini); Culicidae; Stratiomyidae; Tabanidae
Trichoptera: Hydropsychidae; Hydroptilidae; Leptoceridae
(+ many other case-building families)

A major substrate type in large rivers around the world is sand, which may constitute up to 95% of the bed type in lowland rivers (Soluk, 1985). Sand grains settle out of the water column wherever the current is reduced, especially in down-river regions. They do not pack together as do finer silts and clays, and can be resuspended by minor increases in flow. Such sandy substrates represent poor habitats because they hold little in the way of organic matter and are unstable. Despite this, shifting sand supports a limited though characteristic fauna which, besides oligochaete worms and clams, consists of specialized mayflies, dragonflies and dipterans (Barton and Smith, 1984). The mayflies belong to six families (Siphlonuridae, Ametropodidae, Baetidae, Oligoneuriidae, Heptageniidae and Behningiidae) and most species are known from only a small number of localities. Many are excellent swimmers and have various morphological (e.g. slender legs used as anchors) or behavioural adaptations for dealing with shifting substrates. Burrowing is a common behaviour in psammophilic odonates but as this is typical of many dragonflies it cannot be seen as a specific adaptation to this particular habitat. Among the Diptera, the most common sand-dwelling forms are chironomids, especially species of Chironomini. The genus *Cryptochironomus* is very widespread and sometimes is represented by several species in the same section of river (Mordukhai-Boltovskoi, 1979). In fact, most species of the *Harnischia*-complex seem to prefer sandy substrates, and five, in particular, are known only from unstable sand (*Beckidia, Chernovskiia, Cyphomella, Robackia* and *Saetheria*) and have wide geographic distributions (e.g. *Beckidia* has been collected in the U.S.S.R., North America and Africa; Saether, 1977). The larvae of many of these psammophilic chironomids exhibit features thought to enable them to deal with sand and these include: an elongated head and subdivided body segments to bestow greater flexibility; long antennae that may aid "swimming" through sand; long maxillary palps with enhanced sense organs, perhaps to aid the largely carnivorous mode of feeding; reduced but slender posterior prolegs; and a thickened body wall to resist abrasion (Barton and Smith, 1984).

Where the sand has a higher silt content, the chironomid assemblage may shift to one dominated by *Chironomus, Procladius* and other species of *Cryptochironomus*. In places, the flow of water may be constricted forming an area of high current that can expose stones or bedrock. Provided that temperature and oxygen levels are tolerable, this can result in the establishment of rhithronic species in the potamon. Forms that occur in such oases commonly include hydropsychid caddisflies (esp. *Hydropsyche* and *Cheumatopsyche*), heptageniid mayflies, and lithophilic chironomids like some species of *Cricotopus* and *Polypedilum;* the exact composition depends on the proximity of source animals and their colonization abilities. Where extensive, firm substrates exist, black fly and hydropsychid larvae may occur at very high densities.

Where the silt and fine particulate organic matter content of the sand is very high, and where the current is minimal and the turbidity low, aquatic macrophytes grow. This has the important effect of adding a third dimension to the largely two dimensional nature of the typical river bed. Not only does this provide more living space for the benthos but it also allows species to occur that, although able to withstand the temperature and oxygen regimes of the potamon, are unable to deal with silt, perhaps because it clogs their feeding or respiratory apparatus. Macrophyte beds in rivers generally support a rich diversity of insect types similar to that seen in weedy ponds or the sheltered littoral zone of lakes, including mayflies, odonates, hemipterans, beetles, cased caddisflies and dipterans.

At high latitudes, there is a tendency for no true components of the potamon fauna to show up, as many rivers here are short and typically rhithronic throughout their entire length. Running waters in both the Arctic and Antarctic are like this. Conversely, in lower latitudes, for example the Amazon basin, the rhithron is very short relative to the potamon, and rhithronic species may be driven into high altitude regions (see Hynes and Williams, 1962). In temperate latitudes, there is more of a balance between the two zones, but although some river species can survive in the lower temperatures of streams they seldom thrive there. Similarly, relatively few rhithronic species survive the warmer temperatures of large rivers, and then, as we have seen, they do so only if stable substrates are available.

Although in this chapter we have adopted a simple, descriptive framework in which to examine the longitudinal distribution of lotic insects, there are other schemes based on the functional responses of invertebrates to changes along a watercourse. Foremost among these is the River Continuum Concept (Vannote *et al.*, 1980). This concept perceives the headwater region (equated to stream orders 1 and 2 in the Horton [1945] scheme) to be dimly-lit, because of dense riparian vegetation and steep valley sides, to have high slope and to be of fairly constant water temperature. It is often a region of low aquatic primary production, where insects feed on coarse particulate organic matter (CPOM) and its derivatives that mostly come from the breakdown of autumn-shed, riparian leaves (Kaushik and Hynes, 1971). The CPOM is fed on by insect "shredders" (e.g. cased caddisflies and tipulids; see Table 8.2) that reduce the material to faecal pellets and small pieces of leaves, less than 1 mm in diameter (fine particulate organic matter, FPOM), which may become incorporated into the sediment or suspended in the water column. Evidence indicates that shredder species derive most of their nutrients from microorganisms, such as hyphomycete fungi and bacteria, growing on and in the leaf tissue (Barlocher and Kendrick, 1973). Alongside the shredders in the headwater community are "collectors". The latter either trap the FPOM as it passes in the water column (e.g. filter-feeding caddisflies and black flies), or collect it directly from the substrate (e.g. certain mayflies and herbivorous stoneflies). Again, much of the nutrition appears to come from the microorganisms that colonize the particles. Collectors produce faeces that are of much the same size as the particles that they ingest, but minus the microorganisms, and these are returned to the FPOM pool. Because of the oftentimes low primary production in headwaters, the "grazer" and "scraper" members of the benthic community may be poorly represented (Fig. 4.1). Predators are present, particularly in the form of large stoneflies, but dragonflies (especially Aeshnidae) occupy this niche in many Australian headwaters. The community is typically heterotrophic (Production/ Respiration <1).

In the midreaches of the watercourse (generally at the level of 3rd to 4th order streams), there is a shift from heterotrophy to autotrophy (P/R >1). This is primarily due to widening of the river and its valley, reduced shading by bankside vegetation, warmer water, and high nutrient levels derived from upstream. Here, the ratio of CPOM to FPOM is reduced as the latter increases because of immigration from upstream, increased production of periphyton, and the rapid breakdown of dead aquatic macrophyte tissue. Predictably, the importance of shredders in the community is reduced, while collector and grazer species dominate. Predatory insects are present also.

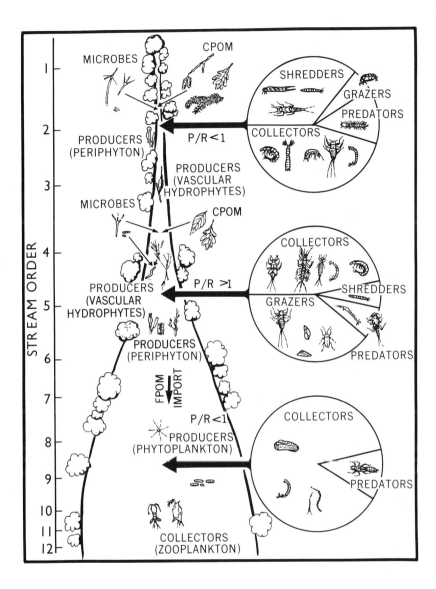

Fig. 4.1   Summary of the longitudinal changes seen in the benthic community
of a watercourse according to the River Continuum Concept (P/R =
production to respiration ratio of the benthic community) (redrawn
after Cummins, 1977).

In the lowermost sections of the watercourse (5th order +), collector species
dominate because of the large amount of FPOM derived both from upstream and
from locally-produced phyto- and zooplankton. The community is again
heterotrophic as a result of reduced primary production because of increased

turbidity and unstable substrates - although different conditions may apply in macrophyte beds. Predators are typically odonates, beetles and hemipterans.

The longitudinal changes in food particle size implicit in the River Continuum Concept may result in a linear succession of species groups within a taxon. An example of this is seen in the Nearctic subfamilies of the net-spinning caddisfly family Hydropsychidae. Three of the four subfamilies have somewhat restricted distributions in the river continuum (Fig. 4.2). The Macronematinae are able to filter very small particles of FPOM from the water column by means of fine-meshed nets; they are therefore well suited to large rivers. On the other hand, the Arctopsychinae spin nets with coarse meshes that are more suited to capturing CPOM and small animals in rapid mountain streams. The Diplectroninae are restricted to cool headwater regions where they also feed on CPOM captured in large-meshed nets. The Hydropsychinae spin nets of intermediate mesh size and are more generalized feeders; predictably, their distribution spans most of the continuum.

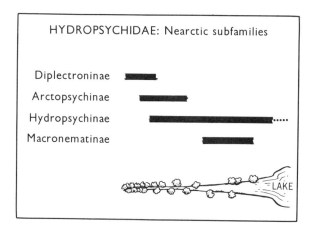

Fig. 4.2   Longitudinal zonation of the Nearctic subfamilies of the Hydropsychidae (redrawn after Wiggins and Mackay, 1978).

Although the River Continuum Concept presents a tidy, apparently widely applicable picture of running waters ecosystems controlled, appealingly, by just a few factors (basically organic matter and stream morphometry), it is not without its critics. The concept fits many watercourses in North America, where it was conceived, and elsewhere in the north temperate zone. However, it is becoming increasingly obvious that it cannot be readily applied to the many lotic systems elsewhere in the world which are of a more indeterminate nature and in which communities are often controlled by local or stochastic events (Williams, W.D., 1988). In particular, the longitudinal community structure of

Australian running waters appears more complicated than that in the Northern Hemisphere as, for example, some Australian streams exhibit considerable spatial and temporal variation in community structure within the same stream order, with no consistent trends (Lake *et al.*, 1986). Statzner and Higler (1986) have advocated a re-evaluation of the effects of the physical characteristics of flow ("stream hydraulics") on the longitudinal zonation of lotic insects.

Microhabitats

Quite apart from the considerable differences in insect distribution between various sections along the length of a watercourse, there are also differences on a smaller scale. Insects may differ from one side of a stream to the other, probably related to differences in flow, oxygen content, accumulations of organic debris, etc. Hydraenid beetles, for example, spend their adult stages near midstream whereas their air-breathing larvae live at the edge. Other, semi-aquatic, insects commonly occur at the margins, particularly larvae of the dipteran families Tipulidae, Ceratopogonidae, Syrphidae, Tetanoceridae and Ephydridae (Pennak, 1953). Dytiscid beetles and larvae of some of the higher dipterans (e.g. Tabanidae, Empididae and Dolichopodidae) may be found in slack water areas created by eddy currents, obstructions to the flow, or small tributaries and, on the surface of these quiet regions, gyrinid beetles and waterstriders (Gerridae, Veliidae) frequently occur.

In slow flowing, tropical rivers, a unique microhabitat is associated with the long (up to 80 cm) roots of floating plants such as the water hyacinth, *Eichhornia,* and bunches of aquatic grasses. In the nutrient-poor, dark-coloured waters of the Amazon Basin, for example, many insects live in these tangled root systems, floating up and down with the severe changes in water level that occur in the region. These large fluctuations in depth together with the flocculent, unstable nature of the river beds make benthic living somewhat tenuous. The aerial portions of these plants provide protection for the adults of insects whose larval stages live in the roots. Most types of aquatic insect found in such rivers are associated with these floating substrates (Junk, 1976).

Streams with deep gravel beds possess an interstitial habitat formed in the mixture of coarse sand, gravel and rocks typical of the rhithron. This *hyporheic* habitat is a middle zone bordered by the surface water above and the true groundwater below. Its fauna, termed the *hyporheos,* thus lives in a habitat in which the boundaries are continually shifting back and forth, rather like the landward and seaward margins of the intertidal zone (Williams, D.D., 1989). The hyporheic community contains elements of the benthos found on the surface of the stream bed together with elements derived from hypogean environments such as groundwater, subterranean waterbodies and the soil. The surface element consists of the larvae of many insect species that seek out this zone as a refuge during their early development. These larvae are usually either long and slender, with a flexible body that allows easy passage among the substrate particles (e.g.

chironomids, elmid beetle larvae, stoneflies and mayflies), or are small and blunt with a hard protective shell or exoskeleton that withstands being crushed as the animal pushes its way through the pore spaces (e.g. elmid adults and helicopsychid caddisflies). Some larvae, such as chironomids, occur as deep as 1m below the bed surface and even hydropsychid caddisflies have been found to 30 cm (Williams, D.D., 1984). As the hyporheic zone has a lateral component, insects are frequently found under the stream bank. In waters that flow over extensive, loosely-compacted floodplain gravels, stonefly nymphs have been recorded up to 50 m from the main channel, and even in aquifers that extend up to 2 km away from the channel (Ward, 1989).

Another interstitial habitat occurs between the sand grains of large rivers. Although this has been little studied, it is assumed that the interstitial flow of water in the potamon is much less than in the hyporheic zone of the rhithron and that conditions most probably resemble those of the *psammon*, the interstitial habitat of sandy lake shores.

Frequently, lotic insects show marked distribution patterns over just a few metres of bed, or even around a single rock. Numerous studies have shown that such microdistributions result from the influence of a variety of factors acting either singly or, more probably, in combination. Foremost among these influences are the nature of the substrate, together with locally induced microcurrents, and the availability of food. For example, high insect diversity has been associated with large gravel (~ 40 mm diameter) compared with small gravel (Williams and Mundie, 1978) and community structure related to the heterogeneity of substrate mixtures (Williams, 1980a). Certain current velocities and water depths (e.g. 75-125 cm/s at a depth of between 20 and 40 cm, in the Tongue River, Montana) have been shown to produce high faunal diversity (Gore, 1978), while the distribution of some insects, for example caddisflies in Hartley Creek, Alberta, has been linked to the algal or moss coatings of rock surfaces (McElhone and Davies, 1983). Many mayflies appear to have preferred positions on rock surfaces. For example Linduska (1941) found several species of *Baetis* and a species of *Epeorus* on the tops of boulders, where the current was 240 cm/s, but halfway down the sides were *Epeorus longimanus* and *Ephemerella doddsi*. A different species of *Baetis* occurred anywhere on the rock surface where the current slowed to less than 120 cm/s, while *Rhithrogena doddsi* occurred where the current was less than 60 cm/s. Under the boulder, where the current was least and debris had accumulated, were found *Cinygmula*, *Rhithrogena virilis*, *Ephemerella inermis* and *E. yosemite*. The positions and densities of some insect species on rock surfaces are known to change on a daily basis (Graesser and Lake, 1984). Other factors known to influence aquatic insect microdistribution include light, temperature, dissolved oxygen, water chemistry, pollution, oviposition habits, territoriality and predation (see Chapter 9).

Floodplains

In regions subject to large seasonal changes in rainfall (e.g. monsoon regions in the tropics) or snowmelt, many rivers cyclically overflow their banks. As the floods subside, extensive shallow lakes and pools are left which, because of high food levels (in the form of decaying, riparian leaf litter) and rapid warming, provide high quality habitats for suitably adapted insects. In the tropics, basins on the floodplain lose their water rapidly by evaporation and seepage, and this concentrates the dissolved nutrients. In regions like the Amazon Basin where the level of nutrients in the rivers is extremely low, the comparatively nutrient-rich floodplain pools often achieve high productivity, both in terms of algae and invertebrates. However, because these habitats are short-lived, they tend to be available only to species with rapid life cycles and good colonization abilities. Insects known to live in floodplain pools include mayflies, odonates, hemipterans, beetles, caddisflies and dipterans, especially mosquitoes, chironomids and tipulids.

Thermal springs

In regions where groundwaters are influenced by geothermal activity, the water issuing from a spring may be warm or even hot. Thermal springs are characterized by year-round high water temperatures and a level of total dissolved solids (TDS) that is usually higher than that of surface waters. Although the unusual water chemistry of thermal springs does not unduly influence the occurrence of aquatic insects, the elevated temperature most certainly does. Water temperatures may be close to boiling at the point of issue, but since no aquatic insect can live above 50°C (and very few can live above 40°C), few thermal springs provide suitable conditions for insects at their source. It is only several metres downstream, where the water has had a chance to cool, that insects occur. However, in a spring in which the annual mean temperature of the emerging water is only 5°C above the annual mean air temperature of the region (the definition of the lower limit of a "thermal" spring), insects may colonize the boil itself. Thermal springs around the world have similar insect faunas, with the greatest diversity occurring in the cooler springs.

Although representatives of all aquatic insect orders, except the Megaloptera, have been found in waters influenced to some degree by geothermal activity, only four orders are commonly encountered: the Odonata, Hemiptera, Coleoptera and Diptera, and each of these is represented by only a relatively few genera. The most common odonates belong to the families Coenagrionidae (esp. *Argia* and *Ischnura*) and Libellulidae (esp. *Erythemis* and *Libellula*); the most common hemipterans belong to the Saldidae and Corixidae; whereas the beetles are typically hydrophilids or dytiscids; and the dipterans are primarily those belonging to the Chironomidae, Ephydridae, Simuliidae and Stratiomyidae (Pritchard, 1991). Thermal spring communities are typically very simple. For example, in a spring in Yellowstone National Park, U.S.A., in which the temperature varied between 30 and 40°C, only five arthropods were present: the ephydrid *Paracoenia bisetosa*, and four predators - two mites, a dolichopodid fly and a parasitic wasp, although the adults of neither of the latter two species lived in the water (Collins *et al.*, 1976). Perhaps the most tolerant and characteristic insects found in hot springs are the Ephydridae, for example *Paracoenia* or *Ephydra* in North America; *Scatella* in Iceland; and *Ephydrella* in New Zealand (Pritchard, 1991).

4.1.2   Standing waters

Lakes

These are large bodies of water surrounded entirely by land, although they may have inflows and outflows connecting them to each other or to rivers. Generally speaking, they have open water and vary in size from over 83,000 km^2, as in the case of Lake Superior, Canada, to only a few metres across. The smallest lakes merge into "ponds" and differentiation at this end of the scale becomes mostly a preference in terminology. However, if we define ponds as those bodies of water which are shallow enough to support rooted plants across their entire width (Reid, 1961), then they can be excluded from the present discussion and dealt with separately, later.

Lake basins vary greatly in both their morphology and genesis. Lake Superior, for example, is the largest in terms of surface area but, with a maximum depth of 307 m, it contains only about half the volume of water (12,000 km^3) found in Lake Baikal, Siberia (23,000 km^3) which, although only 31,000 km^2 in area, is 730 m deep. Such differences impose themselves onto the nature of the habitats available for occupancy by plants and animals. Habitats are influenced also by the way in which a lake is formed, primarily by way of the nature of the sediments created, the depth and water circulation produced, and the nutrients released. Over 70 different types of basin have been recorded but these can be roughly divided into three main categories, based on their geological origin (Murray, 1910).

(1) *Rock basins* are formed through a number of different processes including glaciation, solution, movements of the earth's crust, and meteoric impact. Glacial movement is known to have created basins in solid bedrock by ice scour, to have carved cirque lakes at the heads of glaciated valleys, and to produce paternoster (chain) lakes in the bottoms of valleys. Alternate freezing and thawing of the ground surface by glaciers has resulted in subsidence lakes and the same process continues to produce shallow ponds in the arctic, today; where several such ponds merge, they form a thermokarst lake. Solution lakes are formed in depressions produced by the solution of soluble bedrock, often accompanied by subsidence of the strata underneath. Tectonic lakes are created by movements of deeper portions of the earth's crust, through faulting and uplifting.

Volcanic lakes may be formed in the craters of extinct volcanoes, or where the flow of lava has dammed the end of a valley. In the distant past, meteors have been responsible for creating quite large lakes either through direct impact and explosion, or by the force of compression waves.

(2) *Barrier basins* are the result of sediment deposition processes, typically of glacial debris. For example, where retreating glaciers have left extensive deposits of ground moraine overlying impermeable till, vast areas of very shallow lakes and wetlands are created. Similarly, kettle lakes have been left in many areas by the melting of large masses of ice buried in moraines. Again, where glacial material is deposited at the ends of valleys, water is trapped on the upstream side. Barrier basins also may be formed by landslides.

(3) *Organic basins* are those formed by the actions of living organisms and include lakes created behind beaver dams; those arising from massive growth of vegetation which may block a drainage channel or divert water into a natural depression; river dams initiated by log jams; and coral lakes resulting from raised atolls. Basins engineered by Man will be considered separately.

Lakes that form in organic basins are generally shallow and, being associated with existing organisms and their by-products, tend to have a relatively high nutrient content, making them *eutrophic*. This is often the case for lakes formed in barrier basins, too, as considerable amounts of nutrients may be released from their bottom sediments. Lakes formed from rock basins, in contrast, inherit little in the way of nutrients and tend to be *oligotrophic*.

Deeper lakes are conveniently divided into four main biotic zones, three with respect to the lake floor and expressed in terms of depth, and one encompassing the open water. The three benthic zones are the *littoral*, the *sublittoral*, and the *profundal*, and the open water region is the *pelagic* habitat.

The *littoral zone* extends from the water's edge to the depth at which aquatic macrophytes, through lack of light, can no longer grow. In large lakes, the littoral zone is subject to wave action, and as a result the substrate invariably consists of large rocks, gravel and sand. In many ways, this zone resembles a stream and the larvae of certain primarily lotic insects occur here, for example, caddisflies of the families Hydropsychidae, Sericostomatidae and Helicopsychidae (Wiggins, 1977), heptageniid mayflies, elmid beetles and nemourid stoneflies (Barton and Hynes, 1978; Jónasson, 1978). In smaller lakes and the sheltered bays of large lakes, where macrophytes are less likely to be torn out by waves, and where fine organic debris accumulates, one may find odonates, beetles (especially Gyrinidae, Haliplidae, Dytiscidae and Hydrophilidae), hemipterans, mayflies (Ephemeridae and Caenidae), caddisflies (especially Polycentropodidae, Hydroptilidae, Phryganeidae, Limnephilidae, Lepidostomatidae, Molannidae and Leptoceridae), and dipterans (particularly chironomids) (Edmondson, 1959; Wetzel, 1983; Wiggins, 1977). The chironomids associated with littoral macrophytes are often represented by similar combinations of genera, for example: *Polypedilum, Chironomus, Tanytarsus, Micropsectra, Procladius, Ablabesmyia, Tanypus, Cricotopus* and *Corynoneura* in Lake Sibaya, S.E.

Africa; *Polypedilum, Chironomus, Cryptochironomus, Glyptotendipes, Stictochironomus, Tanytarsus, Procladius, Ablabesmyia, Pentaneura* and *Clinotanypus* in Lake Cassidy, Florida; and *Polypedilum, Chironomus, Cryptochironomus, Dicrotendipes, Stictochironomus, Tanytarsus, Clado-tanytarsus, Tanypus, Procladius* and *Cricotopus* in Lake Kinneret, in the Jordan Rift Valley (Schneider, 1962; Serruya, 1978; Allanson, 1979).

In clean water areas, on large cobbles and boulders, mats of freshwater sponge grow. Larvae of several different insects shelter within the sponge structure and feed on the tissues. Widespread genera include *Ceraclea* (Trichoptera), *Xenochironomus* (Chironomidae) and *Sisyra* (Neuroptera)

The *sublittoral zone* extends from the lower edge of the rooted macrophyte zone to about the level of the upper boundary of the hypolimnion (the cool water lying below the thermocline, the latter being the region of maximum rate of temperature decrease with respect to depth). Here, the bottom deposits are of a transitional kind, grading from those of the littoral to those of the profundal, and showing an increasing accumulation of true bottom deposits on top of the original basin floor. In many ways, the fauna reflects this transition in that some members of both the littoral and profundal insect communities are present, perhaps because of the up and down swing of the thermocline, although diversity is reduced compared with the littoral zone. In some lakes, a few species seem to reach their maximum abundance in this zone (e.g. species of *Stictochironomus, Procladius, Tanypus* and *Cryptocladopelma* in Lake Kinneret; Serruga, 1978).

The *profundal zone* includes all of the lake floor bounded by the hypolimnion. Its sediments are very finely divided soft oozes which cover the floor of the lake sometimes to great thicknesses. Being beneath the hypolimnion, the profundal zone is subject to continuous low temperatures, little or no light, a pH which frequently falls below neutrality, considerable amounts of free carbon dioxide, and complete absence of oxygen (although not in oligotrophic lakes) during part of the summer and late winter (Ruttner, 1953). In addition, in some lakes, there may be an accumulation of methane, hydrogen sulphide, and other gases associated with decomposition. Conditions in the sediment are brought about primarily by the metabolic processes of bacterial life, which control the environmental conditions under which the profundal benthos lives. Frequently, an oxygen gradient is set up across the mud surface, ranging from low oxygen near the mud/water interface, to complete absence in the top few millimetres of deposit. Gradient profile depends on the amount of reducing substances in the sediment, so that in oligotrophic lakes with only small quantities of reducing substances in their sediments and abundant oxygen dissolved in deep water, many species of invertebrates might be expected to occur, although each at low density. In contrast, eutrophic lakes may contain only a fraction of this diversity, but may be quantitatively much richer (Wesenberg-Lund, 1917).

The most common components of the profundal benthos are chironomid larvae, and the abundance of certain species, or species groups, is often characteristic of a particular lake type. Deep lakes, for example, may have only small populations of these midges, or none at all. This is believed to be more a reflection of the weak internal circulation found in many deep lakes rather than of any strict biological limitation of the insects (Barton, 1981). Chironomids have, logically, been used to classify lakes. Thienemann (1925), working on European lakes, and Deevey (1941), working on lakes in Connecticut, U.S.A., recognized several types, for example:

(1) *Chironomus* lakes - these are usually shallow and turbid, and have, in general, oxygen curves characteristic of eutrophic (nutrient-rich) lakes. They are dominated by species of *Chironomus* (Chironominae: Chironomini) in which the larvae typically have two pairs of ventral abdominal gills (Fig. 2.35C); common species are *C. decorus* (in North America), and *C. anthracinus* and *C. plumosus* (in Europe). The culicid *Chaoborus* is often present in open water.

(2) *Tanytarsus* lakes (Chironominae: Tanytarsini) - these are usually deep, oligotrophic (nutrient-poor) lakes which never lack oxygen in deep water. *Chaoborus* tends to be absent.

(3) Mesotrophic *Chironomus* lakes - these have oxygen curves typical of lakes of intermediate nutrient content, and characteristically support species of *Chironomus* that lack ventral abdominal gills.

(4) *Trissocladius* lakes (Orthocladiinae) - these become stratified, but are of inconsistent trophic status.

(5) Dystrophic lakes - these also have variable amounts of nutrients, but they are always high in humic compounds which colour the water brown. They tend to be shallow but can experience oxygen deficiencies in deeper parts. *Chironomus* and *Chaoborus* are often present but their densities are low.

(6) Unstratified, faunistically and limnologically diverse lakes.

Such distinctions have proved useful in comparing lakes within the Holarctic, but in a country like New Zealand, where the chironomid fauna is particularly depauperate, they have little or no value as biological indicators (Forsyth, 1978). The same tends to be true of the littoral fauna, in general. For example, Saether (1975) showed that although lists of littoral dipterans from lakes in both Europe and North America identified some species restricted to oligotrophic systems, those found in eutrophic lakes tended to be more widely distributed and therefore less useful. Further, seasonal differences in littoral species from a given lake are greater than those seen in profundal species (Smock *et al.*, 1981). Shallow lakes therefore fit existing classification schemes less well than deep ones (Harper and Cloutier, 1986).

The *pelagic zone* includes both the surface water, where the animal community is phytoplankton-based (trophogenic), and the water below, which extends down to the bottom ooze. The latter is characterized by low insect diversity and abundant bacteria, and its fauna relies on food in the form of a steady rain of detritus from the water layers above (tropholytic). In ponds, shallow lakes and clear deep lakes, where light penetrates deeply, the tropholytic sub-zone may be missing. Pelagic animals form part of either the *plankton* (passive drifters, or poor swimmers) or the *nekton* (powerful swimmers, capable of vertical and horizontal migrations). For the most part, insects belong to the nekton and forms such as *Chaoborus* are known to make vertical migrations to depths of 10 m or more during the day, returning to the surface at night (Teraguchi and Northcote, 1966). Some chironomid larvae also make large vertical migrations (e.g. *Saetheria* and *Chironomus* in Lake Michigan; Wiley and Mozley, 1978) and chironomid pupae can form an important seasonal component of the nekton.

In addition to the four aquatic zones described above, a few insects belong to a "community" associated with the lake water surface, the *neuston*. Predominant, here, are the Hemiptera (Gerridae, Veliidae and Hydrometridae), Coleoptera (Gyrinidae) and Collembola.

The extent of these biotic zones is related to the origin of the lake basin so that, for example, steep-sided lakes of tectonic origin have small littoral and sublittoral areas, but large profundal and pelagic regions. Conversely, ground moraine lakes have large littoral, sublittoral and pelagic zones, but may lack a profundal zone. Moraine lakes provide another type of habitat suited to specialized insects. This is the interstitial environment of the littoral and sublittoral zones, the fauna of which is termed the. *psammon* (Pennak, 1940). Here, life, as in the hyporheic zone of running waters, is dominated by substrate particle size and many of the organisms are characteristically elongate and flattened, completely filiform, or very small. For the most part, the fauna is dominated by microcrustaceans, but small chironomid larvae may be present also (Williams, 1979).

Very large lakes have unique properties that influence the insects that live in them. Despite differences in local climate and physical and chemical conditions, these large waterbodies have two unique habitats in common: remote offshore regions, and wave-swept surf zones (Barton and Smith, 1984). Relatively few insect species have been able to adapt to life in remote offshore regions, probably because this involves achieving total independence from land. Only a few species of chironomid and chaoborid (especially those belonging to the genera *Chironomus, Tanypus, Procladius* and *Chaoborus*) living in ancient African lakes like Malawi, Victoria and Edward appear to have achieved this (Beadle, 1981). In these species, emergence occurs *en masse* and mating and oviposition occur over the open lake. Emergence is highly synchronized to ensure that these events can be completed within the relatively short time that the adults can remain in flight over the water. An alternative strategy would be development of a fully aquatic life cycle, in which adults might remain totally submerged. Although a very few species appear to have achieved this (e.g. wingless adults of the stonefly *Capnia tahoensis* have been collected on submerged plants at depths of 60 to 80 m in the exceptionally clear waters of Lake Tahoe, Nevada; Jewett, 1963), it is not a solution adopted extensively by aquatic insects. The wave-swept surf zone of very large lakes results in the same dominance by lotic forms, noted earlier, in the exposed littoral zone of smaller lakes. However, the composition of the insect fauna of very large lakes is more similar around the world (Barton and Smith, 1984), with the exception of Lake Baikal where a combination of eradication of the shore fauna by glacial ice and competition from the huge variety of endemic amphipod species is thought to have resulted in low insect diversity.

The insect fauna of very large lakes is particularly influenced by time. For example, the Canadian Great Lakes have a somewhat lean fauna, a consequence of the relatively short time for which they have existed (since the Pleistocene glaciation ~ 10,000 years B.P.). In much older lakes, such as Baikal, Ohrid and Nyasa, there has been sufficient time to allow the evolution of endemic species, such as those belonging to the Baicalinini, a tribe of flightless limnephilid caddisflies unique to Lake Baikal (Kozhov, 1963).

Latitudinal differences are imposed on lentic faunas perhaps more so than on lotic ones, with the exception of the wave-swept faunas of very large lakes. The predominance of the relatively warm-water adapted Chironominae in tropical and temperate lakes, for example, is replaced by that of the more cold-water adapted Orthocladiinae and Diamesinae in Polar lakes (Welch, 1976). Other warm-adapted, lentic insect groups peter out towards the Arctic; there are, for example, no Notonectidae, Corixidae, Gyrinidae or Dytiscidae in lakes north of Brooks Range, Alaska (68°N) (Livingstone *et al.*, 1958).

On a much smaller scale, lake insects are known to be influenced by many of the factors that influence running water insects. The microdistribution of chironomids, in particular, is strongly tied to substrate grain size. In Lake Michigan, for example, *Chironomus fluviatilis* has been shown to be primarily associated with fine sand, *Cryptochironomus* and *Robackia* occur on coarser material, and *Polypedilum, Heterotrissocladius, Hydrobaenus* and *Micropsectra* are strongly associated with very fine sand (Winnell and Jude, 1984). Variation in pH among lakes is also known to affect chironomid assemblages (Raddum and Saether, 1981).

Ponds

Although ponds may be formed in many of the ways described for lakes, though on a smaller sale, the definition proposed earlier sets them apart on the basis of their support of aquatic plants. Macrophytes grow in more or less clearly defined zones and particular groups of insects are associated with each zone.

Closest to the shore are the emergent macrophytes. These are rooted in water-saturated substrate and have their stems and leaves mostly out of the water. They include grasses, cattails (*Typha*), bulrushes (*Scirpus*), grass-like rushes (*Juncus*), a variety of reeds, and wild rice (*Zizania*), among the more cosmopolitan forms, together with locally prevalent species. Some, such as waterlilies (e.g. *Nymphaea* and *Nuphar*), watershields (*Brasenia*), smart-weed (*Polygonum*) and pondweed (*Potamogeton*), have floating leaves.

In deeper water, submerged macrophytes predominate. These are rooted and generally have small, narrow or finely divided leaves, but may have emergent leaves and flowers during their reproductive phase. Examples are: hornwort (*Ceratophyllum*), milfoil (*Myriophyllum*), pondweed (*Elodea*), water-crowfoot (*Ranunculus*), tape grass (*Vallisneria*), and bladderwort (*Utricularia*). Also present may be certain free-floating species some of which have roots that hang down into the water. Many of these are small species (e.g. the duckweeds *Lemna* and *Wolfia*) although, in the tropics, larger forms such as the water hyacinth (*Eichhornia*) occur.

The insects associated with this vegetation include: both Zygoptera and Anisoptera, amongst the Odonata, particularly those families adapted for climbing slowly through dense vegetation, such as the Coenagrionidae and Aeshnidae; certain baetid mayflies; caddisflies of the families Phryganeidae, Limnephilidae and Leptoceridae; and bugs belonging to the families Corixidae, Notonectidae, Pleidae, Naucoridae, Nepidae and Belostomatidae. Also likely to be present are beetles of the families Dytiscidae, Hydrophilidae, Haliplidae, Gyrinidae, Helodidae, Chrysomelidae and Curculionidae; Megaloptera; and Diptera, especially the Tipulidae, Ptychopteridae, Dixidae, Culicidae (Culicinae), Chironomidae (particularly the Chironominae, some of which bore into macrophyte stems), Ceratopogonidae, Stratiomyidae, and a few Tabanidae. Three dipteran families, the Syrphidae, Tetanoceridae and Ephydridae, commonly occur around the water line (Johannsen, 1969).

Some insects live on the undersides of floating leaves. Certain chironomids use them as a base for building their tubes, and many others, like beetles, caddisflies and damselflies, lay their eggs there. Other species live on the tops of floating leaves, for example the waterlily leaf beetle (*Galerucella nymphaeae*) deposits egg masses on the upper surface of *Nuphar luteum* and the resulting larvae feed on the tissues producing irregular trenches in the leaf surface; pupation occurs on the leaf surface. It has been suggested that the unpalatability of these beetles to predatory birds may be due to chemicals, derived from the leaves, that accumulate in these insects' tissues (Otto and Wallace, 1989).

Open areas of pond bottom, consisting mostly of silt and plant debris, are likely to be colonized by caenid mayflies, chironomids and dragonflies, especially the Cordulegasteridae, Petaluridae and Gomphidae.

Several important pond types characterize different parts of the globe. *Arctic tundra ponds*, for example, are formed by the melting of ice. They are quite shallow and are isothermal from spring to fall, with no real thermocline. When these ponds are ice free, dissolved oxygen is close to saturation, but under snow and ice they become sealed off from the air and, for 6 to 7 months steadily lose oxygen. Benthic insect diversity is low, but populations can be large, and chironomids predominate (up to 95% of the community; Oliver, 1968). Most arctic chironomids have at least a two-year life cycle with emergence occurring throughout much of the ice free stage (Stocker and Hynes, 1976), although different species emerge at slightly different times in the same pond. Eggs deposited in the summer develop to second instar larvae, then overwinter, develop to fourth instars in the second summer, overwinter again, and finally emerge in the third summer. During overwintering periods, larvae become frozen into the ice. Many coenagrionid damselfly nymphs are able to survive the winter frozen in the ice of *aestival ponds* (also known as sloughs) which are shallow, semi-permanent ponds that occur at high altitudes or latitudes (Daborn, 1971). Various other pond types hold water only intermittently and these will be considered later.

Marshes and peatlands

Most shallow ponds represent stages in the transition from aquatic habitats to terrestrial ones. This change is brought about, primarily, by the accumulation of organic matter faster than it can be degraded. These stages are represented in Fig. 4.3. As the water regime and vegetation change, so does the insect fauna. At a certain point in this transition, ponds contain water for only part of the year, or not every year; they are then classed as temporary aquatic habitats (see later section).

Fig. 4.3   Transitional steps from a permanent aquatic habitat to terrestrial and wetland habitats (after Wetzel, 1983).

Wetlands are found in landscapes that are neither fully terrestrial nor fully aquatic, particularly where there is a moisture surplus or poor surface drainage, or adjacent to water bodies where waterlogged conditions are maintained. In such areas, the groundwater table remains near or above the soil for most of the growing season and the dominant vegetation is adapted to wet environments (Zoltai, 1980). Local climate is clearly an important contributing factor.

As with all other water bodies, attempts have been made to classify wetlands. In the broadest sense, there are five classes: bogs, fens, marshes, swamps and shallow open waters. *Bogs*, are peat-covered wetlands in which there is a general lack of nutrients and where the vegetation is influenced by a high water table (e.g. cushion-forming *Sphagnum* mosses are common, as are heath shrubs) and acid water (pH < 4.5). *Fens* also represent waterlogged peatlands but, here, the pH of the water is circumneutral (5.5 to 7.5) and tends to be rich in mineral nutrients. Grasses, sedges, brown mosses, shrubs and even trees dominate the vegetation. *Marshes* represent areas of wet mineral soil,

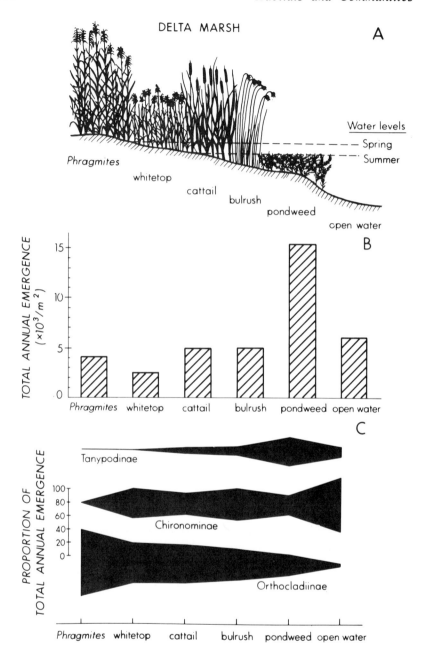

Fig. 4.4 Microdistribution of chironomids along a moisture and vegetation gradient in a typical prairie marsh, Delta Marsh, Canada: (A) elevational and vegetation gradients; (B) total annual chironomid emergence from each microhabitat; (C) proportion of the total emergence represented by each chironomid subfamily (redrawn after Wrubleski, 1987).

sometimes incorporating a shallow layer of peat. They are characterized by emergent vegetation, particularly rushes, reeds and/or sedges. As they are typically flooded by adjacent water bodies, they are usually rich in nutrients. *Swamps* mainly consist of dense growths of large shrubs and trees with, on occasion, only a scattering of peat-forming plants (Zoltai, 1987).

The communities of insects found in bogs, fens and marshes, although not well studied, include species belonging to most of the major aquatic groups. However, some groups, such as the Plecoptera and certain dipterans (e.g. Simuliidae), appear to be absent, while others, like the Ephemeroptera, are poorly represented. Fens are the least studied peatlands, but preliminary surveys suggest that they contain a rich insect fauna, especially in terms of chironomids, hemipterans and beetles. Besides these three groups, bogs and marshes also contain odonates, caddisflies (especially Phryganeidae and Limnephilidae) and a variety of other dipterans, notably culicids, tabanids and ceratopogonids. Most species known from Canadian peatlands and marshes are aquatic generalists, particularly those from marshes, as these habitats typically continue from lake margins and are colonized by species common to a number of different shallow aquatic habitats. Apparently, few species are restricted to wetlands, apart, perhaps, from bogs where certain insects (e.g. dragonflies) appear to be specially adapted to the unique conditions found there (Danks and Rosenberg, 1987).

Microdistributions are evident in the insects of some wetlands, and these are thought to be related to vegetation changes linked to water level. In Delta Marsh, Canada, for example, the microdistribution of chironomids, based on adult emergence sites, has been shown to vary among major vegetational communities along a gradient from deeper open water to drier land (Fig. 4.4).

4.1.3 Salt waters

Seashores

Most marine insects live in the intertidal zone (i.e. between the high and low tide marks), especially on rocky shores or associated with detached and decaying seaweed on the strand-line of sandy beaches. Although diversity can be high, it is only a few orders that have colonized these habitats in any numbers, most notably the Diptera, Coleoptera and Collembola. Physical features of the intertidal zone, such as powerful waves, currents, and alternate but cyclical exposure to water and air, have forced many insects to reside buried in sand or mud, or hide under seaweed or in crevices. Thus marine beetles, for example, are typically benthic in habit and most are specialized members of largely terrestrial families, primarily the Carabidae and Staphylinidae. Those families adapted to fresh water, with the exception of those that can live in intertidal rockpools, have few marine species. In particular, there are no marine representatives of the Elmidae, Dryopidae or Psephenidae, other than a single species of dryopid which occasionally enters brackish water (Doyen, 1976).

In contrast, most marine dipterans belong to families that commonly occur in fresh waters: Chironomidae, Ceratopogonidae, Culicidae, Tabanidae, Ephydridae and Tipulidae. Only three subfamilies of chironomid are found in the sea, the Chironominae, Orthocladiinae and Telmatogetoninae, and these are represented by about 12 genera (~ 50 species) which are primarily restricted to the intertidal zone; a very few species are found in the open sea. The larvae of marine chironomids, unlike those of freshwater species, lack anal gills and, in the pupa, the prothoracic respiratory horns are lacking. Further, in the adult, the antennae, palps and eye facets are much reduced and, in some, the wings, halteres and legs may be modified or degenerate. Alongside such morphological modifications, may be behavioural ones, particularly those associated with locomotion, reproduction and emergence (Hashimoto, 1976).

Three genera of Ceratopogonidae are common in marine environments: *Culicoides* (represented by approximately 50 - 60 marine species), *Leptoconops* (~ 9 marine species) and *Dasyhelea* (whose larvae are poorly known); and the three occur in all major regions of the world. Adult ceratopogonids ("no-see-ums", "sandflies", "punkies") are blood-suckers and, as well as being a nuisance, some may transmit pathogenic organisms. Marine species are reported to be especially troublesome, even to the point of jeopardizing the regional economy on certain coasts (Linley and Davies, 1971). They occur in a variety of habitats including the intertidal areas of sheltered beaches, moist supralittoral sand, where streams and rivers flow over beaches, wet mud in mangrove swamps, saltmarshes and tidal flats, and crab burrows (Linley, 1976).

Marine culicids belong to nine major genera, seven of which also contain species found in fresh water. Their habitats include temporary and permanent pools in saltmarshes and mangrove swamps, crab holes, and supralittoral rockpools kept brackish by sea spray (O'Meara, 1976). As in the case of the ceratopogonids, saltwater mosquitoes may transmit diseases, and this is true also for some of the tabanids whose larvae develop in the soil of saltmarshes, brackish pools and tidal overwash areas (Axtell, 1976), and which may loosely be called "marine" insects.

The Ephydridae is one of the most diverse and widespread families of the more advanced dipterans, having species in almost every type of shallow aquatic and semi-aquatic habitat. Some 36 genera live in marine environments ranging from rocky shores and sandy beaches to mangrove swamps and tidal and estuarine marshes. Within each of these areas, species often live in specific microhabitats characterized by certain features of substrate, tidal level or vegetation (Simpson, 1976).

A family of dipterans specializing in life on the rafts of decaying seaweed deposited on the strand-line, is the Coelopidae, or kelp flies. Since they have a preference for brown kelps, their distribution is largely restricted to temperate and subarctic shorelines where these algae grow. Adults are attracted to the smell of the seaweed and colonize newly stranded plants within hours of deposition. They are able to breed even throughout the winter in temperate regions, as the eggs are kept from freezing by the heat from the decay process which is accelerated by the actions of the larvae. On ice-covered beaches, temperatures are known to reach 40°C over a distance of 10 to 20 cm into the seaweed mass. Life cycles are often completed in 2 to 4 weeks. Although the coelopids are the main inhabitants of stranded seaweed, two other families of Diptera may be found, the Dryomyzidae and Muscidae (Dobson, 1976).

Some hemipterans are associated with seashores. Shore bugs belong primarily to the family Saldidae (although there are a few species of Ochteridae), whose members mostly live in fresh water. Saldids show behavioural rather than morphological adaptations to a marine existence, and even though they may survive periods of submersion in salt water, they basically appear to be terrestrial bugs with little in the way of adaptations to permit respiration under water (Polhemus, 1976). Typical habitats include the littoral zone, saltmarshes and mangrove swamps. Many species of Gerridae are able to extend their habitat ranges to include the surfaces of coastal waters, particularly marine pools, estuaries and brackish lagoons. There are also coastal Veliidae, in the tropics, particularly in the sheltered waters of lagoons and mangrove swamps. Some 12 genera of Corixidae have been recorded from saline waters, but these are restricted to shallow waters with suitable oviposition sites, and frequently are limited by high salinity. Only species of *Trichocorixa* are found, regularly, in salinities exceeding 30 parts per thousand (ppt) (Scudder, 1976).

Two other hexapod groups commonly occur on the shore: the Collembola, typically found in crevices or amongst algae on rocky shores (although there are some that live interstitially between the sand grains on open beaches); and the Thysanura which tend to be restricted to rocks and crevices around the high-water mark.

Some other insect groups occur in brackish water tidepools, in which the salinity is below that of full strength sea water. These include caddisflies (e.g. *Limnephilus tarsalis, L. affinis* and *Oecetis* sp.) and dragonflies (e.g. *Aeshna*) (Williams and Williams, 1976). Around the coasts of New Zealand and southeastern Australia, larvae of the caddisfly *Philanisus plebeius* live in tidepools after hatching from eggs which develop in the coelomic cavity of the starfish *Patiriella regularis*. It is thought that the female deposits her eggs through the papular pores on the aboral surface of the host, and that the larvae may leave by this same route or through the wall of the stomach (Winterbourn and Anderson, 1980). The significance of this unique early development is not understood and merits further study.

Open ocean

The only insects that inhabit the open oceans are hemipterans. Some 44 species of gerrid are now known to live on the surface of the oceans, and all belong to the genus *Halobates*. The greatest diversity occurs in the Indo-Malaysian and Australasian region, which is thought to be the evolutionary centre of the genus.

Thirty-nine species have coastal distributions, usually in sheltered localities, but five live on the open ocean, including the middle of the Pacific. The latter are all found in tropical or subtropical waters, where the winter temperature does not fall much below 15°C, and where distribution and abundance seems to be regulated by surface water temperatures. *Halobates micans* is the only cosmopolitan species, being found in all three major oceans (Pacific, Indian and Atlantic). Endemism appears to be high, particularly among the coastal species (Cheng, 1985).

One of the most fascinating questions concerning the distribution of insects is: given the unparalleled success of insects in both terrestrial and freshwater environments, why are they so poorly represented in the sea? There are in excess of 30,000 species of insect that live in fresh waters, but the number that can be called "marine" amounts to only several hundred. Surprisingly, few entomologists or marine biologists have attempted to resolve this paradox. Some hypotheses have been proposed, particularly in the 1920s and 1930s - some of which still tend to be quoted today, but these, in our opinion, have not been examined rigorously. The primary explanations are: (1) that physical barriers, such as marine currents, tides and waves, prevent effective colonization; (2) that chemical barriers, particularly high salinity but also the low calcium content of sea water, do likewise; and (3) that before the insects were in a position to begin colonizing the sea, the highly successful and already diversified Crustacea had become so well established that, in effect, most available marine niches were full (Buxton, 1926; Pruthi, 1932; Usinger, 1957). It is clear that purely physical obstacles like waves (and to some extent tides) have been overcome by insects living in the surf-zone of very large lakes, whereas currents pose little problem to most lotically-derived insects. Although this was conceded to some extent, early on, the view was held that, in combination, such obstacles still acted as a formidable barrier, unsurmountable by the vast majority of freshwater insects (Buxton, 1926). It is thought by some (e.g. Norris, 1991) that water level fluctuation in the intertidal zone is too much and too often for many of the air-renewal mechanisms employed by freshwater insects (e.g. siphons and plastrons) to be effective. However, there are many coastlines throughout the world where tidal amplitude is only of the order of about 1 m (e.g. Bermuda), creating conditions comparable with those produced by seiches on the shores of large lakes where insects can be abundant. Potential marine endophytic breathers (see Chapter 6) are thought to face problems of contending with a radical change in host plant group.

In a more recent analysis of the chemical barriers (Cheng, 1976), it has been suggested that the physiology of the muscular or nervous systems of an insect is so specialized that it cannot tolerate the physicochemical conditions found in the sea; alternatively, the osmotic regulation and submarine respiration needed involves the evolution of such different physiological adaptations that only a handful of insects have been successful in attaining both goals. Nevertheless, the necessary adaptations to deal with full strength sea water are seen in a number of different aquatic groups, for example the Collembola, Diptera, Hemiptera, Coleoptera and even Ephemeroptera (Pruthi, 1932), and the ability to deal with brackish water is seen in additional groups (e.g. Odonata and Trichoptera). Since at least a few species from virtually all the major insect orders occurring in fresh water clearly have managed to adapt to the physiological demands necessary for living in salt water, water chemistry does not seem to be too great an obstacle.

The idea that the crustaceans colonized marine habitats first and have since repelled the majority of insect invaders seems unlikely. Natural selection is not a static process and it is therefore unlikely that the occupancy of many marine niches has remained the same since the Cretaceous, when many freshwater insects were established. On land and in freshwater benthic habitats, it is the crustaceans that are in the minority, in some cases no doubt, because of the superior competitive abilities of insects. A seed of contention, however, is perhaps sown by the fact that insects are not as abundant in ancient Lake Baikal, coincident with a huge array of endemic amphipod species, as they are in other large lakes.

Another aspect, related to the occurrence of insects in very large lakes, is that relatively few species inhabit remote offshore regions. As we have noted, living at such large distances from shore necessitates development of total independence from land (Barton and Smith, 1984). This is a trait not common among freshwater insects and one that seems to be linked to an ability to congregate and mate during the short time that adults can remain in flight over the water. Even so, some freshwater insects have overcome this by developing a fully aquatic life cycle in which there is no need for the adults to emerge from the water. This also appears to have been achieved by a marine chironomid (Bretschko, 1982). Thus even spatially-related impediments to typical insect life history structure, that might be created in the oceans, have been overcome by at least one species of chironomid.

Since the majority of marine insects are found in what have been termed "bridging habitats" (e.g. estuaries, saltmarshes, the intertidal zone and mangrove swamps; Cheng, 1976), it could be argued that we are currently looking not at the end of an evolutionary pathway leading freshwater insects into marine habitats, but at early steps in the journey.

Inland saline lakes, ponds and springs

The argument that insects might have difficulty in attaining independence from land seems plausible enough for the oceans, but what is the situation in small, inland saline waters?

Saline lakes and ponds occur in a number of different parts of the globe, but especially in Australia, northern and central Africa, the Middle East and western North America. The amount of salt dissolved in these habitats is far more variable than is found in the sea. Inland saline waters are usually defined as having a salinity of > 3.0 ppt although concentrations may reach 350 ppt, ten times the concentration in standard sea water. Although the ions most commonly encountered are sodium and chloride, magnesium, calcium, sulphate, carbonate and bicarbonate ions may also be present, depending on local geology.

Faunal diversity in saline lakes and ponds tends to be negatively correlated with habitat salinity. However, this relationship is not clear-cut and may be confounded by the fact that many such water bodies are temporary in nature, in which case diversity may be linked also to habitat predictability (Williams, W.D., 1984). Thus, in addition to having to contend with initially high salt levels, insects in temporary saline water bodies have also to deal with water loss from their habitat which, of course, compounds ionic stresses through evaporation. It may be that such adversities counteract any tendencies towards higher diversity that the small size (relative to the oceans) of many inland saline water bodies might bestow. In practice, relatively few insects appear to be able to cope with these adversities and thus higher diversity occurs at lower salinity in permanent waters.

Among the insects recorded from inland saline waters are: dragonflies (e.g. *Ischnura* and *Enallagma*), hemipterans (especially corixids), beetles (e.g. *Enochrus*, Hydrophilidae; *Hygrotus* and *Necterosoma*, Dytiscidae), caddisflies (Leptoceridae), pyralid moths, and dipterans (especially chironomids, mosquitoes, ceratopogonids and ephydrids) (Bayly and Williams, 1966; Hammer *et al.*, 1975; Tudorancea *et al.*, 1989). Ephydrids (*Ephydra* sp.) and ceratopogonids (*Culicoides variipennis*) have been collected in lakes with salinities in excess of 200 ppt (Williams, W.D., 1985). In magnesium carbonate - sodium bicarbonate dominated saline lakes in British Columbia, Canada, Cannings and Scudder (1978) were able to identify distinct chironomid associations according to salinity and productivity levels. A *Cricotopus abanus* - *Procladius bellus* association prevailed in the lowest salinities (40 to 80 μmho/cm), whereas in conductivities between 400 and 2,800 μmho/cm a *Glyptotendipes barbipes* - *Einfeldia pagana* association predominated. In the most saline lakes (4,100 to 12,000 μmho/cm) a *Tanytarsus gracilentus* - *Cryptotendipes ariel* association was evident.

Rare amongst inland saline habitats are salt springs. Few studies have been made of the insects of these habitats but indications are that the fauna consists mainly of halophilic species known from other saline environments such as the intertidal zone, saline lakes, salt marshes and brackish water. Among the insects recorded from natural salt springs on Saltspring Island, British Columbia, were chironomids (*Thalassosmittia marina*), ephydrids (*Ephydra* spp.) and the hydraenid beetle *Ochthebius*, within the pools, and collembolans, saldid bugs and the carabid beetle *Bembidion* around the margins of the spring. Most of these insects were present throughout the year suggesting life history patterns that take advantage of the relatively benign and constant environmental features present in this particular type of saline habitat. Notable absentees from this particular spring were staphylinid beetles, a diverse and abundant group on seashores (Ring, 1991).

4.2   Temporary habitats

Temporary ponds and streams are widespread and are essentially natural bodies of water that experience a recurrent dry phase of varying duration. From a faunistic point of view, it is the cyclical nature of the drought that is important, because some permanent water bodies may also become dry in exceptional years. In the latter case, however, much of the fauna may be exterminated, whereas the faunas of temporary waters are adapted to survive such conditions (Williams, 1987). As we have noted, differences between the insect faunas of permanent standing and running waters may be considerable, but because many temporary streams flow slowly and invariably form pools as they dry, there is a greater similarity between the inhabitants of temporary lotic and lentic water bodies.

In general, temporary water bodies exhibit much greater amplitudes of both physical and chemical parameters than are seen in most permanent aquatic habitats. In temporary streams, for example, the current may range from torrential proportions during spring floods to zero in the pools that remain after the flow has stopped. Water temperature can be highly variable, for example, in temporary ponds in temperate regions, from 0°C under winter ice to approaching 40°C on a hot day in summer. The latter is very near the thermal death point of most insects and under such conditions insects seek refuge by burrowing under debris on the pond bed or, in the case of adult beetles and bugs, by flying away to find cooler conditions. Because these habitats tend to be shallow, temperatures can change markedly even on a diel basis. Water chemistry fluctuates markedly, too. Large quantities of decaying organic matter on the bed, combined with high temperatures and extensive patches of respiring algae, can very quickly deplete dissolved oxygen levels. Further, as the volume of water is reduced by evaporation and retreat of the groundwater table, there is a progressive concentration of the ions in the water. In ponds with large algal biomasses, removal of dissolved carbon dioxide through photosynthesis may result in significant increases in pH.

Flow control in temporary streams is dependent on several factors, foremost amongst which is the balance between precipitation and the infiltration capacity of the stream bed and surrounding soil. The rate of evaporation may be important also, particularly during low flow conditions. The degree and rate of descent of the groundwater table and the permeability of the bed substrate are important in so far as they control the formation and duration of the pool stage. These pools support certain species of insect that are not normally able to live in a lotic environment and, at the same time, they invite colonization from many purely lentic forms that use these habitats as temporary sites for breeding, for example many of the Coleoptera and Hemiptera (Fernando and Galbraith, 1973). Bankside vegetation may affect the moisture retention ability of the drying stream or pond bed by controlling permeability of the surface substrate and by providing a humid canopy over the bed.

Because of the changing nature of the physical and chemical environment seen in temporary waters, the faunas often show seasonal succession. For example, the following succession of insects was recorded in a small temporary stream in southern Ontario, Canada (Williams and Hynes, 1977a):

(1) *Stream stage* (fall-winter) - Plecoptera, Trichoptera and Diptera (some Tipulidae, Simuliidae and Chironomidae);

(2) *Pool stage* (spring) - Ephemeroptera, Odonata, Hemiptera, Coleoptera and Diptera (other Tipulidae, Culicidae, Ephydridae, Syrphidae and Psychodidae);

(3) *Terrestrial stage* (summer) - Coleoptera (Heteroceridae, Staphylinidae and Scarabaeidae), Hymenoptera (Formicidae) and Diptera (Sepsidae, Sphaeroceridae and Ceratopogonidae).

Comparison of the insects found in temporary ponds around the globe shows that several groups consistently predominate: the Hemiptera, Coleoptera, cased Trichoptera, Odonata, Chironomidae, Culicidae and Ceratopogonidae. Temporary water species within these groups invariably show special features of either their physiology or life cycle, which make them successful in these habitats and allow them the means to colonize them (Williams, 1987). A recent study of a temporary pond in eastern Australia (Lake *et al.*, 1989) has shown that in addition to the ubiquitous taxa listed above, a stonefly (*Dinotoperla bassae;* Gripopterygidae) and two species of aquatic moth (Pyralidae and Oecophoridae) are capable of surviving in a temporary, standing water environment. Three species of stonefly (Nemouridae) are also known from temporary ponds in North America (Lehmkuhl, 1971) but, in general, the Plecoptera are absent from these habitats.

Rainpools

In many parts of the world rain collects in small depressions in bedrock. Because these pools tend to be shallow and are often very exposed to wind and sun, they are highly ephemeral in nature, particularly in the tropics and subtropics. Nevertheless, many support aquatic insects, although diversity is low. On isolated hillsides in tropical Africa, McLachlan and Cantrell (1980) identified three types of such pools that differed in depth and hence life span. The longest-lived pools (several weeks) were populated by larvae of the chironomid *Chironomus imicola* which are not able to tolerate much in the way of water loss. Pools that existed for only a few days were inhabited by the ceratopogonid *Dasyhelea thompsoni* which survived buried in the mud at the bottom of the pool when the water evaporated. Pools in which water was present for only 24 hours contained larvae of the chironomid *Polypedilum vanderplanki* which are extremely tolerant of desiccation and can lose up to 99% of their tissue water, yet recover on rehydration when the pools next fill. Most of these pools support monospecific populations which perhaps underlines the rigours of such habitats. *Dasyhelea* and two species of chironomid have been found, along with aquatic beetles and hemipterans, in similar but longer-lived rainpools in southwestern Australia (Bayly, 1982). Large, longer-lived rainpools in the tropics can sometimes support dragonflies, gerrids and veliids (Hynes, 1955).

Phytotelmata

This term embraces a number of very small aquatic habitats associated with living plants. The most common are: treeholes, pitcher plants, bromeliads and inflorescences. Certain localities, particularly in the tropics, may contain large numbers of phytotelmata and thus, potentially, a regular supply of aquatic habitats. However, the fact that the insect communities which develop in individual phytotelmata can never reach equilibrium suggests that they represent habitats of low stability, justifying their categorization as temporary habitats.

Treeholes are formed where rainwater collects in depressions in the bark of a tree - usually at main branch forks or behind outgrowths of bracket fungi or scar tissue. Treehole communities have a simple structure, which appears to be remarkably similar across the globe, and include only two main insect orders: the Diptera (typically mosquitoes, chironomids, ceratopogonids, tipulids and syrphids) and Coleoptera (pselaphids, mould beetles; and scirtids, marsh beetles - especially *Prionocyphon*). Most of these insects feed saprophagously on decaying plant matter trapped in the holes, and competition appears to be avoided by temporal segregation of some species and the presence of food refuges among others (Kitching, 1983).

As in treeholes, dipterans are the most frequent inhabitants of the fluid-filled leaf chambers of *pitcher plants*. Three insect species are known from *Sarracenia purpurea*, the common pitcher plant of glacial peat bogs in North America. Only newly-opened pitchers attract and capture insect prey and these slowly decompose as the pitcher ages. The three insect inhabitants consume this material at different stages in the decomposition process, which produces a seasonal succession in the pitcher fauna. Larvae of *Blaesoxipha fletcheri* (Sarcophagidae) feed on freshly-caught prey floating on the surface of the pitcher fluid, whereas larvae of the mosquito *Wyeomyia smithii* filter-feed on the partially decomposed material in the water column, and larvae of the chironomid *Metriocnemus knabi* feed on the remains that collect on the bottom of the pitcher chamber (Fish and Hall, 1978). In Indo-Malaya, the predominant pitcher plants belong to the genus *Nepenthes* and these support assemblages of insects (chiefly mosquitoes) and other organisms that live either in the pitcher fluid or on the upper walls of the chamber. Complexity of the food web varies considerably in different species of *Nepenthes,* and this is believed to be related to the degree of spatial and temporal isolation of pitcher species, the size of the local pool of species capable of living in pitchers, and the diversity of *Nepenthes* present, locally (Beaver, 1983).

Bromeliads are abundant in the tropics and subtropics of the New World. For example, some 2,000 species are known from tropical and warm temperate regions of the Americas, where they range in altitude from sea level to over 4,000 m. Tank bromeliads have enlarged leaf bases that overlap tightly and collect rainwater. Some species are long-lived (> 20 years) and thus provide relatively long-term aquatic habitats that are colonized chiefly by insects, microcrustaceans and amphibians. Although the volume of water contained in a single, large bromeliad is seldom larger than 1.3 litres, it has been calculated that, in the cloud forests of Columbia, the total volume of water contained in leaf axils amounts to more than 50,000 litres/hectare (Sugden and Robins, 1979). Again, it is the Diptera that dominate these habitats. Most abundant are mosquitoes, but chironomids, ceratopogonids, psychodids, tipulids, syrphids, stratiomyids, tabanids, muscids and phorids are common also. Other insects include beetles (especially the Scirtidae = Helodidae), bugs (Veliidae), one species each of stonefly and caddisfly, and many odonates, particularly damselflies (Fish, 1983). Bromeliads growing in forested areas (dendrophilous bromeliads) collect nutrients leached by rainfall from tree canopies and from plant debris, and have communities that are detritus-based. In contrast, those growing at the tops of forest canopies, or on cacti in deserts, (anemophilous bromeliads) collect windborne nutrients and tend to support algal-based communities (Frank, 1983).

Other parts of large plants can retain sufficient water to support small aquatic communities, besides leaf axils. *Heliconia*, for example, is a genus of largely Neotropical plants with large *inflorescences* in which each flower bract may hold around 8 to 9 ml of water. *Heliconia caribea* in the northern, lowland tropical rainforests of Venezuela is known to harbour up to 15 species of insect (Machado-Allison *et al.*, 1983). Of the eight most abundant species, three are mosquitoes (belonging to the genera *Wyeomia, Culex* and *Toxorhynchites*), one a syrphid, one a richardid, and one a stratiomyid. Food input to the community appears to consist of decomposition products of the flowers and the inner-wall of the bract.

Dung

Several semi-aquatic species of insect live in dung, and there is a distinct similarity between some of the taxa found here and in the summer terrestrial fauna of temporary ponds (Williams, 1987). Cow dung produced in summer, for example, when the animals are grazing on fresh grass, is quite liquid (73 - 89% water; Hammer, 1941). Common inhabitants include the larvae of a number of hydrophilic dipteran families (especially the Sphaeroceridae, Tipulidae, Sepsidae, Ceratopogonidae, Empididae, Stratiomyidae and Chironomidae) as well as beetles belonging to the Hydrophilidae (scavenging waterbeetles), Histeridae (hister beetles) and Staphylinidae (rove beetles) (Laurence, 1954; Merritt, 1976). The habitat changes with age, and this leads to a seasonal succession of species. A species of ephydrid (*Allotrichoma livens*) is known to breed in pig droppings in muddy pens (Bohart and Gressitt, 1951). In Britain, larvae of *Eristalis* (Syrphidae) live in tanks of liquid manure. In Hong Kong, chironomid larvae (*Chironomus* spp.) are grown commercially on chicken manure for use as live food for aquarium fishes and fish fry under aquaculture (Shaw and Mark, 1980).

The larvae of many dipterans inhabiting dung show respiratory features in common with aquatic and semi-aquatic species.

4.3   Man-made habitats

Most of the insect habitats considered thus far are formed naturally. There is, however, a group of commonplace habitats that are anthropogenic.

Reservoirs are man-made lakes, and might be expected to have lake-like insect faunas. However, their mode of operation prevents this. Management practices also have a profound effect on outflow streams. Unlike most natural lakes, water allowed to flow out of a reservoir is frequently taken from the hypolimnion. Summer water temperatures of the receiving stream are thereby kept low and the water may be chemically unusual, containing little oxygen at first, but many ferrous, manganous and sulphide ions, and bacteria (Symons *et al.*, 1964). Retention time of the water may be short, which again affects the chemical composition of the outflow. Because of wide fluctuations in water level, an extensive "intertidal" zone is created on the banks of both the reservoir and the stream. This zone, unlike its natural marine counterpart, is not inundated at regular intervals and thus most insects (apart from chironomids) seem unable to colonize it.

When reservoirs are first filled, a considerable adjustment takes place in the fauna. Initially, the benthos consists of remnants of the terrestrial and semi-terrestrial faunas (including some tipulids) together with obligate and facultative rheophiles derived from the impounded stream. Within a few weeks to months, populations of lentic insects (particularly chironomids - Chironomini) develop and the terrestrial and obligate rheophile components are lost (Paterson and Fernando, 1970). As reservoirs age, they assume some of the insect components of shallow lakes, and chironomid diversity can become particularly high. However, while *Procladius culiciformis, Endochironomus albipennis* and *Tanytarsus lugens* appear to characterize the littoral faunas of reservoirs in the Palaearctic, none of these species is known from natural lakes in the region (Rosenberg *et al.*, 1984). As in natural lakes, dominant chironomid groups can be correlated with the trophic status of a reservoir, for example Tanypodinae in eutrophic reservoirs and Orthocladiinae in oligotrophic ones.

The effect of a reservoir on the insects of the outflow stream may be considerable. For example, in Canada, Spence and Hynes (1971) showed that the downstream effects of a flood control reservoir were comparable to those resulting from mild organic enrichment; Plecoptera disappeared, and certain mayflies (*Baetis* and *Caenis*) increased in number, while others (e.g. *Stenonema*) became scarce. Populations of chironomids, simuliids, elmid beetles and net-spinning caddisflies increased downstream, associated with an increase in available detritus and plankton coming from the reservoir. In addition, the submerged support structures of the dam wall served as ideal attachment sites for many of the filter-feeders. Prevalence of filter-feeders, and especially caddisflies, below dams is a worldwide phenomenon (e.g. *Cheumatopsyche* and *Amphipsyche* in South Africa; *Hydropsyche, Cheumatopsyche, Macrostemum* and *Chimarra* in North America; and *Amphipsyche* in Java; Chutter, 1963; Boon, 1984).

Stream canalization, as part of the process of drainage alteration, also tends to create ideal attachment sites and flow regimes (fast and laminar) for filter-feeding insects, as well as reducing the variety of microhabitats through removal of substrate, littoral areas and macrophytes.

Techniques in modern *agriculture* have created specialized aquatic habitats for insects. Excavation of farm ponds, for storing rainwater in dry areas or to store slurry, is a common practice. In the case of the latter, levels of phosphate, ammonia and nitrate can be extremely high, removing all but the nuisance species of insect such as mosquitoes and other pollution tolerant dipterans (e.g. Syrphidae, Scathophagidae and Ephydridae). Cleaner farm ponds may become intermittently dry, as the water is drawn off for livestock, and may develop faunas similar to those found in natural temporary ponds. Farm ponds, together with temporary rainpools on agricultural land, often support high populations of mosquitoes. Some pools are selected over others and it is thought that these are: (1) chemically more attractive to ovipositing females; (2) adjacent to plant cover which provides suitable resting places for gravid females; and (3) closest to cattle herds, a convenient source of blood (Dixon and Brust, 1972).

Clearing of land, the proliferation of ploughed fields over pasture, and the use of wire fences in place of earth-bank hedgerows changes local drainage patterns, especially in areas of clay soil, so that many temporary streams and runoff channels are created where previously the uncultivated land would have supported a much smaller number of permanent streams and rivers. A desire to control this increased surface runoff, as well as to make marshy areas agriculturally productive, has led to the practice of laying drainage tiles. These subterranean lotic systems have been found to support collembolans, chironomids and several types of aquatic beetle (Hydrophilidae, Scirtidae and Staphylinidae; Williams, 1976).

Rice fields in the tropics are shallow, nutrient-rich, man-made lakes, that attract many species of insect. Records from rice fields in north eastern Thailand include: a baetid mayfly (*Cloeon*); odonates (Agrionidae, Libellulidae, Gomphidae); and a great diversity of hemipterans (Belostomatidae, Nepidae, Ranatridae, Pleidae, Mesoveliidae, Gerridae, Hydrometridae, Notonectidae, Corixidae, Hebridae); beetles (Gyrinidae, Dytiscidae, Hydraenidae, Hydrophilidae, Hydroscaphidae, Curculionidae, Carabidae, Staphylinidae); and dipterans (Chironomini, Tanypodinae, Culicidae, Chaoboridae, Dolochipodidae, Empididae, Ephydridae, Ceratopogonidae, Sphaeroceridae, Stratiomyidae, Tabanidae) (Heckman, 1979).

Roadside ditches represent another common type of temporary aquatic habitat that contains water during spring runoff and after heavy rain. In the

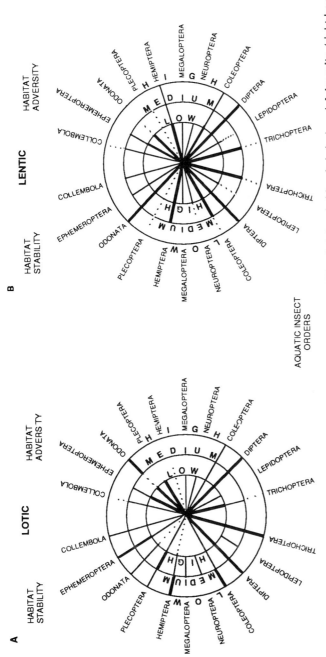

Fig. 4.5   Distribution of the aquatic insect orders along two habitat axes: **Stability** (left-hand side of circles) and **Adversity** (right-hand side of circles). (A) Lotic habitats - highly stable habitats are represented by permanent, cold, freshwater springs; medium-stable habitats = permanent streams and rivers; low stability habitats are represented by temporary streams; high adversity habitats are represented by hot springs, saline springs and waters of extreme pH; medium-adverse habitats include polluted streams and rivers, and estuaries; low adversity habitats encompass clean, cool, fresh, running waters, in general. (B) Lentic habitats - high stability = very high salinity, permanent ponds and lakes; medium stability = temporary ponds; low stability = phytotelmata; high adversity = very high salinity, extreme pH, gross pollution and petroleum ponds; medium adversity = very large lakes, brackish water, mild pollution; low adversity = clean, freshwater lentic habitats of small or moderate size (the width of the distribution lines is proportional to the importance of the taxon in a particular habitat type, broken lines indicate that the taxon is poorly represented in that habitat type).

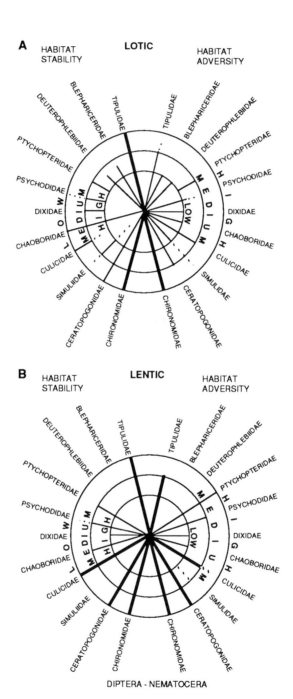

DIPTERA - NEMATOCERA

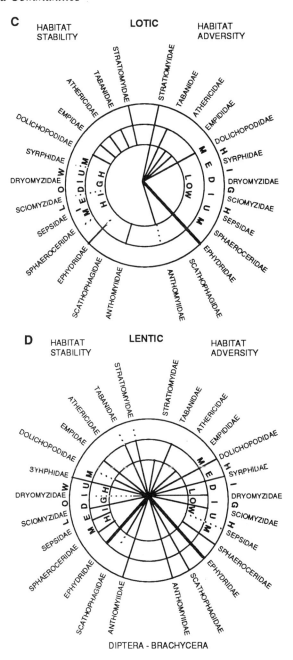

Fig. 4.6   Distribution of the aquatic and semi-aquatic Diptera along two habitat axes: **Stability** (left-hand side of circles) and **Adversity** (right-hand side). (A) Nematocera in lotic habitats; (B) Nematocera in lentic habitats; (C) Brachycera in lotic habitats; (D) Brachycera in lentic habitats (see Fig. 4.5 for descriptions of typical habitat types).

interim they often become choked with fallen leaves or support luxuriant growths of grasses and weeds. Despite the brief aquatic phase, which may be lotic, lentic or a mixture of both, they contain species of Collembola, Corixidae, Gerridae, Hydrophilidae, Dytiscidae, Chironomidae, Tipulidae and Culicidae, most of which have short larval stages and adults with good powers of dispersal.

Within *urban environments*, aquatic insects become established in a variety of habitats such as eaves troughs (semi-aquatic Coleoptera and Diptera), non-chlorinated backyard swimming pools (Coleoptera, Hemiptera and Chironomidae), bird baths (Coleoptera and Diptera) and water butts. Parthenogenetic species of *Paratanytarsus* frequently become a major problem in water supply pipes (Langton *et al.*, 1988). In the tropics, man-made aquatic microhabitats harbour important vectors of disease. For example, the yellow fever mosquito *Aedes aegypti* breeds in water that collects in rainbarrels, cisterns, tin cans, jars, iron cooking pots and old motorcar tyres. In the third world, *A. aegypti* breeds in the large water-filled pots kept inside houses and this gives rise to urban dengue.

The trickling filter beds of *sewage treatment plants* create a good aquatic habitat for certain pollution tolerant insects. Here, the depth of the habitable zone is great, it is well aerated, there is a constant supply of both soluble and fine particulate food, and daily and seasonal temperature variations are buffered. Predominant insects are chironomids, psychodids and syrphids, together with collembolans (Lackey, 1949). Further aspects of the effects of pollutants on aquatic insects are covered in Chapter 11.

Helaemyia petrolei (Ephydridae) is known to breed in pools of crude *petroleum* and *waste oil* in California and Cuba. The larva feeds on insects that become trapped on the surface film and respires through posterior spiracles set on elongate siphons which extend to the liquid surface (Thorpe, 1931).

4.4   Summary

It is clear that aquatic insects occupy a great diversity of habitats, with some groups being more prevalent in certain habitats than others. Identification of patterns from within this melange will undoubtedly allow significant ecological, phenological, behavioural and physiological questions to be raised which, when answered, will lead to a greater understanding of the factors that have produced this diversity. To this end, we have summarized the distributions of all the aquatic and semi-aquatic orders along two habitat axes (Southwood, 1988). The first axis is habitat stability, which ranges from highly stable aquatic systems such as permanent, cold, freshwater springs and permanent ponds and lakes, to unstable systems such as temporary streams and phytotelmata. The second axis is habitat adversity and ranges from benign habitats (clean, cool, freshwater lotic and lentic water bodies) to ones in which conditions for life are close to the survival limits of insects (e.g. hot springs, very high salinity, extreme pH, gross pollution and petroleum ponds). It must be understood that these models are tentative and that there are bound to be some subjective allocations of taxa to habitat type, particularly where data are lacking or ambiguous. Nevertheless, some relationships are undeniable, thus fulfilling the primary purpose of the exercise which is to highlight major trends rather than to create discourse over minutiae.

In terms of aquatic insects, generally, in running water habitats, mayflies, caddisflies and dipterans are well represented in habitats of differing stability (Fig. 4.5A). However, caddisflies and mayflies deal less well with increasing habitat adversity than with decreasing habitat stability. In standing waters, dipterans cope well with variation in both stability and adversity (Fig. 4.5B), and odonates cope well with decreasing habitat stability - more so than in lotic habitats. Groups that appear not to have been able to adapt to decreasing lotic habitat stability include the Megaloptera, Neuroptera and Lepidoptera; they are also absent from highly stable lotic habitats. Few taxa, apart from the Diptera and Odonata, are well represented in adverse lotic habitats, and this is mirrored in lentic habitats where, although a number of groups can tolerate moderate adversity, only the Hemiptera and Diptera tolerate high adversity.

Although, as a group, the Diptera cope well with conditions in most aquatic habitats, there are marked differences in adaptability at the family level. In running waters, tipulids and chironomids deal particularly well with variations in habitat stability (Fig. 4.6A,C). In adverse lotic habitats, however, it is the chironomids and ephydrids that predominate. Ephydrids are abundant, too, in adverse lentic habitats and in highly- or moderately stable ones (Fig. 4.6 D). Families that fare particularly well across both the stability and adversity axes, in lentic waters, are the Chironomidae, Ceratopogonidae and Culicidae (Fig. 4.6 B). Other brachycerans occur in highly adverse (grossly polluted) standing waters, notably the Dolichopodidae, Syrphidae, Sphaeroceridae, Scathophagidae, Anthomyiidae and Sepsidae.

Some of the features that allow insects to adapt to such a range in habitat types are discussed in the next two chapters.

5   LIFE HISTORY AND HABITAT

Information on life histories is fundamental to most ecological studies of freshwater insects. However, we still lack many of the life history data critical to the progress of contemporary research on the structure and function of aquatic communities and ecosystems. Observing patterns of life history may lead to testable hypotheses on causal mechanisms and the reasons for the variation seen among species.

5.1   Life history theory

Most definitions of life history tend to include all aspects of an animal's general biology. Because physiology, behaviour, evolution and phylogeny, for example, all impinge on the nature of a life history, broad definition is to be expected even though it makes conceptualization of life history theory cumbersome (Butler, 1984). Additionally, confusion frequently exists in terminology between "life history" and "life cycle". Butler's (1984) definition of life cycle as the sequence of morphological stages and physiological processes that link one generation to the next, is a sound one. It emphasizes that the cycle components are essentially identical and consistent in sequence (e.g. egg, larva, pupa, adult, egg) for all members of a species. A life history, however, consists of the life cycle and qualitative and quantitative information on the variable factors associated with it which can differ among individuals or populations. Thus a population of a species in which the life cycle is repeated twice in one year is considered to have a bivoltine life history.

The variety of life history patterns seen in aquatic insects is great. What factors promote such diversity? Broadly speaking, there are two categories: (1) *intrinsic factors* such as physiology, behaviour and morphology which tend to restrict life history traits within certain genetically determined ranges; and (2) *extrinsic factors* such as temperature, nutrition, photoperiod and other biota (predators, prey, hosts, parasites) present in the environment whose effects, too, tend to be range limited. It is often enormously difficult to attribute life history patterns to either intrinsic or extrinsic cause. However, the life cycles of some aquatic arthropods exhibit features that have deep phylogenetic roots (e.g. hemimetaboly in the exopterygote insect orders). In species or species groups where certain features of the life cycle diverge from the phylogenetic norm for the group, environmental factors may be invoked (Butler, 1984). Certain phenotypic traits have arisen in association with specific ecological conditions so that the behaviour, morphology and physiology of an organism can be related

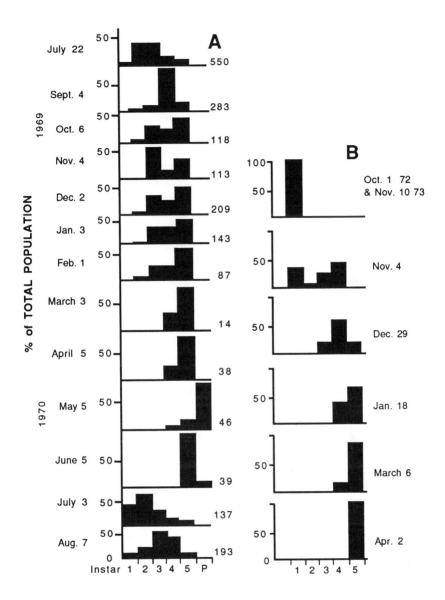

Fig. 5.1   Examples of the size distribution of the larvae of two species of Trichoptera with univoltine life cycles from lotic habitats: (A) *Cheumatopsyche oxa* in the Eramosa River, Ontario (P = pupal stage; the numbers to the right of each histogram represent the total numbers of individuals caught/unit sampling effort); (B) *Ironoquia punctatissima* in Moser Creek, Ontario - a temporary stream (redrawn from Williams and Hynes, 1973; Williams and Williams, 1975).

to adaptations for the acquisition and use of resources (food, habitat, mates). Characteristics of resource use are genetically controlled and are selected for according to which features best promote the proliferation of the genes that code for them. Thus genotypes determine phenotypes but the evolution of the genotype is constrained by the physiology of the phenotype because the processes and structures involved in the acquisition of limited resources are themselves limited (Sibly and Calow, 1986). "Phylogenetic drag" or "inertia" ensures that all options are not available simultaneously. Consequently, trait expression will vary among taxa (Wilson, 1975; Southwood, 1988).

Hynes (1970) proposed that the life cycles of lotic invertebrates could be divided into two main types: (1) *Non-seasonal*, in which the cycle spans either more, or much less, than one year with overlapping generations, and in which all stages of development are present at all times; (2) *Seasonal*, which is subdivisible into slow or fast seasonal, but in which a distinct change in the mean body size of members of a cohort can be observed over time (Fig. 5.1A,B). The majority of temperate lotic species tend to be univoltine with seasonal development, whereas in those exhibiting non-seasonal cycles wide size distribution usually can be traced to semi-, bi- or multi-voltinism. Wide size variation in univoltine species is, however, poorly understood and it seems particularly common in species from coldwater springs and in species from freshwater habitats in very cold climates (e.g. Iceland, Gislason *et al.*, 1990). Poor synchronization of annual life cycles, together with extended periods of hatching and emergence, is also prevalent in lotic insects in New Zealand and Australia (Hynes and Hynes, 1975; Winterbourn, 1978; Towns, 1981), and possibly in South American montane species (Illies, 1969). Winterbourn (1978) interpreted this trait in terms of "ecological flexibility" and concluded that this together with low species differentiation may be the rule in the Southern Hemisphere. It has been proposed that the prevalence of steep, unstable streams and the unpredictability of the physical environment in these regions may have promoted selection for opportunism, while, at the same time, the risks associated with non-seasonal emergence or larval cohort loss may have been spread (minimized) by the evolution of poorly-synchronized life histories or the maintenance of this primitive, generalized trait (Winterbourn *et al.*, 1981). However, linking poor synchronization of life cycle with unpredictability of lotic conditions does not account for the prevalence of this trait in highly stable habitats such as coldwater springs.

Southwood (1988) identified two approaches to considering life history strategies: (1) a "bottom up" approach in which one starts with basic observations on the traits and tactics of a variety of species and then attempts to arrange these traits into strategies by theoretical considerations of various combinations of fitness under a suite of conditions characteristic of a particular habitat; (2) an approach (top down?) in which broad categories of selection are identified, representing selection forces that successful organisms would respond to by adopting a particular set of characters. Either approach has merit in the

Table 5.1 Correlates characteristic of extreme r-, K- and A-selection (from Greenslade, 1983).

	Selection type		
	r	K	A
Properties of the habitat:			
Favourability	Variable	High	Low
Predictability	Low	High	High
Community attributes:			
Diversity	Low	High	Low
Interspecific competition	Occasional, can be intense	Frequent, often diffuse	Rare
Investment in defence mechanisms:	Low	High	Low
Specialization:	Low	High	Low
Population or species attributes:			
Capacity for dormancy	Variable	Low	Variable
Vagility	High	Intermediate	Low
Geographical distribution	Wide	Restricted	Variable
Parthenogenesis	Variable	Low	High
Life span	Short	Intermediate	Long
Maturity	Early	Intermediate	Late
Rate of development	Rapid	Intermediate	Slow
Fecundity	High	Intermediate	Low
Population density	Very variable	More constant, near carrying capacity	Variable, below car. capacity
Rate of increase	High	Intermediate	Low
Density dependence	Weak at low dens.; strong & overcomp. at high density	Moderate, compensating at high density	Weak
Key factors	Adult losses: mortality & migration	Juvenile mortality; variation in fecundity	Mortality at all stages; variation in fecundity & rates of development

study of aquatic insects, although the dearth of life history studies from certain types of aquatic habitat often makes the bottom up approach less workable.

Life history theory frequently explains observed traits as the products of single causal systems. Three theories currently prevail: (1) the deterministic view based on the prediction of r-, K- and A-selection (MacArthur and Wilson, 1967; Greenslade, 1983; Table 5.1); (2) the so-called "bet-hedging" hypothesis (Murphy, 1968; Schaffer, 1974; Stearns, 1976); and (3) the balanced-mortality hypothesis (Price, 1974). Data on the reproductive strategies of four species of lacebug (Heteroptera: Tingidae) contradict the idea of unilateral control, suggesting that covarying life history traits evolve in response to a suite of selective pressures, many of which interact. Further, different combinations of traits may be equally adaptive in particular environments. Thus, although a specific habitat may select for traits predicted by r, K and A, similar selective forces may produce more than a single life history solution. In the case of lacebugs, for example, *Corythuca marmorata* lives in an environment typically reserved for r-strategists, but exhibits classic K-selected traits (Tallamy and Denno, 1981).

In the ***deterministic view***, three types of selection are thought to operate depending on the nature of the environment: r-selection occurs in unpredictable, uncrowded environments where traits of early maturity, many small young, reduced parental care, and larger reproductive effort prevail; K-selection occurs in predictably favourable environments that are typically crowded and where the common traits are late maturity, multiple broods, a limited number of young, increased parental care and small reproductive effort; and A- or adversity selection typically is found in severe, but stable and predictable environments, and favours conservation of adaptive traits (see Table 5.1; MacArthur and Wilson, 1967; Greenslade, 1983). However, because many of these traits were originally established for vertebrates, some, such as the degree of parental care, are inappropriate for aquatic insects.

The ***balanced-mortality*** model, unlike the r- and K-models, does not distinguish between density-dependent and density-independent mortality. Thus when risk to a species in a particular environment is high, the theory predicts that fecundity will be high in order to offset increased mortality (Price, 1974). Risk can take many forms. Predation, parasitism and seasonal availability of food are three major biotic forms but harsh physical and chemical conditions, typically associated with fluctuating environments, also are important. For springs, for example, balanced-mortality theory would predict low fecundity for species not under high risk from biotic factors whereas, in ponds and lakes, fecundity should be high.

Bet-hedging theory predicts optimization of reproductive effort, rather than maximization, in fluctuating environments (Stearns, 1976). This contradicts r-K selection theory. Production of a single, large batch of eggs ("big-bang strategy") in a variable environment, where larval survival is uncertain, runs the risk of total loss of the next generation. A bet-hedging species should be iteroparous (i.e. reproduce two or more times during its lifetime) or have a flexible life history strategy (Baird *et al.*, 1987) to ensure that the young of at least some egg batches survive. In stable, coldwater permanent springs, semelparity (one reproductive event/lifetime) should, predictably, prevail but for insects in temporary ponds and streams, either iteroparity or semelparity (but with staggered egg hatching) would be a safer "bet"; the latter being characteristic of temporary water mosquitoes.

5.2   Life history traits and their relationship to habitat type

One method of comparing the biology of different faunas is to characterize features of their respective habitats, particularly in terms of the selective pressures that they exert on populations. MacArthur and Wilson (1967) proposed that habitats could be categorized according to whether they promote r- or K-selection. r-selection favours rapid population growth in unsaturated, "unstable" environments, whereas K-selection favours competitive ability and predator avoidance in saturated, more "stable" environments. Other factors, both in the environment and of the organism, correlate with these two forms of selective pressure (Table 5.1). K-selected species therefore should fare better in stable habitats such as permanent coldwater springs, whereas r-selected species are better suited to living in temporary streams (Williams, 1987). This relatively

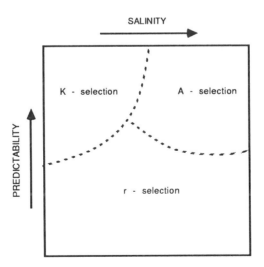

Fig. 5.2   Habitat templet showing the distribution of r-, K- and A- selection according to habitat predictability and salinity (redrawn from Williams, W.D., 1985; the broken lines delimit the approximate areas in which each type of selection predominates).

simple dichotomy is made more complex by inclusion of another strong variable in aquatic habitats, namely salinity. A habitat templet originally developed for staphylinid beetles in tropical montane forests (Greenslade, 1972) has been applied to lentic waters in the semi-arid regions of Australia (Williams, W.D., 1985). The two axes of the templet (Fig. 5.2) are the extent of predictability of the habitat and salinity level. Plotted are the presumptive distributions of the three broad types of selection: K (interactive selection); r (exploitative selection); and A (adversity selection). Note that as one moves towards more predictable (stable) habitats, K-selected species replace r-selected species but that along a gradient of stable habitats in which salinity differs, K-selected species are replaced by A-selected species. However, salinity is by no means the only strong habitat variable in aquatic habitats. For example, the salinity axis in Fig. 5.2 could be replaced by temperature and one would predict the same replacement of K-selected species by A-selected species on, for example, a gradient from coldwater springs to hot springs. It has been argued, however, that no organism is entirely r-, K-, or A-selected; generally, a balance is struck that maximizes the adaptive value of features drawn from each type of selection (Pianka, 1970).

Other models underline the importance of habitat in determining life history strategies. Southwood (1988) showed that the four main models, when arranged with their axes in a common orientation, made similar predictions on the relative importance of different tactics and features of organisms in the communities, despite their different origins (Fig. 5.3). How closely can these models, none of which is at substantial variance with r-K theory when its unidimensional character is recognized, predict the faunas of specific waterbodies? As an example, let us consider the case of springs.

Permanent, coldwater springs are habitats characterized by low disturbance and low adversity - assuming that low water temperature is not an adversity to lotic organisms, many of which have their ancestry in cold, headwater streams (Hynes 1970). According to the predictions in Fig. 5.3, most of the fauna should be K-selected, illustrating the properties outlined in Table 5.1 and Fig. 5.3A. The community should show intermediate to high diversity, and consist primarily of sessile grazers and filter feeders with narrow specialized niches due to heavy competition. Productivity and individual growth rate should be high. Survival rate may be low, length of life, time of maturity and fecundity intermediate and population densities near carrying capacity. Incidence of dormancy should be low, as should that of parthenogenesis. Invertebrate predators should be rare and vagility low to intermediate (Williams, 1991).

Permanent, hotwater springs represent habitats that are stable but continuously harsh and therefore of high adversity. As we have seen, such habitats should support primarily A-selected rather than K-selected species. The models (Fig. 5.3) predict low diversity with few herbivores but a large number of invertebrate predators, low growth rate and survival of individuals, and low productivity. Physiological tolerance should be high but mortality should occur at all stages. The life cycle should last a long time with late maturity. Dormancy

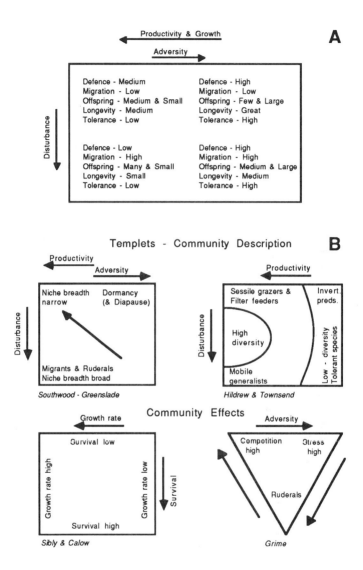

Fig. 5.3   (A) The relative importance of tactics in different habitats as predicted by various templet schemes; (B) The predictions of some community features from the different templet models (the term "ruderal" refers to species with strategies geared towards habitats with low stress but high disturbance; Grime, 1977). The models have been given a common orientation with the least disturbed, most productive habitats at the top left; redrawn from Southwood, 1988; original citations are given in the references.

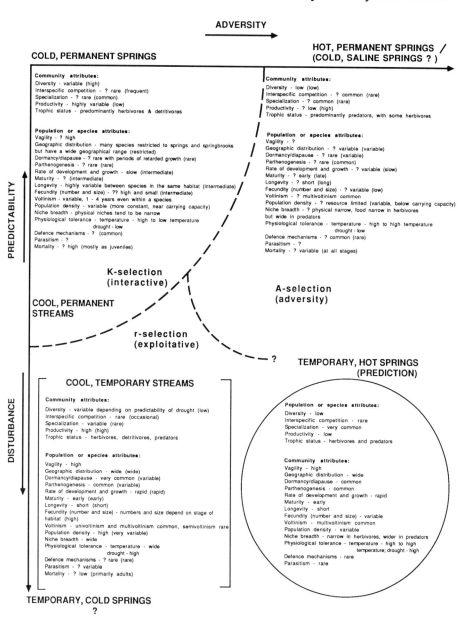

Fig. 5.4 Habitat templet showing various attributes of insect communities, populations and species in different types of running waters (primarily springs). Habitats are arranged along axes of degree of stability (disturbance/predictability) and adversity. Attributes are based on findings in the general aquatic literature. Question marks indicate uncertainty in available data (usually only one or two studies existing) or lack of study of particular habitat types. Approximate selection type boundaries are indicated by the broken lines (after Greenslade, 1983; and Southwood, 1988). Attributes in brackets are those predicted by Greenslade (1983); see Table 5.1.

and diapause should be common, parthenogenesis and defence high but vagility low.

The somewhat limited data on life history traits of spring-dwelling species, in the general literature, show most support for the deterministic view, based on the predictions of r-, K- and A-selection. Specifically, the attributes of both cold and hot permanent spring faunas do conform to those of K- and A-selected species, respectively, but there are instances of non-conformity. Notable examples of the latter concern productivity, competition and specialization in hot and cold springs, and parthenogenesis, time of maturity, longevity, defence mechanisms, and control of population density in hot springs (Fig. 5.4). In permanent springs, species do not seem to employ bet-hedging techniques to optimize their reproductive effort (although this may be appropriate in temporary springs), nor does the balanced-mortality model appear to apply although, again, it might be employed by species in temporary springs or hot springs.

Species living in *temporary coldwater springs* can be considered to occupy habitats that appear to be unstable but which exhibit a predictable cyclical return to stable conditions, for example each springtime. At first sight we might expect their faunas to be primarily r-selected, similar to species found in cool, temporary streams (Fig. 5.4; data from Williams, 1987), although environmental conditions are far from stable in the latter. It is not inconceivable, however, that some species might show K-selected traits provided that they possess some adaptation to survive the dry period (e.g. dormant immature stage or long-lived or moderately vagile adult). Unfortunately, there are no data available on the life histories of species from temporary coldwater springs. To our knowledge, there have been, similarly, no faunal studies on *temporary hot springs*. However, the attributes of their faunas might reflect both those of temporary, cold spring species and those of permanent, hot springs and they would occupy a position at the extremes of both disturbance and adversity axes. Community, population and species attributes are predicted in Fig. 5.4 and are largely self explanatory. The faunas of *cold saline springs* would probably exhibit attributes similar to those of hot, permanent springs (high on the adversity axis), except for physiological tolerance of high temperatures shown by species in the latter.

None of the various theories and hypotheses covered in the preceding sections is without its problems, and attempts to force a universal application are seldom successful. However, their usefulness lies as guidelines for thinking about the mechanisms underlying observed life history traits. Based on this kind of reflection, and in concert with constantly updated knowledge, new generations of models with even better fit to the data will evolve.

5.3   The influence of temperature

Growth and other biological processes in insects are particularly dependent on environmental temperatures. Aquatic insects are no exception, and the temperature characteristics of a waterbody influence both properties of individuals and communities. For example, the low species diversity typical of springs may be a consequence of stable temperatures which reduce thermal niches (Ward and Stanford, 1982). Because many insects have become well adapted to living within specific temperature ranges, it has been suggested that the pattern of temperature change is more important in regulating their life cycles than the absolute temperature attained (Hynes, 1961).

The temperatures of temperate, polar and even tropical streams typically fluctuate (Harper, 1981; Awachie, 1981). These fluctuations are primarily associated with seasonal changes, for example winter to spring to summer to fall, or dry to wet and, for the most part, are highly predictable. Superimposed on these changes are usually smaller and less predictable diel changes that result from changes in factors such as intensity of solar radiation, amount of cloud cover and rainwater inflow. The thermal history to which a species has been exposed shapes responses at the level of the individual, population and community, and these responses are exhibited on both short-term ("ecological") and long-term ("evolutionary") time scales (Ward and Stanford, 1982).

The effects of temperature on the life histories of aquatic insects have been reviewed by Sweeney (1984) and features that bear on running waters, and on springs in particular, are summarized in Table 5.2.

For the most part, insect body temperatures reflect those of the environment, be it air or water. A few insects appear to be capable of a limited degree of body temperature control, generally through some behavioural mechanism. Among aquatic insects, however, this has only been shown in the larvae of a species of cranefly (Kavaliers, 1981). In most insects, though, an increase in environmental temperature results in a corresponding increase in body temperature which increases the animal's metabolic rate; lowered environmental temperature may cause a decreased metabolic rate. This relationship holds only between certain upper and lower temperature thresholds which are species specific but which can be modified according to the previous temperature experience of the insect (acclimation; Wigglesworth, 1965). Many features of the life history of an aquatic insect are directly related to environmental temperature, particularly development and growth of the larva, size of adult and fecundity (Vannote and Sweeney, 1980). In many bi- or multivoltine species, development is faster in the summer generation(s) (Ross and Merritt, 1978), because duration of the larval stage can decrease with increasing temperature (Brust, 1967). However, in some species, instar duration can become independent of temperature past some particular threshold (Nebeker, 1973). Growth of the mayfly *Baetis rhodani* proceeds at the same rate in both thermally constant and fluctuating streams despite a difference in degree days (Fahy, 1973). A particularly detailed study of two lotic mayfly species has revealed complex relationships between temperature and a number of developmental parameters (Söderström, 1988). Growth rate of *Parameletus minor* was maximal between 10.8 and 19.8°C, whereas develop-

Table 5.2   Summary of the effects of temperature on the life histories of aquatic insects (Based on Ward and Stanford, 1982; and Sweeney, 1984).

Egg development

- Duration of egg stage often is inversely related to temperature in non-diapausing species and may be described by a power function (Humpesch and Elliott, 1980);
- Development time is more variable at lower temperatures (Elliott, 1972);
- Hatching success often is high at intermediate temperatures but falls near the upper and lower thresholds (Brittain, 1977) even though the optimum constant temperature for egg development most often occurs near the upper lethal limit. Such optima may be the result of selection for rapid development in variable temperature environments (Howe, 1967);
- Eggs may hatch sooner under diel cycles of temperature than in habitats with constant temperatures (Humpesch, 1978);
- Eggs of some species show flexibility in response to the different temperature regimes among habitats - this may result in locally-adapted populations (ecotypes) (Humpesch, 1980);
- Diapause is an adaptation enabling survival through unfavourable conditions (including cold weather) and the duration of all stages of diapause is affected by temperature (Bohle, 1972). Low or seasonally decreasing temperatures frequently terminate diapause (Snellen and Stewart, 1979).

Larval development

- At higher temperatures, larvae usually develop faster because of decreased time spent in each larval instar or this combined with a reduction in the number of instars (Ross and Merritt, 1978);
- The total number of degree days required for complete larval development appears to be constant, regardless of rearing temperature, in some aquatic species but not in others (Konstantinov, 1958; Sweeney and Schnack, 1977);
- Temperature influences growth through its effects on ingestion and assimilation rates of food. Generally, ingestion rate increases with increasing temperature but a levelling out or even a decrease may occur as temperature thresholds are approached (Gallepp, 1977; Iversen, 1979; Lacey and Mulla, 1979). In some species, assimilation efficiency decreases as ingestion rate increases (McCullough, 1975) although this has been little studied;

Table 5.2 (contd)

- Many species may grow and remain active at water temperatures close to 0°C (Ulfstrand, 1968);
- Differential temperature-growth responses may promote temporal niche segregation (Vannote and Sweeney, 1980);
- Seasonal temperature changes may synchronize life histories (Corbet, 1957).

Pupal development / adult emergence

- Increasing water temperature results in early emergence, whereas decreasing temperature delays emergence, in both lab (Sweeney and Vannote, 1981) and field studies (Lehmkuhl, 1974; Wise, 1980);
- Temperatures that are optimal for high growth rates may prove to be suboptimal for growth efficiency, success of emergence or adult longevity (Heiman and Knight, 1975);
- In a single habitat, a successional pattern of emergence frequently results because of temporal separation of even closely related species that exhibit differential responses to temperature (Brittain, 1978). Between-year variation in habitat temperature may alter the time of emergence, but it does not alter the sequence in which species emerge (Hynes, 1976).

Adult size and fecundity

- Considerable intraspecific variation exists in size among pupae and adults of many aquatic species which seems related to the fact that growth and duration of development are not affected to the same relative degree by the same change in water temperature (Mackey, 1977; Vannote and Sweeney, 1980);
- Seasonal temperature differences may produce large adults in winter - spring cohorts compared with smaller summer cohorts in multivoltine species (e.g. mayflies; Wise, 1980). It is not clear whether these differences are promoted through direct changes in physiology and/or developmental processes, or through indirect changes such as alteration of food type or amount, or habitat;
- Fecundity is influenced by temperature because of its positive correlation with adult body size (Clifford and Boerger, 1974), thus deviations from optimal thermal conditions may well affect the competitive potential of populations;
- Adult longevity is, in part, determined by the thermal experience of the immature stages and also by the air temperatures to which the adults are exposed (Humpesch, 1971; Nebeker, 1971).

---

ment time was shortest at 14.6°C, adult size and fecundity were maximal between 5.9 and 10.8°C, and survival rate peaked at 5.9°C. Growth of *P. chelifer* was greatest between 14.6 and 19.8°C, development time was again shortest at 14.6°C and survival was greatest at 5.9°C, but adult size was greatest at 10.8°C and fecundity was highest between 10.8 and 14.6°C.

Quite apart from having an effect on development, temperature is implicated in the timing of life cycles of arthropods, enabling critical stages to be optimally phased with important environmental events such as the appearance of specific foods, factors favouring successful dispersal, or the availability of hosts in parasitic species (Vannote and Sweeney, 1980). Several models have been proposed through which synchronization of life histories could be brought about by temperature: Corbet (1957) showed that in species of damselfly that overwinter in more than a single instar, the thermal threshold for growth is positively correlated with the size of individual nymphs (see also Lutz, 1968). Adult emergence is synchronized because the smaller nymphs resume growth earlier in the spring than larger nymphs. Sweeney and Vannote (1981) proposed that synchrony of emergence in the stream mayfly *Ephemerella subvaria* is achieved when rising water temperatures in the springtime exceed a threshold above which nymphal metabolism and/or the production of hormones are altered. The latter promote the development of adult tissues, and are produced as long as this threshold is maintained. Although emergence is synchronous, a wide range of adult sizes reflects the wide size range of nymphs present in the population at the time the nymphs were thermally stimulated. Although species such as *E. subvaria* have precise temperature thresholds at which certain physiological processes are turned on (or off), other species have been shown to respond to thermal summation (Ross and Merritt, 1978; Ward and Stanford, 1982). More commonly, winter temperature decrease may cause accumulation of an insect population in a particular diapause stage. Time of entry into diapause thus varies among individuals, but life cycles are synchronized when development resumes in springtime (Danks, 1987). Dormancy, with temperature as a regulator, is thus an important synchronizer of insect life cycles.

Because temperature influences the timing of insect life cycles, it can determine the frequency with which life cycles are repeated; in other words, it can affect voltinism (Ward and Stanford, 1982). A population in which the individuals grow quickly and emerge early may be able, if there are no genetic constraints or resource limitations, to complete one or more additional cycles annually, if the environment remains favourable. Such species usually grow in cohorts among which there is little size variation. In species which exhibit a wide range in larval size, as often occurs in springs (e.g. Iversen, 1976), cohorts may become split resulting in a range of life cycle duration. This is most conspicuous when some individuals in a cohort reach the adult stage and emerge within one season but the rest do not and delay their emergence until the next season, adding another year to the life cycle (Carter, 1980; Butler, 1984). On occasion, emergence of the second part of the cohort is not at precisely the same time as the first part of the next cohort in the following season which results in life cycles of differing duration within the same species in the same habitat. Examples include 14 and 22 months in the lake mayfly *Hexagenia limbata* (Flannagan, 1979) and one or two years in the lotic caddisfly *Sericostoma personatum* (Elliott, 1969).

Where populations of the same species are found in streams with different temperature regimes, the life cycles may be timed differently: for example, *S. personatum* took two and three years to complete its life cycle in two Danish spring/springbrook systems with temperature ranges of 0.5-20.5°C (Fonstrup Baek) and 2.4-13.6°C (Rold Kilde), respectively (Iversen, 1980). Iversen (1973) pointed to the obvious inference that differences in temperature accounted for the difference in length of life cycle between the English (Wilfin Beck; annual temperature range 2.0-21.0°C) and Danish (Rold Kilde) populations. However, he concluded that the temperature differential was too small to explain so great a contrast and suggested that poor and patchy food in the Rold Kilde (scattered, decaying beech leaves) was a more likely cause of slower growth in the latter. However, several other records of the life cycle of *S. personatum* show that the temperature effect is most important. If life cycle duration is plotted against the maximum water temperature of the habitat (accumulated degree days would be a better descriptor but such detailed data are not given in these studies), a negative relationship is indicated.

Hynes and Hynes (1975) showed that two species of stonefly in Australia, *Riekoperla rugosa* and *R. tuberculata*, were univoltine in warm water streams but semivoltine in colder streams. These authors concluded that this was an effect of temperature and that such lack of seasonal rigidity in growth and emergence is important to species living in variable and uncertain habitats. They further hypothesized that such flexibility of life cycle, imposed by the erratic nature of Australian lotic waters, has reduced the number of niches available on that continent because single species can adopt one or more options in their life history. Décamps (1967) similarly suggested that the one-, two-, or three-year life cycle seen in the caddisfly *Rhyacophila evoluta*, in the Pyrénées, may be related to habitat temperature as, in this region, it is found in both cold springs and streams with wide temperature ranges. The duration of the life cycles of both the stonefly *Pteronarcys dorsata* and the damselfly *Ischnura elegans* increases with increasing latitude (Hilsenhoff and Narf, 1972; Barton, 1980; Corbet, 1980; Lechleitner and Kondratieff, 1983) and, in other species, with increasing altitude (Ward and Stanford, 1982). However, many other environmental variables may change with latitude or altitude and thus most field evidence for the control of voltinism by water temperature is largely correlative or circumstantial.

In North American hot springs, the intervention of a short-day induced diapause in the fall extends the life cycle of the dragonfly *Argia vivida* to one year, preventing the emergence of cold-sensitive adults in winter, even though the springs provide enough day-degrees to support a bivoltine life history. Interestingly, this diapause of *A. vivida* was present also in a population from a non-thermal stream (annual temperature range 5.0-20.0°C) where the life cycle took three years (Pritchard, 1991). Pritchard concluded that the lack of differences observed in temperature responses between populations of *A. vivida* from different thermal regimes is evidence that a micro-evolutionary response to temperature is rare in thermophily.

Further studies of the life cycles of species like *Sericostoma personatum* are necessary before temperature can be definitely implicated as a controlling factor in the field. Manipulative experiments would be particularly powerful tests of the relationship. This serves to underline the fact that insufficient data exist on even such a well-known species as *S. personatum* to make definite conclusions concerning basic environmental influences.

5.4   Other environmental factors

It is often difficult to separate the effects of temperature from those of food; both can change an organism's metabolic processes and rates. In some species, environmental temperature seems to be more important in determining growth rate and body size than seasonal changes in food quantity and quality, and the limits imposed by food quality may be exceeded if the temperature rises. On the other hand, food quality also can overide temperature effects (Scriber and Slansky, 1981).

Many studies have shown that changes in diet of lotic invertebrates can influence growth rate and thus voltinism, size at maturity, and fecundity (e.g. Söderström, 1988), and, indeed, many species shift from herbivory or detritivory to predation in later instars (e.g. Siegfried and Knight, 1976), or the reverse may occur (Winterbourn, 1974). There are also differences in the quality of food collected by different techniques. For example, most filter feeders are unselective and thus their diets may vary more than species that scrape patches of algae off rock surfaces (Hynes, 1970). The food potentially available to freshwater insects ranges from that which is easily assimilated (e.g. algae and animal tissues) to that which requires considerable processing before yielding any nutrients (e.g. wood). Species belonging to different functional feeding groups may therefore be expected to exhibit different life history traits. For example, the life cycles of many shredders are synchronized, in temperate regions, to the input of riparian leaves in the autumn, and these insects show their major growth period in the late fall and winter. The diet of scrapers similarly may be influenced by the seasonal availability of various species of periphyton. Populations of black flies filtering particles at lake outflows may have faster growth than populations of the same species occurring downstream where food particles tend to be more dispersed. Variation in the quantity (prey density) of a predator's food may influence its generation time (Carlsson *et al.*, 1977). Reduced food levels have been shown to prolong the length of the larval period in predaceous hydrophilid beetles (Hosseinie, 1976), and to delay pupation or adult emergence and to produce smaller adult black flies (Colbo and Porter, 1979). It is difficult to isolate the exact effect of food, unless it is done in the laboratory or in carefully controlled field experiments (Anderson and Cummins, 1979). Pritchard (1976) has pointed out that detrimental effects of poor diet will be most noticeable in multivoltine species because univoltine or semivoltine species frequently have a period of arrested development that may be shortened to accommodate an increase in the feeding interval.

In the field, photoperiod tends to covary with environmental temperature, so the relative influences of these two factors may be difficult to separate. One explanation of the persistence of seasonal cycles in some species from springs, despite relative uniformity in water temperature, is that photoperiod (either as day-length or its rate of change) acts as a time-setter. Thus temperature and photoperiod may interactively control ontogenetic events (Hynes, 1970). However, in the case of *Argia vivida*, the effect of photoperiod appears to overide the influence of temperature (Pritchard, 1991). Khoo (1968), working on the stonefly *Capnia bifrons*, showed that although emergence of nymphs was normally stimulated by a rise in temperature it could be induced also under winter temperatures if nymphs were exposed to long days. In this species, there is a marked decrease in adult size as the emergence season progresses, perhaps because of abbreviated nymphal growth and reduction in the number of instars. Tardy individuals therefore emerge in time to mate before the end of the season (Khoo, 1964). Interestingly, in the Arctic, chironomid emergence periodicity in the summer seems not to be affected by light but is controlled instead by temperature which, at high latitudes, impinges significantly on all aspects of the life cycle (Danks and Oliver, 1972).

Dissolved oxygen concentration also may influence insect life cycles through its effect on growth. In Lake Esrom, Denmark, growth of profundal benthos was depressed by oxygen depletion in summer but increased coincident with the spring and autumn overturns (Jónasson, 1972). Similarly, in Lake Memphremagog, Canada, depressed growth of the chironomid *Chironomus anthracinus* occurred during a period of low oxygen concentration at the mud - water interface in August (Dermott *et al.*, 1977).

The influence of a major, biotic environmental factor, predation, on insect life histories is examined in Chapter 9.

6 MORPHOLOGICAL AND PHYSIOLOGICAL ADAPTATIONS

6.1   Problems of living in or on water

As we have seen, the hexapods, as a group, evolved on land. It is clear that they were aided in this by many of the adaptations to terrestrial existence inherited from their ancestors, the terrestrial myriapods. The Collembola appear to have evolved only partially towards independence from water as, for example, amongst other things their continued reliance on cutaneous respiration confines them to aquatic or, at least, damp habitats. Insects, on the other hand, have become adapted to some of the most harsh and dry environments on the planet. Having spectacularly overcome the severe morphological and physiological problems involved with disassociation from free water (as opposed to that contained in food), it might seem odd that some insects have returned to aquatic environments. The reasons for this are not clear. However, exploitation of unoccupied food niches in fresh waters, perhaps those associated with riparian macrophyte litter (especially that derived later alongside the evolution of the flowering plants in the Cretaceous), is one possibility. Those aquatic insects present earlier, in the Permian, may have fed on algae or the litter from more primitive multicellular plants (e.g. mosses, ferns, cycads or gymnosperms) or, like the Odonata, have been predaceous. Whatever the reason, the fact that mayflies were among the earliest of insects indicates that the invasion of fresh waters began not too long after the establishment of terrestrial hexapods. Since that time, there have been several other, probably independent, invasions by different insect groups.

Foremost among the problems particular to insects living in water are those concerned with respiration and osmoregulation. In addition, feeding in water presents an array of problems not found on land. For example, aquatic insects have to contend with continual dilution of their food source as readily assimilable sugars and other useful soluble organics are leached into the surrounding water. Contents of cells broken in the process of being gathered or chewed may similarly be lost and this may have prompted the development of sucking types of mouthparts (Anderson and Cargill, 1987). There are also some special problems that are habitat specific, such as living in flowing water or in waterbodies that dry up. Some parallels are to be seen with other terrestrial animal groups some of whose members have taken up an aquatic existence, for example birds. Compared with the latter case, however, it could be argued that the insects have completed the transition more successfully as the vast majority of aquatic insects can still fly, although crossing the air - water interface is potentially hazardous for small organisms.

Retention of flight is doubtless linked to maintaining adequate powers of dispersion, particularly between small, isolated bodies of water and, in running waters, it may serve to compensate for downstream displacement in the immature stages (see Chapter 7). However, immature insects, along with some aquatic adults, are also quite mobile, exhibiting a variety of locomotory mechanisms suited to moving through a liquid. These include flattening of limbs into oar-like structures, often with fringes of setae to increase surface area still further, as in many adult beetles and hemipterans; elongation of larvae combined with worm-like or a thrashing/wriggling motion, as is seen in many dipterans and mayflies; possession of stout claws enabling safe crawling across rock and plant surfaces (e.g. stoneflies and caddisflies); and more exotic methods such as forcing jets of water from the rectum as in anisopteran dragonflies.

6.2   The water surface

The air - water boundary acts like a resilient membrane and many aquatic insects use its surface properties to their advantage. The better-known examples include pondskaters, water measurers and whirligig beetles. These are members of the neuston community which live at the water surface. Such insects are physically supported by upward forces resulting from surface tension and buoyancy, provided that this combination is not exceeded by the downward force due to the weight of the insect. This relationship is described by the equation: $W = F + B$; where W (the weight of the insect) = m g (body mass x acceleration due to gravity); where B (buoyancy) is the upward thrust due to displacement of water by any part of the body below the water surface; and where F is the component of the surface tension force acting vertically upwards. F is perhaps

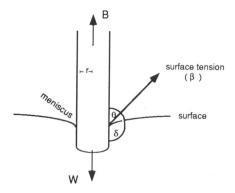

Fig. 6.1   The forces acting on a cylindrical insect limb as it rests in the water surface (after Guthrie, 1989).

the most difficult value to determine, but if we consider the example of a cylindrically-shaped insect limb resting vertically in the water surface (Fig. 6.1), $F = 2\pi$ r ß cos θ. The surface of the limb is water-repellent (∂ is > 90°) resulting in a convex meniscus. The length of the line of contact of the meniscus with the insect's cuticle is equal to the circumference of the limb, in this case a cylinder with r equal to the radius of the limb. The upward force, derived from surface tension acting around the cylinder, is 2π r ß (where ß is the surface tension for clean water = 73 dyn/cm at 20°C), acting at an angle θ to the vertical (Alexander, 1968; Guthrie, 1989).

Although some gerrids may weigh up to 15 mg, neustonic insects in general tend to have lightweight bodies combined with features that spread their weight over as large an area of water surface as possible. Gerrids contact the water surface only with the large flattened tarsal segments on each leg and the flattened tibiae on the metathoracic legs, the rest of the body being held clear of the water. The tarsal segments are covered with fine, velvety hairs which protect the cuticle against wetting and the claws are raised so as not to pierce the surface film. Using the mesothoracic limbs to push rhythmically against the water surface, the trailing metathoracic limbs as a rudder, and the prothoracic limbs to detect the ripples made by prey, the insect can glide very effectively and quickly across the water. On occasion, gerrids do accidentally break through the surface and fall into the water. If this occurs, they swim back to the surface with the assistance of air bubbles trapped among their body hairs and must rest on some solid object while their leg-hairs dry off.

A similar "pondskater-type" action is seen in males of the marine chironomid *Pontomyia*. Males skate across the surface film, in search of the vermiform and structurally degenerate females, supported on the tips of both the stout mesothoracic legs and the longer metathoracic legs. The wings are modified to form paddles which flick at the air adjacent to the water surface. The prothoracic legs skim the surface in an "outrigger" fashion and adjust the pitch of the insect's body so that the wings can operate effectively (Norris, 1991).

Adult water beetles are typically covered by a thin waxy layer which is water repellent. Such a strong hydrophobic property results in a high contact angle between the water and the body surface and, consequently, a significant upward force due to surface tension. High contact angles are found in many of the diving beetles (Dytiscidae) and also in aquatic bugs such as the corixids and notonectids. These insects are thus highly buoyant and have to swim strongly downwards in order to submerge. When they take to the air, the water repellent property of the cuticle substantially reduces the amount of muscular effort required to break free from the surface film and water droplets simply fall off as the insect becomes airborne (Guthrie, 1989). *Gyrinus*, the whirligig beetle, lives primarily on the water surface with the ventral portion of its body, the cuticle of which has slightly hydrophilic properties, dangling through. Its dorsal cuticle is waxy and so gains some support from surface tension because it is hydrophobic. The unique swimming posture of adult gyrinids has led to modification in some of their sense organs. Most obvious is the division of the compound eye into an upper and lower section, adapted for seeking prey on and below the water surface, respectively.

As body size decreases, the area of contact with the water surface needed to support the insect becomes less. In the case of the Collembola, for example, the animal balances on the tips of the legs and the collophore organ. Small insects wishing to enter the water may have difficulty in breaching the surface film. The only way that the adults of some of the smaller species of hydrophilid beetle can do this is by crawling down emergent plant stems. A novel use of the properties of surface tension is seen in some of the smaller riparian staphylinid beetles. If, while hunting, these beetles fall onto the water surface, a secretion from the tip of the abdomen lowers the surface tension causing the insect to be propelled towards the bank at speeds of up to 70 cm/s (Jenkins, 1960).

Many benthic species run the risk of being trapped on the underside of the surface film should they come too close. However, at certain times in their lives, the immature stages of most benthic species have to come to the surface in order to emerge to adulthood. Certain adaptations protect them from being trapped during this transition. For example, the pupal caddisfly cuts its way out of its silken cocoon on the substrate and swims to the surface using specialized limbs. Once it has broken through the surface film, the adult emerges using the pupal exuvium as a floating platform and thus avoids being wetted. Use of the pupal skin in this manner is particularly common in the Diptera, and mosquito and chironomid adults frequently rest on the cast skin for some time while they expand their wings. Black flies are unusual in that the adult rises to the surface, fully expanded, enclosed in a bubble of gas secreted beneath the pupal skin. Once the skin is split, the adult can fly away immediately. Many other holo-metabolous aquatic insects (especially the Megaloptera, Neuroptera, many brachyceran Diptera and most Coleoptera) pupate on land and this requires the final instar larva to crawl out of the water and establish some form of pupal cell, or chamber, in damp soil. The final instar larvae of many hemimetabolous insects also climb out of the water, on rocks or vegetation, prior to eclosion.

6.3   Dealing with currents

One of the most influential factors in running waters is the force of the current. At a low level of resolution this has resulted in the vast majority of lotic insects being benthic in habit, as being planktonic risks getting washed out of areas of optimum habitat and being nektonic involves the expenditure of large amounts of energy to maintain station. At a high degree of resolution, it is clear that current has promoted a wide spectrum of insect adaptations to both lessen the chances of physical dislodgement and take advantage of some of the benefits derived from water flow.

Mechanisms for reducing physical dislodgement of animals from substrate surfaces in running water have been studied for over 60 years (e.g. Hubault, 1927). Early on, it was demonstrated that as one approaches the substrate, current decreases until, some 1 to 3 mm from the surface, there is virtually no current. This is the so-called *boundary layer*, the thickness of which decreases with increasing current speed (Jaag and Ambuhl, 1964). This discovery, combined with observations that many insect larvae living in fast water are dorso-ventrally flattened, led strong support to Steinmann's (1907) theory that flattening was an adaptation to avoiding dislodgement, by enabling insects to take refuge in the boundary layer. It has been pointed out, however (e.g Hynes, 1970), that flattening can serve other functions, such as allowing insects to squeeze under rocks or preventing them from sinking into silty substrates in slower water. Baetid mayflies, for example, cope well with strong currents even though they typically stand raised on their legs with their abdomens trailing free in the water. Their bodies, although not flattened, are aerodynamically streamlined with their widest section about 36% of the distance from the head to the tail and tapering to a point at the rear. This "fusiform" shape offers the least resistance to passing fluids (Bournaud, 1963) and is the one commonly approximated by many pelagic fishes. Even so, at very high currents, baetids crouch down closer to the boundary layer (Hynes, 1970). In torrential water, several ventral sucker-like devices aid adhesion such as those seen in the larvae of the Blephariceridae (Diptera) and certain mayflies (e.g. *Ephemerella doddsi*) (Fig. 6.2).

Other morphological adaptations for resisting displacement include prominent claws, friction pads, incorporation of heavy "ballast" stones into the cases of caddisflies (e.g. as in *Silo*, *Goera*, *Agapetus* and *Glossosoma*) and the use of silk for physical attachment to the substratum (as in simuliids, some chironomids, and both cased and caseless caddisflies).

In psephenid beetles (water pennics), not only does the streamlined form of the larva minimize drag forces and allow the larva to maintain station with the minimal expenditure of energy, but the insect also makes postural adjustments that reduce the risk of being swept away as current increases. As Reynolds number increases, transition from laminar to more turbulent flow occurs and this begins to affect even the boundary layer[1]. However, even under such conditions, a thin laminar or viscous sublayer persists next to the substrate. The shape of the psephenid larva modifies this laminar sublayer, increasing its thickness to encompass the larva. At a very high Reynolds number, turbulence reduces the thickness of the laminar sublayer to a point where the possibility of boundary separation at the surface of the larva arises and it is in danger of being swept off the substrate. At this stage, it has been observed that larvae of *Sclerocyphon*

---

[1] Reynolds number relates the factors that determine whether the flow will be laminar or turbulent. It may be calculated from the relationship: $R = \partial vd/\mu$; where ∂ is the density of the fluid, v is the velocity of the fluid, d is the diameter of a body placed in the flow, and μ is the viscosity of the fluid.

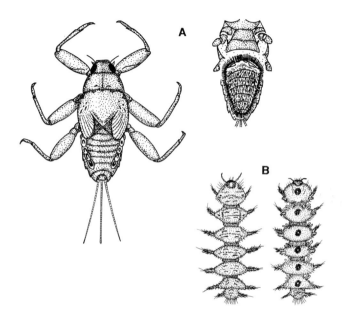

Fig. 6.2   (A) Dorsal and ventral views of the mayfly nymph *Ephemerella doddsi* showing the ventral "sucker" formed from a ring of outwardly-directed hairs and an inner area of backwardly-directed hairs; (B) dorsal and ventral views of a blepharicerid (Diptera) larva showing the segmentally arranged ventral suckers (redrawn after Hynes, 1970).

aquaticus modify the nature of the boundary layer by applying suction. This is done by making precise adjustments to the spaces between the lateral laminae on the abdominal segments (Fig. 2.29H). Small amounts of water from the boundary layer are allowed to pass through these spaces and under the larva, creating a negative pressure that pulls the larva down onto the substrate (Smith and Dartnall, 1980). Thus, in water pennies, a combination of morphological (streamlining) and behavioural (boundary layer control) adaptations allows movement across substrates even in fast turbulent streams.

Recently, the advent of new technology has enabled a more quantitative analysis of streamlining in aquatic insects to be made. Using Laser Doppler Anemometry, Statzner and Holm (1982, 1989) have shown that the water velocities around the bodies of benthic insects, and probably the forces acting upon them, are much more complicated than suggested by the boundary layer concept. These authors found that all three caddisflies tested (*Anabolia*, *Micrasema* and *Silo*) experienced a downstream drag force when exposed to currents ranging from 7.9 to 40 cm/s. In addition to drag, larvae had to contend with at least four other physical factors acting on them simultaneously: *diffusion* through the boundary layer, *corrasion* (abrasion by suspended solids), *friction* and *lift*. Statzner and Holm concluded that (1) the physical impact of these five factors depends to a large extent on the Reynolds number of the larva (being affected by flow separation which is related to the ratio of body length to height and the slope of the posterior contour) and therefore must change as the insect develops and grows; and (2) a simultaneous, effective morphological adaptation to all five factors would probably be physically impossible, particularly as it would have to change from life at a low Reynolds number (as in the case of small instars) to life at a high Reynolds number (final instar larvae).

One of the major benefits of having a dynamic, unidirectional environment is that of having food pass by. In addition to the more conventional methods of food gathering found in most environments (grazers, detritivores, carnivores, etc. see Chapter 8) many lotic insects have developed specialized techniques to capture this suspended material. Filter-feeding insects belong to only three of the aquatic insect orders, the Ephemeroptera, the Diptera and the Trichoptera. The material itself is a mixture of organic and biofilm-coated, inorganic particles, ranging in size from <1 μm (e.g. bacteria and fine clay) to >1 mm (e.g. fragments of plant tissue and coarse sand), depending on the force of the current. The nutritional content of this suspended matter is not as high as some other food types (e.g. algae or animal tissue) but the amount of energy required to collect it is potentially less. With assimilation efficiencies in the range of 2 to 20%, suspended detritus needs to be ingested in bulk.

Filtering techniques range from the simple to the elaborate. Among the simplest are dense fringes of setae, either on the legs or mouthparts. Nymphs of the mayfly genus *Isonychia*, for example, have hairs on their prothoracic legs which are held out into the current. Trapped particles are combed off, to form a food bolus, by brushes on the mouthparts. Nymphs of another mayfly genus *Tricorythus*, from Africa, have hair fringes on the outer margins of their mandibles. Together with hairs on the labial palps, these form a funnel the base of which is directed towards the mouth (Hynes, 1970). Some species of brachycentrid caddisflies anchor their cases to the substrate, facing upstream, and then spread their legs at the case entrance; arrays of short setae on the long meso- and metathoracic legs collect particles which are then wiped off and passed to the mouth by the shorter prothoracic limbs (Fig. 6.3).

Black fly larvae, apart from those of a few primitive genera, have a pair of highly modified mouthparts which form cephalic fans. These organs are retractable and are located between the antennae and the mandibles. When expanded they form a filtering apparatus of considerable surface area. Although the smallest openings in the fans measure about 35 μm across, larvae are capable of collecting particles in the range of 0.091 to 350 μm, the majority being < 30 μm in diameter. This is possible because the fans are coated with a non-sulphated acid mucosubstance containing sialic acid. This is similar to the mucosubstances produced by many filter-feeding marine invertebrates and which have particle-adsorbing properties (Ross and Craig, 1980). The mucus is thought to be spread over the fan rays by a sweeping motion of the mandibular raking

Fig. 6.3   Cased larva of *Brachycentrus* (Trichoptera) using its meso- and
metathoracic legs to capture suspended food particles (redrawn after
Merritt and Wallace, 1981).

bristles after food and old mucus have been cleaned off the rays by the mandibles.
Black fly larvae typically position themselves on smooth rock surfaces in areas
of fast, laminar flow. Attached to a pad of silk threads by their minute posterior
hooks, so that their posterodorsal surface faces upstream, the larvae rotate their
bodies 90 to 180°, longitudinally. This causes the adoral surface of the cephalic
fans to face the water flow. The larvae, not being particularly streamlined, are
deflected from the vertical (the angle of deflection being related to water velocity)
so that one fan is closer to the substrate while the other is located near the top of
the boundary layer. Velocity profiles around larvae of *Simulium vittatum* have
shown that body shape and feeding stance create two vortices. One of these
remains in the lower boundary layer, while the other rises up the downstream
side of the body and passes through the lower fan. The latter entrains particulate
matter from the substrate which the larva then filters (Fig. 6.4). The upper fan
filters water only from the top of the boundary layer (Chance and Craig, 1986).
The position of adjacent larvae can enhance feeding through the mutual influence
of flow.
       Larvae of many species of free-living caddisfly (e.g. Psychomyiidae,
Polycentropodidae, Philopotamidae, Hydropsychidae, Stenopsychidae and

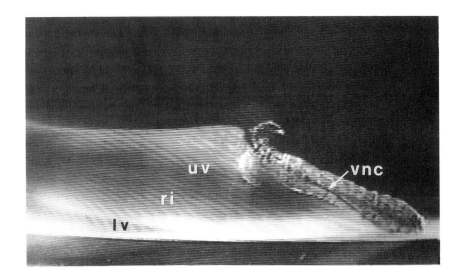

Fig. 6.4   Larva of *Simulium vittatum* (Diptera) capturing particulate matter from
the lower boundary layer (brought up by the upper vortex; **uv**) in its
lower cephalic fan. **Lv** denotes the lower vortex; **ri** is the region of
interaction between the vortices; **vnc** is the ventral nerve cord which
appears twisted because of the rotation of the larva's body needed to
attain the filtering posture (from Chance and Craig, 1986).

Ecnomidae; see Table 2.11) filter particles in the water by means of nets
constructed from silk. Mesh size and configuration vary greatly among species
and are related to the type of particle exploited. Coarse meshes (0.5 x 0.6 mm)
are typically spun in fast currents by species of *Arctopsyche* (Hydropsychidae),
which frequently eat small invertebrates, whereas nets with openings as fine as
0.5 x 5.5 μm are spun by the larvae of *Dolophilodes distinctus* (Philopotamidae)
which feed on FPOM in quieter waters (Wallace and Malas, 1976a,b). Associated
with the different mesh sizes are morphological structures on the larvae to
construct the different net designs and to gather and process the food trapped. For
example, in *Arctopsyche* the prothoracic legs are spiny and are used for subduing
prey whereas, in *Dolophilodes*, the highly setose labrum and maxillary brushes
are used for removing fine particles from the net meshes. There is evidence to
suggest that in the nets of some species, particles adhere to the silk threads rather
than solely being retained by a size-sieving method. Mesh size increases,
proportionally, as caddisfly larvae grow. This is not surprising as the dimensions
are related to the sizes of the various bodyparts used in construction of the net.

    Net architecture varies considerably in the Trichoptera. The simplest takes
the form of shallow funnels stretched between adjacent substrate particles, as is

typical of many hydropsychids. The more complex involve convoluted structures some of which may be partially buried. The net of a Brazilian species of *Macrostemum*, for example, is deeply embedded in the sandy beds of woodland streams. The entrance, constructed of sand and silk, is in the form of a curved cylinder, rising about 3 cm above the stream bed and opening into the current (Fig. 6.5A). The net is located roughly 1 cm below the bed surface and in front is a vertical sand-grain shaft which forms the larva's retreat. Posterior to the net is a vertical tube with a horizontal opening just level with the bed. Water is driven through the system by pressure differences between the two openings, the principle of the Pitot tube. Placing a net in the centre assures capture of any suspended particles (Sattler, 1963). Modifications on this basic design are to be seen in other species of *Macrostemum* throughout the world (Fig. 6.5B). As we saw in Chapter 4, specialization of larvae on certain food particle size classes may lead to marked longitudinal zonation in filter-feeding trichopterans.

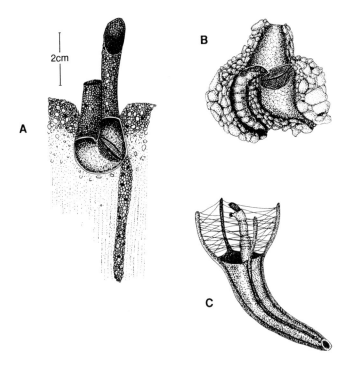

Fig. 6.5   (A) Retreat and feeding net of a species of *Macrostemum* from Brazil; (B) larval retreat and feeding net of *Macrostemum zebratum* (Trichoptera: Hydropsychidae) from Canada; (C) larval retreat and feeding net of the chironomid genus *Rheotanytarsus* (Diptera) (redrawn after Hynes, 1970; Wiggins, 1977; Merritt and Wallace, 1981).

Among the Diptera, larvae of the stream-dwelling chironomid genus *Rheotanytarsus* also employ a filter-feeding technique. The larva builds a tube-like shelter attached to the substratum, made from particles of silt bound together with a sticky, silk-like saliva (Fig. 6.5C). The end attached to the substrate has a small opening while the other end, which is curved up into the water, ends in a wide mouth surrounded by two to five slender projections. Between these, the larva strings several threads of saliva which trap passing detritus. Periodically, the larva emerges from its shelter and eats both the threads and attached particles. New, clean threads are then produced (Merritt and Wallace, 1981).

Filter-feeding methods involving the production of silk or similar secretions on a regular basis are not without their costs. Few studies have attempted to evaluate the efficiency of such techniques. However, Dudgeon (1987) examined the energy budget associated with net construction in the polycentropodid *Polycentropus flavomaculatus*. When, in the laboratory, larvae were forced to rebuild their nets daily, they lost more weight than larvae forced to rebuild every second day. Although net allocation (net weight/larval weight) remained constant, the weight of the nets of larvae disturbed daily decreased over successive constructions. Whereas the major energetic expense of undisturbed larvae was routine metabolism, that of the daily rebuilders was silk synthesis. Construction behaviour was linked to food capture, with larger nets (not meshes) being built at times of greater food availability. The energy gain (food intake - the combined costs of metabolism and silk production) was greater in larvae fed *ad libitum* compared with larvae fed every second day. Presumably, in nature, the larvae arrive at some sort of optimum net size and replacement time based on the food available at specific sites.

6.4   Respiration

Aquatic insects have dealt with the problem of respiration in several different ways. Broadly speaking, they can be divided into two categories: *aeropneustic* and *hydropneustic*. Insects belonging to the first group primarily use oxygen from the atmosphere, whereas those belonging to the second group extract oxygen dissolved in the water.

Aeropneustic insects can be further divided into three subgroups: (1) surface breathers that remain permanently in contact with the atmosphere; (2) surface breathers that periodically contact the atmosphere; and (3) insects that obtain their oxygen from the stems of aquatic vascular plants.

Larvae of the Syrphidae (Diptera) are good examples of insects that remain more or less permanently in contact with the air. They do so via a pair of posterior spiracles located at the end of a telescopic breathing tube. Hydrophobic hairs on the end of the tube keep the spiracles open and prevent water from entering. Syrphid (rat-tail maggot) larvae typically live in very shallow water, as the breathing tube will not stretch beyond about 5 cm. Larvae of some of the

Ephydridae (shore flies) respire similarly.

Insects that periodically come to the surface to breathe include mosquito larvae and pupae, beetle larvae and adults, and the mature and immature stages of many hemipterans. Culicine mosquito larvae have a pair of spiracles at the tip of a sclerotized "siphon", whereas anopheline mosquitoes have no such extension and the spiracles are borne on the dorsal side of abdominal segment 8 (Fig. 2.33). These spiracles lead directly into the insect's tracheal system, an internal network of air-filled tubes. This network is formed from invaginations of the cuticle, in which larger tubes (*trachea*) down to 2 - 5 μm in diameter, gradually give rise to smaller tubes (*tracheoles*) less than 1 μm in diameter which end blindly in close contact with individual body cells. The end of each tracheole is fluid filled and it is across this surface that gas exchange takes place. Here, there is a self-regulating mechanism which ensures that areas of tissue that are metabolically very active can be preferentially supplied with more oxygen. This system evolved in the terrestrial ancestors of aquatic insects and is the one that persists in both groups today.

Once mosquito larvae have renewed their air supply, they can dive in search of food. As they do this, the ring of hydrophobic hairs surrounding the spiracles is compressed to prevent the entry of water. The time for which mosquito larvae and pupae can remain submerged is limited by the amount of air carried in the tracheal system but also depends on the activity level of the insect and the temperature of the water. In adult beetles and many hemipterans, submergence time is greatly extended by carrying an additional, ex-tracheal, supply of air.

Adult gyrinids and dytiscids hold an extra store of air under their elytra. When dytiscids rise to replenish this store, their naturally buoyant abdominal region ensures that they break the water surface at the correct angle to allow air to be taken into the sub-elytral cavity and into the abdominal spiracles - especially the enlarged posterior pair. Although hydrophilid and hydraenid beetles also store air under their elytra, the method of renewal is different. Hydrophilid adults, for example, rise to the surface head first, tilting slightly to one side. The water surface is penetrated with the aid of hydrofuge hairs borne on the tip of a specially modified antenna which establishes an air channel to the sub-elytral air store. These antennae are typically short and club-shaped and some of their previous sensory functions have been transferred to the palps. Some hydrophilids are capable of storing additional air on their general under surface, trapped by a dense pile of hairs, and replenished through the antennal channel via special hair tracts (Evans, 1977). Entrapment of air by regions of unwettable hairs or scales is common in the adults of many bugs and beetles and the film of air thus formed is known as a *plastron*.

There are basically two types of plastron found in aquatic insects. The *macroplastron* acts as an additional reservoir of breathable air, but also allows some oxygen to be extracted from the water through its action as a physical gill. This is the type typically found in large active beetles like the dytiscids and hydrophilids and, characteristically, it gradually decreases in size over time since

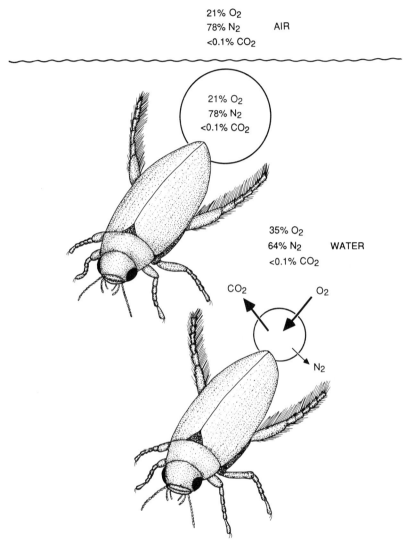

Fig. 6.6   Plastron respiration in an adult beetle.

the last visit to the surface. It functions as follows. When the insect dives, the proportions of the major gases held in the plastron bubble are the same as in the atmosphere (~ 78% nitrogen; 21% oxygen; and a little carbon dioxide; Fig. 6.6). As the oxygen begins to be used up, the partial pressure of nitrogen in the bubble increases and, in order to maintain an equilibrium, some of the oxygen in the water diffuses into the bubble; carbon dioxide produced during respiration is quickly removed from the bubble because it is highly soluble. Although nitrogen is far less soluble, over time it diffuses out into the water with the result that the bubble volume gradually decreases and eventually the insect must surface to renew it. However, over the life of the bubble, the insect obtains a greater total amount of oxygen (approximately 8 times) than was contained in the original bubble. The hairs maintaining the macroplastron are relatively long (~ 200 µm) and flexible and as the gas bubble is used up, they may become clumped and collapse. The time that a bubble lasts is dependent on a number of factors including the activity level of the insect, the temperature of the water, the amount of oxygen that it contains, and the amount of organic pollution present. During periods of ice cover, some normally very active water beetles (e.g. the Colymbetinae) are known to hibernate out of water, while others (e.g. *Dytiscus*) diapause in the substrate where sufficient oxygen to sustain them diffuses across the cuticle surface (Galewski, 1971; Landin, 1976).

Some insects, such as dryopid beetles, have a layer of much shorter, more densely-packed hairs which forms a *permanent plastron*. The latter is more resistant to collapse. *Aphelocheirus*, a naucorid bug found in the palaearctic, has one of the most efficient permanent plastrons. It consists of very short hairs (5-6 µm high) with a density of about 4.3 million hairs per mm^2, capable of withstanding a pressure of some four atmospheres before collapse. The spiracles connect with the plastron through small pores located along a series of canals in the cuticle and the canals are lined with hairs to prevent the entry of water. Under normal environmental conditions, the amount of oxygen supplied to *Aphelocheirus* via this system is sufficient to allow the insect to remain permanently submerged (Chapman, 1975). Naucorids are known sometimes to ventilate the ventral portion of their plastron by rowing their hind legs under the abdomen (McCafferty, 1981). The permanent plastrons seen in adult elmid beetles are believed to be supplemented, on occasion, by bubbles of oxygen produced by photosynthesizing benthic algae and collected by fringes of setae on the beetle's legs.

The pupae of many dipterans and beetles, and the larvae of sphaeriid and hydroscaphid beetles, which are associated with the margins of water bodies in which the level of water is likely to vary greatly tend to have complex, cuticular networks surrounding their spiracles (Hinton, 1968). These structures resemble columns supporting a perforated canopy, under which a film of air is trapped which functions as a physical gill. This system seems ideally suited for both obtaining oxygen when the insect is submerged and for retaining water when it is beached.

Insects that obtain their oxygen from within the aerenchymal tissues of aquatic plants have spiracles that are modified for piercing and holding on to stems. Although not a commonly-adopted technique, it is seen in a number of dipterans (e.g. the ephydrid *Notiphila*, the syrphid *Chrysogaster* and several genera of mosquito: *Mansonia*, *Taeniorhynchus* and *Coquillettida*) and also in the chrysomelid beetle *Donacia*. Many of these insects are capable of cutaneous respiration which seems to act as a backup system. In regions where there is macrophyte die-back, in winter, the metabolism of endophytic breathers appears to be shut down (Eriksen *et al.*, 1984).

Hydropneustic insects all basically respire cutaneously, but can be divided into groups based on various types of accessory cuticular organs (e.g. gills) that enhance the process. Many do not have functional spiracles. Hydropneustic insects tend to be more common in running waters or well-aerated lentic waters where their distributions may be closely linked to available oxygen.

In the larvae and pupae of small hydropneustic insects, like chironomids, simuliids and chaoborids (plus the Collembola), gaseous exchange occurs by diffusion through the general body wall. The air in their tracheae contains both oxygen and nitrogen and, as the oxygen in their tissues is consumed, it is replaced by oxygen diffusing in from the water. This restores the normal oxygen-nitrogen ratio of the tracheal air, and resembles the functioning of the plastron "gill". Simple cuticular respiration of this sort is sufficient to supply the oxygen needs of only small, relatively inactive, insects. It is possible that the system may be made more efficient by: (1) using a more effective oxygen carrier in the haemolymph (e.g. some chironomids possess haemoglobin) but this has not

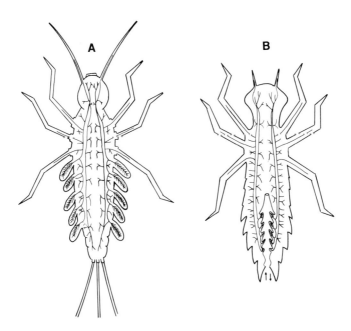

Fig. 6.7   Two different types of closed tracheal systems in aquatic insects: (A) with abdominal tracheal gills as seen in a mayfly nymph; (B) with rectal tracheal gills as seen in a dragonfly (Anisoptera) nymph (after Wigglesworth, 1965).

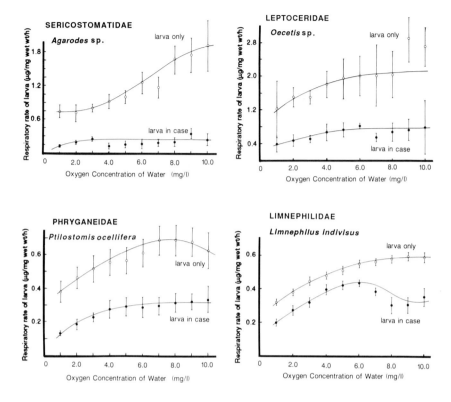

Fig. 6.8   Oxygen uptake curves of four species of cased caddisfly collected
from lentic or temporary habitats (from Williams *et al.*, 1987; mean
values ± 1 SE are shown).

been well studied; and (2) expanding the peripheral tracheal network just under
the cuticle.

Larger and more active aquatic insects have developed membranous
outgrowths from the body wall that increase the effective surface area for gas
exchange (Fig. 6.7A). In many larvae, these *gills* now serve as the primary sites
for respiration. Gills are of two basic types: (1) thin, flat, plate-like structures
commonly rich in tracheae; and (2) filamentous structures that are more fleshy
and tube-like, and which may be branched and not necessarily direct outgrowths
of the tracheal system (McCafferty, 1981). Plate-like gills are typically found in
mayfly nymphs (on the first seven abdominal segments) and in damselfly
nymphs (as three large leaf-like structures at the end of the abdomen). In
dragonfly (Anisoptera) nymphs, they take the form of folds in the lining of the
rectum (Fig. 6.7B) and water is drawn rhythmically in through the anus to
irrigate them. Filamentous gills are typically seen in stonefly nymphs, as single
or branched structures around the bases of the legs or the basal abdominal

segments. In larval Trichoptera, gill filaments are usually located on abdominal segments, although they may also be on the thorax in some families (e.g. the Hydropsychidae). Not all of the gill-like structures seen in aquatic insects are actually used for respiration, some may be associated with protection, osmoregulation or burrow maintenance.

Sometimes, respiratory efficiency is enhanced by behaviour. For example, when stressed by low oxygen levels, stonefly nymphs frequently embark on "push-ups", mayfly nymphs twitch their gills and dragonfly nymphs increase the frequency of water intake into their rectums; these are all means of increasing the circulation of water around the gills. In caddisflies, the elaborate tubular case constructed around the abdomen is thought to aid respiration in many species, although it may act as camouflage in others. Larvae of species in which the case appears to enhance respiration show a lower, more uniform, respiration rate and a greater tolerance of low oxygen levels when in their cases (Fig. 6.8). In these instances, the cases seem to be designed to aid respiration by optimizing, rather than maximizing, the amount and rate of oxygen that the larva can extract from the water. This might be of particular advantage to species living in habitats where oxygen levels in the environment are seasonally variable and, indeed, many of the species in which the case does bestow an advantage are found in warm lotic, lentic or temporary habitats (Williams *et al.*, 1987).

6.5 Osmoregulation

Osmoregulation refers to the processes that maintain specific concentrations of internal salts which are necessary in order to allow basic biochemical reactions to occur. This involves careful control of the amounts of water and ions both entering and leaving the body. Early terrestrial insects had to overcome major problems of water loss but this is clearly not a problem for their freshwater descendants, in fact quite the reverse. The haemolymph of freshwater insects is hypertonic to the water in their environment and, consequently, water passes into the insect, largely through the cuticle. Some aquatic insects are more permeable than others, primarily because of differences in cuticle structure, specifically in the nature of the epicuticular waxes. The cuticles of adult water beetles and bugs (Heteroptera), for example, are relatively impermeable compared with those of many aquatic larvae and pupae which tend not to have a lipid layer (Chapman, 1975). High cuticle permeability also puts aquatic insects at risk from loss of salts which may be leached out into the environment.

Excess water is removed from the body through the production of copious amounts of dilute urine and along with this go large quantities of dissolved ammonia, a major nitrogenous waste product of at least some aquatic insects. It is not clear whether ammonia is formed in these insects by direct deamination of amino acids (nitrogen excretion in its most primitive form), or whether it comes from the secondary breakdown of other excretory products (Gilmour *et al.*, 1970).

Salts are partially retained by reabsorption in sections of the gut and in the rectum. In the rectum, Na and Cl ions are known to be resorbed through an active process, and perhaps this is the case for K^+ also. Resorption is regulated against the composition of the haemolymph (Shaw and Stobbart, 1963) but as this system is not entirely effective, freshwater insects have developed supplementary systems for the active absorption of ions from the surrounding water. Some salts are also replaced through the diet.

Active absorption of ions from the environment occurs across specialized regions of the body. Odonate nymphs and the larvae of many species of caddisfly and Diptera have patches of integument that are well supplied with cells that specialize in ion uptake. These patches are known as *chloride epithelia* and are typically located on ventral or dorsal parts of the abdomen in caddisflies, adjacent to the anus in dipterans, and within the rectal chamber of odonates. All these areas are actively ventilated by the larvae, and the rate at which this is done, together with the size of the patches, is related to the ionic content of the water - both increase with decreasing salinity.

The nymphs of stoneflies and mayflies, as well as the adults of some heteropterans, have specialized *chloride cells* scattered over certain parts of their bodies and these are also involved in the active absorption of ions from water. Although more generally distributed on the bodies of bugs, chloride cells are typically found on the gills and sides of the body in mayflies, and on the gills, intersegmental areas and lateral and ventral parts of the abdomen of stoneflies. The density and degree of ventilation of chloride cells change in reponse to the ionic strength of the surrounding water in the same way as seen for chloride epithelia (McCafferty, 1981). In a species of *Callibaetis* (Ephemeroptera) which lives in temporary ponds, the number of chloride cells has been shown to decrease as the pond dries up and, consequently, as the ionic content of the water increases (Wichard and Hauss, 1975).

Anal papillae are also known to be sites of active ion uptake. These are elongate, thin-walled evaginations of the body most typically seen in dipteran larvae. In mosquito and chironomid larvae (Figs 2.33B; 2.35C) they are more or less permanent structures although their size varies according to the salinity of the surrounding water (they are larger in more dilute solutions). Black fly larvae have papillae formed from evaginations of the rectal wall that protrude from the anus. Larvae of *Aedes*, *Culex* and *Chironomus* are known to be able to absorb sodium, potassium, chloride and phosphate through their papillae (Chapman, 1975). Structures analogous to anal papillae include, for example, the protrusible anal "gills" found in helodid (Coleoptera) larvae and the non-tracheated processes on abdominal segment 9 in gyrinids (Crowson, 1981).

The mechanisms for active ion absorption given above are only those that are known or understood. In general, the precise means by which many aquatic insects absorb ions is unknown. Freshwater insects seem well able to cope with a drop in the salt concentration of their surrounding water, but most do not fare as well when faced with an increase in salinity (McCafferty, 1981). This may be because in hypertonic (salty) media the insect's haemolymph soon becomes isotonic with the medium and osmoregulation ceases. Unlike terrestrial and marine insects, most freshwater insects seem incapable of producing urine that is hypertonic to their haemolymph (Chapman, 1975). However, as we saw in Chapter 4, at least six of the aquatic insect orders contain one or more (and sometimes many) species that have managed to overcome the problems of living in salt water. Faced with such evidence of repeated, independent, solutions to living in salt water, can the absence of mass colonization of marine habitats by insects really be attributed to a problem of osmoregulation?

Marine insects tend to have haemolymph that is hypotonic to their surroundings and they are thus subject to loss of water through osmosis. They lose more water in the form of urine but some of this, as in terrestrial insects, is retrieved in the rectum. Water is replaced by drinking the surrounding medium and absorbing only the water through the walls of the midgut (Chapman, 1975). Excess salts from both this process and from those taken in as part of the insect's food are excreted in very hypertonic urine (Tones, 1978). Species that can switch between fresh water and marine habitats have part of their rectum (or, in some cases, the ileum) that specializes in water reabsorption and another part that is adapted for ion reabsorption (McCafferty, 1981).

6.6   The Chironomidae and Ephydridae - special cases?

Comparison of the habitat ranges of aquatic insects (Chapter 4) identified the dipteran families Chironomidae and Ephydridae as being well able to cope with habitats characterized by high adversity and also, particularly in the case of the chironomids, by low stability (Fig. 4.6A-D).

Chironomid larvae are found in all types of aquatic and semi-aquatic habitat from the most ephemeral (temporary rainpools) to the most permanent (the sea). As a group, chironomids range over huge environmental gradients of temperature, pH, oxygen concentration, salinity, etc. They are also capable of living and, in some cases, thriving in a variety of anthropogenically-polluted waters (see section 11.1). What has enabled this particular group of nematocerans to become so successful? Intuitively, it must be a result of an enormous spectrum of morphological, physiological, phenological and behavioural adaptations ranging from the ability of *Polypedilum vanderplanki* to recover after losing up to 99% of its tissue water to the modifications in locomotion, reproduction and emergence seen in marine chironomids. However, despite such intriguing diversity, comprehensive and detailed study of these adaptations is, for the most part, lacking.

The same is less true for the ephydrids as at least some aspects of their physiological adaptations to extreme environments are known. For example, the ability of *Ephydra cinerea* to deal with varying levels of salinity is known to be through maintenance of its own haemolymph osmotic pressure regardless of whether the larva is in distilled water or a 20% sodium chloride solution (almost 6 times the strength of sea water; Stobbart and Shaw, 1974). The osmotic pressure generated in the haemolymph of *E. cinerea* has been estimated to be equivalent to 20.4 atmospheres (Nemenz, 1960). A variety of other adaptations is seen in specific habitats. These include: endophytic respiration (from plant roots) by the larvae of *Dimecoenia spinosa*, which lives in the anoxic muds of salt marshes; the ability of eggs, larvae and pupae to withstand the high temperatures associated with hot springs (e.g. *Scatella thermarum*, up to 48°C); the temporary air store carried beneath the wings of adult *E. hians* when ovipositing or feeding under water; the varied surface microsculpture of significance to the respiration of eggs and enabling those of some species (e.g. *Paracoenia fumosa*) to remain submerged but respire directly in the air via a snorkel; and the adaptability of the larval mouthparts for a range of foods (Simpson, 1976). In some of the most adverse habitats, such as hot springs, the larvae and adults eat the same food (Tuxen, 1944). This is thought to be a consequence of reduced interspecific competition, and may serve to keep the adults close to the larval habitats, which may be sparsely distributed. Although interest has been generated in the adaptability and abundance of selected species of ephydrid in extreme habitats, comparatively little is known of the natural history of the family as a whole (Simpson, 1976). For example, of the approximately 347 Nearctic species, the feeding habits of only 20 are known (Foote and Eastin, 1974).

It is evident that much more basic biology, set against a clear phylogenetic background, will have to be gathered before the factors responsible for the unparalleled success seen in both the ephydrids and chironomids can be properly understood.

7   POPULATION BIOLOGY AND DYNAMICS

7.1   Dispersal and colonization

There has been considerable confusion and re-definition of the various terms associated with insect movement (Johnson, 1966). For present purposes, *colonization* is taken to be the end product of a series of steps in an individual insect's behaviour. These steps are (based on Fernando, 1958): (1) *dispersal*, which provides the basis for colonization; (2) *location* of a new habitat; (3) *selection* of a new habitat; and (4) *colonization* itself, which represents a period of residence in the new habitat during which the insect may feed and/or reproduce. These steps have evolved as a result of two types of stimuli: *proximate* factors, such as environmental temperature and photoperiod; and *ultimate* factors, such as food or suitable oviposition sites.

Adult aquatic beetles, for example, are known to respond directly to proximate stimuli by adaptations of behaviour or physiology. Changes in water or air temperature, modification of light intensity or duration, rain, or a combination of these factors cause many beetle species to disperse. They leave the water and take to the air from where they may locate new water bodies. It is believed that internal physiological states, such as ovarian development in the female, or hunger, can initiate dispersion. In the spring, flying female *Helophorus brevipalpis* (Hydrophilidae) have larger oocytes than non-fliers, and fliers' guts tend to be empty (Landin, 1980). Habitat location, selection and colonization are responses to ultimate stimuli, primarily the presence and availability of suitable food and substrate (in the case of benthic species). Ultimate factors may not be directly linked to the proximate factors that initiated dispersion, but the former are often "anticipated" as being favourable, either for the adult insect or its offspring, because of the evolution of seasonal dispersal (Fernando, 1958).

In temperate and polar regions, dispersal of adult insects from water bodies has a distinct seasonal basis. This may be evident also in the tropics where cyclical events occur in the physical environment, for example in areas influenced by monsoons or predictable droughts. Dispersal to new habitats seems more prominent in temporary water bodies than in permanent ones. In Britain, *H. brevipalpis* has two dispersal peaks. The first is in the spring (April - May) and the second, which is much larger, occurs from the middle of June to the end of August (Fig. 7.1). The spring dispersal consists primarily of females carrying mature eggs which are deposited in other ponds. The later dispersal is non-reproductive and consists of adults, including those produced from eggs laid in

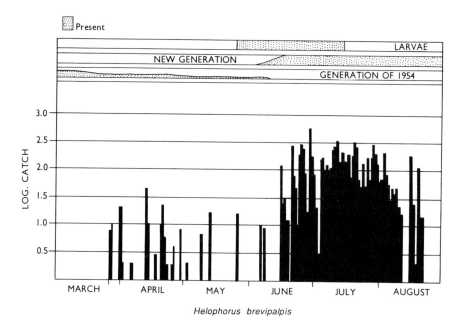

Fig. 7.1   Logarithm of the catches of *Helophorus brevipalpis* in the field
experiment in Wytham during 1955, using six artificial habitats (4 ft x
4 ft), three glass traps (5 ft x 3 ft), and three glass traps (2 ft x 3 ft) - a
daily census was made from March to September (redrawn from
Fernando, 1958).

the spring, apparently redistributing themselves among suitable, available
habitats. In addition to seasonal flight patterns, many aquatic insects exhibit a
diel pattern. Again, this has been studied best in beetles. In Sweden, *H.
brevipalpis* disperses bimodally (Landin, 1968), with the number flying in the
evening being 3 to 4 times that in the morning. Temperature is thought to be
the controlling factor because of a fixed threshold that must be exceeded before
flight can be initiated; for *H. brevipalpis* this lies between 11 and 15°C. High
winds and light levels, together with air temperatures over 25°C, may play a role
in depressing flight in this species (Fig. 7.2).

     Flight terminates after habitat location and selection, with either
colonization by the adult insect itself, as is often the case in the Coleoptera and
Hemiptera, or egg laying by the female. Habitat location and selection processes
may involve one or more of the insect's senses. In corixid bugs, for example,
sight is important, and flying adults orient themselves at a specific angle to the
incident light. The insects home-in on reflections from water surfaces, perceived
as a result of this orientation, and may make their final selection on the basis of

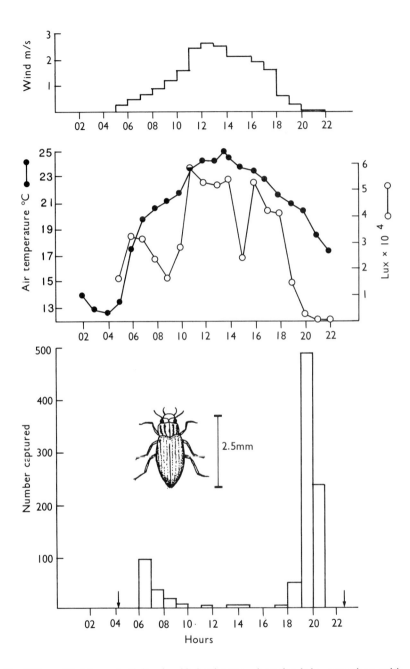

Fig. 7.2   Flight periodicity in *Helophorus brevipalpis*, together with air temperature, light intensity and wind velocity on 22 July, 1965 (wind velocity measured from 5 am to 10 pm; sunrise 3.12 am, sunset 8.34 pm; redrawn from Landin, 1968).

the amount of organic matter present in the habitat (Macan, 1939; Popham, 1953). Pajunen and Jansson (1969) observed that adult brackish water corixids could discriminate between pools of different size, moving from shallow pools where they live in spring, to deeper pools in late autumn, prior to over-wintering.

Habitat selection/oviposition stimulation processes have been studied in some detail in mosquitoes, and are known to be influenced by a variety of environmental factors including degree of water movement, alkalinity, salinity, the amount of shade, background colour and emergent vegetation. Most remarkable is the process by which the females of many insects that inhabit temporary waters select suitable ponds and streams when they are dry. By what mechanism is the female assured that the dry bed upon which she lays her eggs will later become an aquatic habitat capable of sustaining aquatic larvae? Females seem to be attracted by special features of the bed that are associated with water. Female dragonflies, for example, are thought to be attracted to stands of *Typha* (cattails) which tend to grow where the ground is susceptible to seasonal flooding (Corbet, 1963). Gravid female mosquitoes may be drawn to the moisture gradient emanating from the saturated soil beneath a dry bed, or to the scent of rotting aquatic vegetation (Williams, 1987).

The processes considered above are all examples of *active* dispersal and colonization mechanisms. These are the primary ones employed by aquatic insects, although they may differ in scale among different groups. For example, those insects possessing good powers of flight, such as the beetles, hemipterans, odonates and certain dipterans, are able to range farther afield than poor fliers like the mayflies and stoneflies. That is not to say, however, that the latter are any less effective dispersers, just that they are more likely to colonize adjacent areas of the same water body rather than a different one. Running water insects, for example, although they may disperse to nearby drainage basins, more often lay their eggs in the same or another section of the stream or river from which the females emerged. Scale is an important factor in considering dispersal and colonization dynamics. They range from long-term, long-distance events involving exchanges between continents, which are the domain of the biogeographer (e.g. Illies, 1965), through mid-range events involving the colonization of local, newly created habitats (e.g. Williams and Hynes, 1977b), to the colonization of adjacent habitat units (e.g. Williams, 1980a), each of which may be an integral part of the life history tactics of a particular species (Sheldon, 1984).

Some aquatic insects are dispersed by *passive* means, on the wind, and this may explain the wide geographical range of some weak fliers. For example, some chironomid species are too short-lived and too weak to disperse actively over large distances, and large numbers have been recorded far out at sea (Bowden and Johnson, 1976). An extreme example is the 400 to 500 km migration of species in the *Simulium damnosum* complex associated with monsoon winds in West Africa. Convection currents have been known to carry *Simulium meridionale* to an altitude of 1520 m (Crosskey, 1990).

Newly emerged insects frequently provide a means by which other aquatic organisms, such as mites, ostracods, algae and bacteria, are passively transported between habitats (Revill *et al.*, 1967).

Thus far, we have considered only adults but immature stages may also relocate, though for reasons other than reproduction. Nowhere is this better illustrated than in the case of running water insects. It is now well established that the immature stages of most lotic species of insect, together with some aquatic adults, drift downstream with the current at some stage in their lives (Waters, 1972). Some do so with such regular periodicity that it is thought to be initiated intentionally by the insect rather than by accidental dislodgement. As a result, several categories of drift have been proposed: (1) *behavioural drift* which includes insects that either actively enter the water column, for example to escape from a predator, or are dislodged as a consequence of an increased period of general activity on rock surfaces, as in foraging; (2) *distributional drift* which represents a method of dispersal, and is particularly common in newly-hatched insects, or even eggs masses (Williams, C.J., 1982); (3) *catastrophic drift* which is usually associated with flood conditions during which insects are physically removed from the bed, *en masse* ; and (4) *constant* or *background drift* which represents the low numbers dislodged accidentally but irrespective of any diel periodicity. There is evidence that insects drift for at least one of these reasons, but some authors have warned about the dangers of confounding the study of drift through excessive terminology (e.g. Brittain and Eikeland, 1988).

Drift is, unequivocally, a very important aspect of life in running water, particularly in terms of the redistribution of benthic insects in response to competition for food and space. It has been shown that up to 82% of the invertebrates colonizing experimentally denuded areas of stream bed arrive via the drift (Townsend and Hildrew, 1976), and estimates of the percentage of the benthos occurring in the drift at any given point in time range from 0.004 to 0.13% (Rutter and Poe, 1978; Williams, 1980c). Based on field experiments, Townsend and Hildrew (1976) calculated that, on average, 2.6% of benthic invertebrates change their position each day by entering the drift, and that, for some individual species, the proportion is much higher (e.g. 14% for the caddisfly *Plectrocnemia conspersa*). It has been proposed that as the predominant means of movement among suitable microhabitats, drift provides a reasonable model to account for the patchy distribution of benthic species so typical of the beds of rivers and streams (Minshall and Petersen, 1985).

Entry into the drift is clearly due to a variety of different reasons, some of them endogenous, some of them exogenous. Species of the mayfly genus *Baetis* are well-known drifters in streams all over the world. In laboratory experiments, Kohler (1985) showed that *Baetis tricaudatus* actively enters the water column at night, after abandoning patches of stream bed when habitat quality, primarily assessed as food availability, falls to some threshold level. Drift by insects is typically nocturnal and while some species show a "bigeminus" pattern of one large peak just after sunset and another, smaller, one just before sunrise, other species show the converse "alterans" pattern (Müller, 1966; Cowell and Carew, 1976). There is also a seasonal component to drift. Typically, in temperate regions, the rate of drift is low during the winter and high in summer whereas, in the tropics, differences are less marked (Williams, 1980d).

Termination of drift by individuals is thought to be largely a function of behaviour rather than of morphology (Walton, 1978). Mayfly nymphs, for example, which are common in the drift, have been observed to swim downwards and re-attach to the substrate. Drift response has been shown to vary according to a number of features in the stream environment including photoperiod, water temperature, discharge and the presence of large predators such as trout (Williams, 1990), and these may govern the drifting process in a hierarchical fashion (Statzner *et al.*, 1984). Several studies have attempted to measure the distances drifted by insects. This appears to be a function of species, stage in the life cycle, water depth, the presence and frequency of pools as well as of some of the other environmental factors discussed above. For example, first instar larvae of the caddisfly *Potamophylax cingulatus* are known to drift more than 10 times as far as fifth instar larvae (Otto, 1976). With such a complexity of influential factors it is not surprising that distance estimates are wide ranging, from 0.5 to 19.3 m at low current velocities to several hundred metres during floods (McLay, 1970; Neves, 1979).

Over the period of their larval life, it has been calculated that certain insects may drift as much as 10 km (Hemsworth and Brooker, 1979). The sheer numbers of insects being displaced downstream together with the distances involved has made researchers speculate about the possibility of headwater regions becoming denuded. It is known that many larvae are positively rheotactic and tend to crawl against the current, but the distances involved seem generally insufficient to compensate displacement through drift. Müller (1954) proposed that compensation is achieved primarily through the upstream flight of adult females which lay their eggs in headwaters. There is evidence for such upstream flights in a few species (e.g. Madsen and Butz, 1976) but more research is necessary before this hypothesis can be accepted fully. In some cases the numbers of females flying upstream is only a few percent more than those flying downstream. Of course one female may lay several thousand eggs (e.g. up to 4,500 in *Baetis rhodani*; Elliott and Humpesch, 1980) and this may be sufficient to maintain upstream populations of larvae.

Besides dispersing up and down the length of streams, immature insects are known to move across the bed, even into the interstices of the bank, and also vertically and obliquely within the bed materials themselves. Larvae and aquatic adults living in the hyporheic zone are known to colonize areas of denuded stream bed from below (Williams, 1984). For some species, movement into and out of the hyporheic zone is part of their life cycle. In the Speed River, Ontario, Canada, for example, the chironomid *Cladotanytarsus* is univoltine. Eggs of a new generation begin to hatch in mid-August and early instar larvae are found

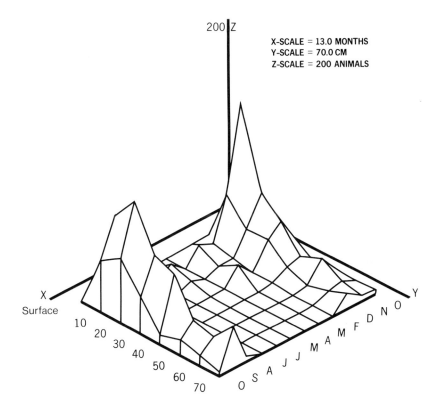

Fig. 7.3   Annual distribution of *Cladotanytarsus* spp. within the bed substrate of the Speed River, Ontario. Plot of time vs. depth vs. number of animals (after Williams, 1981).

interstitially at this time (Fig. 7.3). This is either the result of a downward dispersal of small larvae coincident with a drop in stream surface water temperature, or settling of eggs into the interstices as a consequence of gentle agitation by the current. Larvae overwinter at various depths down to 70 cm before dispersing upwards in March as the surface water temperature rises. The majority of larval growth takes place in the interstices where the food is fine particulate detritus. Pupation occurs on the stream bed surface and adults begin to emerge in April.

Because life histories differ among aquatic insect species, even those occurring in the same habitat, there is frequently a difference in the timing of their dispersal and/or colonization phases. This, alongside potential seasonal differences in environmental influences, results in changes in the composition of species within communities over time. Species succession is just as much a part of aquatic insect communities as it is of terrestrial ones, and has been studied both experimentally and descriptively, especially in streams. Often the causal factors are simple and obvious, sometimes they are less so. For example, the succession of insects seen in Moser Creek, a small temporary stream in Ontario (see Chapter 4), was linked to the changing nature of the physical and chemical environment, resulting from the gradual loss of water and the increase in air temperature. On the other hand, changes in the species of chironomids in a newly formed lake in England might at first sight have been attributed to sediment distribution, whereas, in fact, larvae of *Tanytarsus gregarius* were competitively displaced by final instar larvae of *Chironomus plumosus* (Cantrell and McLachlan, 1977).

7.2   Coexistence and competition

Whether competition significantly affects the population dynamics and community structure of living organisms has been a matter of discussion for approximately 100 years (Hardin, 1960). Surprisingly, this concept did not receive formal recognition until Gause's (1934) contention that, "As a result of competition two similar species scarcely ever occupy similar niches, but displace each other in such a manner that each takes possession of certain peculiar kinds of food and modes of life in which it has an advantage over its competitor". Niche, as used here, referred to the role of an animal in its environment and its relationship with its food and enemies (Elton, 1927). Gause's concept of competition was modified by Hardin (1960) in the *Competitive Exclusion Principle*, which is briefly stated as, "complete competitors cannot coexist". Hardin suggested three pre-conditions for competitive exclusion to occur: (1) two non-interbreeding populations must be sympatric (i.e. they occupy the same spatial/temporal dimension); (2) these populations must occupy the same niche; and (3) one population (A) must multiply faster than the other population (B), thus eventually A will displace B, which will become extinct.

Although the above description of habitat exclusion and species extinction is intrinsically appealing, it is also paradoxical. How is it that so many apparently antagonistic pairs of animals persist together without one driving the other to extinction (Connell, 1980)? What mechanisms stabilize negative interactions thus allowing species to persist? MacArthur (1972) inverted Hardin's Competitive Exclusion Principle to ask how different two species have to be to coexist sympatrically? Traditionally these questions have been addressed through studies on the guilds of birds and mammals (Lawton and Hassell, 1981), whereas more recent investigations have focused on aquatic insects (e.g. caddisfly larvae - McAuliffe, 1984a; dragonfly and damselfly nymphs - Wissinger, 1989; and stonefly nymphs - Peckarsky and Penton, 1985).

To demonstrate that coevolution is shaped by competition is difficult, and requires that three essential points be addressed (Connell, 1980): (1) that divergence in resource use has occurred between species; (2) that competition was the causal mechanism; and (3) that the divergence has a genetic, and not simply a phenotypic, basis.

To address point number **one**, that divergence has occurred, requires that observations be made before and after contact between putative competitors, which can be done most effectively using fossil evidence (Eldridge, 1974). Caddisflies are particularly well-suited for this purpose as palaeoecological studies show that their sclerotized body parts preserve well in sediments, the larvae cannot fly (thus their remains indicate local assemblages), they are distributed on almost every habitable land mass (Wiggins, 1977), and habitat selection by extant species is highly specific (Williams, N.E., 1988). Although fossil caddisflies are used to make inferences about past climates and habitats (assuming that extant species are reasonable analogues of their fossil relatives - i.e. the principle of "uniformatarianism"; Birks and Birks, 1980), stratigraphic analyses could also be used to assess divergence over time.

It is more difficult to address point number **two**, that the causal mechanism affecting divergence was competition. This hypothesis may be supported using field experiments which, for example, have been used to demonstrate that aquatic insects do compete for food and space (e.g. Lamberti *et al.*, 1987; Dudley *et al.*, 1990). Non-experimental approaches may be used to compare insect populations in sympatry and allopatry, but it is very difficult to attribute differences in resource use to competition. To establish that other variables (e.g. current speed, pH, oxygen concentration, substrate compostion, etc.) have not affected observed differences is extremely difficult. However, complementary field and laboratory studies may be used to make a logical argument that differences in the use of resources may be due to competition. For example, Reznick and Endler (1982) used this method very effectively to assess differences in the life histories of guppies (*Poecilia reticulata*) between sections of stream above (fish predators absent) and below (predators present) waterfalls.

A procedure used to assess the effects of competition in sympatrically occurring species is to construct resource utilization curves along some resource axis and to calculate the degree of niche overlap (Fig. 7.4). If the overlap is below a theoretical threshold (Hutchinson, 1959) this supposedly enables both species to coexist. However, this method has several inherent problems: (a) if the overlap in resource use is greater than should theoretically occur, then the argument can be made that the resource measured is not the one over which species compete; (b) in reality, niche space is multidimensional, yet to measure these dimensions is nearly impossible (Pianka, 1975); (c) acceptable levels of niche overlap are usually derived theoretically, with no sound empirical basis; (d) simply sharing resources does not necessarily mean that species compete, as the resources may not be limiting (Menge, 1979); and (e) to measure the availability of resources is highly problematic. This final point is especially applicable to aquatic insects as, for example, food may exist in interstices that deny access, or it may exist in areas where the risk of predation is high thus forcing a trade-off.

To demonstrate point number **three**, that evolution has actually occurred

[4]

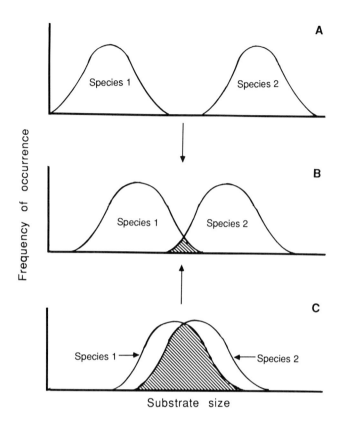

Fig. 7.4   Resource use curves: (A) depicts a situation where mid-sized substrata are underused. Situation (C) would result in significant competition. Natural selection would favour situation (B), depicting a "trade-off" between resource underuse and competition scenarios.

(i.e. that divergence has a genetic basis), breeding experiments and/or general "genetic marker" techniques (e.g. DNA fingerprinting) may be used. This latter approach usually involves taking a sample of blood (haemolymph) or tissue from the insect, and isolating genetic markers which may be used to infer relatedness of individuals. If, for example, phenotypically similar species A and B were paired, and ethograms were used to establish that species A was a superior forager, fingerprinting could be used to determine, (a) whether subsequent generations of species A survived disproportionately as compared with species B, and (b) if the behaviour persisted over multiple generations of species A.

Debate over the types of mechanism capable of regulating species assemblages has yielded three distinct schools of thought (Grossman *et al.*,

1982). The first school, the ***determinists***, suggests that assemblages are generally at equilibrium and that competitive exclusion is avoided through processes of resource partitioning (Schoener, 1974), predation on competitive dominants (thus negating their ability to exclude inferior competitors: Paine, 1974), or non-linear competitive hierarchies (Buss and Jackson, 1979). Determinists maintain that competition is or has been a force affecting species coexistence (Diamond, 1978; May, 1978), and that regardless of its overtness at any point in time competition did shape past events - hence, the influence of the "ghost of competition past" (Connell, 1980). The second school of thought, the ***stochasticists***, believe that the environment is sufficiently unstable that equilibrium is rarely reached (Chesson and Werner, 1981), and that species assemblages are controlled by differential responses to unpredictable environmental changes. These stochastic events reduce population density to levels at which competitive exclusion is not realized, and resources become available in an unpredictable manner. The third school of thought is a hybrid of the first two, and adheres to the ***"competitive crunch"*** hypothesis of Wiens (1977). The suggestion here is that competition mediates coexistence, following long periods of docility interrupted by aperiodic (once every two to 20 generations) environmental perturbations. Between crunch periods, when resources are not limiting, presumably behavioural and morphological characteristics of species become more variable, whereas the reverse is expected during and immediately following competitive bottlenecks.

It is naive to suggest that species assemblages are regulated by one process (Grossman *et al.*, 1982). It is probably more realistic to consider all three theories of community regulation as part of a continuum, with determinism and stochasticism occupying opposite poles, and the competitive crunch hypothesis lying somewhere in between.

7.2.1 Effects of disturbance and predation on population dynamics and competition

Disturbance There are several types of disturbance (i.e. stochastic) processes that may affect the population dynamics of aquatic insects. For example, in lentic systems, the direct effects of wave action along shorelines during storms (Sousa, 1979), together with the indirect effects of pounding by floating materials, often result in aquatic insects being buried or dislodged from the substrate (Lubchenco, 1986). In lotic systems, seasonal or aperiodic flooding is the most prominent disturbance mechanism (Sagar, 1986) and, as such, much research has focused on the manner in which floods affect competition, primarily in two groups, larval caddisflies and black flies (e.g. Feminella and Resh, 1990; Hemphill, 1988, 1991). These insects have been used as models to test competition theory since their field densities periodically escalate, they can be spotted readily, optimal locations to build retreats may periodically be limiting (suggesting potential competition for food and space), and they behave naturally under laboratory conditions.

In a small Montana stream, McAuliffe (1984a) examined ways in which competition and disturbance interacted to affect species evenness, a structural characteristic of aquatic insect communities. He initially established that the sessile caddisfly larva *Leucotrichia pictipes* was territorial, and that it aggressively excluded conspecifics and other species from its foraging territory. When *L. pictipes* was experimentally removed from an area, higher densities of other insects like *Baetis*, *Glossosoma*, and *Simulium* ensued; competition with *L. pictipes* limited their distributions and abundances within otherwise suitable microhabitats. However, when small stones overturned during floods, *L. pictipes* density decreased significantly, thus allowing the faster colonizing species to make use of this ephemeral resource. On larger stones unaffected by floods, no such species response was observed.

Initial studies by Resh *et al.* (1984) and Lamberti *et al.* (1987) suggested differential effects of disturbance factors (droughts, floods) on intraspecific competition within populations of the algivorous caddisfly, *Helicopsyche borealis*. These observations led to a study (Feminella and Resh, 1990) in which the influence of hydrologic disturbance on competition between *H. borealis* larvae was examined. A series of enclosure experiments with caddisfly densities which corresponded to mild, moderate and harsh wet-season hydrologic regimes (high, intermediate and low larval densities, respectively), indicated that an increase in larval density had a negative impact on food (measured as algal biomass using chlorophyll *a* content), the proportion of larvae that pupated, and with pupal size. Additionally, at high densities (indicative of the mild hydrologic regime), adult size (male and female) and fecundity were lower than observed at intermediate and low (moderate and harsh regimes, respectively) densities. Feminella and Resh suggested that, during most years, storms usually reduced late-instar caddisfly densities to levels below that necessary for competition to occur, thus countering competitive effects realized during low-flow conditions. However, in drought years, a lack of wet-season storms exacerbated competitive interactions during all stages of growth, thus reducing individual fitness. As a counterpoise to reduced individual fitness, adult densities would have been higher as a result of the relatively mild conditions (no disturbance events) experienced during larval development. Thus, the reproductive capacity of the population would not have been depressed.

The relative importance of natural disturbance and competition on populations of the caddisfly *Hydropsyche oslari* and the black fly *Simulium virgatum* was considered by Hemphill (1991) across seasons and sites (upstream/downstream) in a Californian stream. Caddisflies and black flies were naturally restricted to the upper surfaces of embedded boulders, and higher black fly densities were found in upstream rather than downstream sites. Winter storms and floods tended to reduce densities of both species, and summer droughts decreased the availability of habitable space; thus, competition was thought to vary seasonally. To test this hypothesis, a series of quadrats was maintained at upstream and downstream sites for one year. By late winter, storms had lowered hydropsychid densities, thus the effects of *H. oslari* on *S. virgatum* were minimal. Spatial differences in competitive interactions were shown by relatively little competition in downstream reaches where black flies were rare, whereas high densities of both species upstream resulted in significant competition for space. Analyses also indicated that the ability of species to recover after disturbance was inversely related to their ability to withstand competition (although *S. virgatum* could be easily excluded from a site by *H. oslari*, it recolonized more rapidly). In sum, competition and disturbance accounted for up to 60% of the reductions in the numbers of individuals present in this stream. From these two studies it is clear that the relative importance of competition may be negatively correlated with disturbance, which itself may vary over time.

Predation Competitive interactions may be affected indirectly by predators as they: (1) reduce densities of competing species (Roughgarden and Feldman, 1975; Jeffries and Lawton, 1984; Kotler and Holt, 1989); or (2) influence habitat partitioning among prey. Holt (1984) proposed that competition and predation are sufficiently co-dependent to be considered as commensurative components in a unified theory of the ecological niche. Evidence from studies with aquatic insects supports this thesis.

To determine if the dragonfly nymphs *Libellula lydia* and *L. luctuosa* interact as competitors and/or predators and prey, Wissinger (1989) manipulated density, species composition, and size range of nymphs in a series of artificial ponds. In treatments containing similarly-sized nymphs, growth rate showed a density dependent response during spring, when resources tended to be limiting, yet no such response was observed in the fall, when more resources were available. Survivorship was not directly affected by competition during either season. In treatments containing disparate size classes of nymphs, mortality was significantly higher (mostly on smaller nymphs) than in treatments containing nymphs of uniform size. Wissinger suggested that, in nature, predators (both odonates and other invertebrates) may reduce nymphal density, thus allowing release from competition, and eventual increases in density. When nymphal density is sufficiently high, predators may "switch" to the more readily available prey, thus lowering prey density, again affecting competitive release. This "feedback" relationship between predatory and competitive processes may facilitate coexistence of ecologically similar species.

Competitive interactions between species may change as a function of their response to a common predator. This was evident in a study by Kohler and McPeek (1989) using the mayfly *Baetis tricaudatus* and the caddisfly *Glossosoma nigrior*, both of which competed for periphyton found on the upper surface of substrate. Countering the benefits of feeding on periphyton was an associated high risk of being consumed by benthivorous sculpins (*Cottus bairdi*). Manipulations of sculpin presence/absence and food abundance, resulted in significant decreases in the number of *B. tricaudatus* found on the upper surfaces of the substrate, and the percentage of time spent feeding there. In contrast, *G. nigrior* (a less vulnerable species) showed no behavioural response to sculpins. Wissinger suggested that these different responses to the predator may have benefited one species (*G. nigrior*) at the expense the other (*B. tricaudatus*).

Predators also may interact positively in such a way that their total prey consumption is increased, a process which Soluk and Collins (1988) called "facilitation". A series of laboratory experiments demonstrated that the total number of mayfly nymphs (*Ephemerella subvaria*) consumed by predaceous sculpins (*Cottus bairdi*) and stoneflies (*Agnetina capitata*), held together in aquaria, was significantly higher (2x), relative to combined totals for each when they were held apart - presumably because of a reduction in the availability of enemy-free space (Fryer, 1986) when both predators were present. In contrast, a repeat of the experiment, using *Baetis tricaudatus* as prey, yielded the opposite result. Few studies have examined facilitation processes, yet these findings demonstrate a potentially significant process by which predators may affect aquatic insects; thus further study is needed.

The above examples, and others (e.g. Mittlebach, 1984, 1988; Mittlebach and Chesson, 1987), lend support to Holt's (1984) suggestion that competition and predation should be considered as part of a continuum of factors which affect the population dynamics of interacting species.

7.2.2   *Interference and exploitative competition*

There are two general types of competition: (1) interference competition, also referred to as contest competition; and (2) exploitative competition, also referred to as resource or scramble competition. Interference competition involves organisms behaving aggressively towards one another when contesting a resource, the supply of which may or may not be limiting. Aggressive actions may be expressed as encounter conflict, territoriality, allelopathy (i.e. antibiotic interactions between plants), or overgrowth (MacIsaac and Gilbert, 1991). Exploitative competition involves organisms depleting resources which are in limited supply (antagonistic interactions are absent) (Tilman, 1982). Both types of competition may be either interspecific (between species) or intraspecific (within species).

Many studies have focused, from largely exclusive perspectives, on discerning whether interference or exploitative forces affect the population dynamics of aquatic insects. Interference is usually assessed on observational data of agonistic encounters between either con- or allospecifics. However, as Dudley *et al.* (1990) suggest, the incidence of such behaviour does not prove that these interactions have important consequences for individuals, populations, or community dynamics. Exploitation is usually tested by holding a resource variable constant (e.g. food concentration) while simultaneously altering the frequency of encounters between potential competitors and checking for differences in some response variable(s) (e.g. growth rate). In only a few studies has the relative importance of both exploitative and interference competition on aquatic insect population dynamics been assessed (e.g. McAuliffe, 1984a; Anholt, 1990). The following sections will focus on aquatic insects and: (1) interference mechanisms; and (2) evidence that exploitative competition may be a significant force affecting population dynamics.

Mechanisms of interference One of the most notable displays of overt aggression in aquatic insects occurs in coenagrionid damselflies. Nymphs will often rest on perches (macrophytes) in close proximity to preferred feeding sites (Baker, 1980), where they will contest "ownership" of the perch. Detailed observations of agonistic behaviours were made by Baker (1981) by placing two *Coenagrion resolutum* nymphs onto opposite ends of a suspended, horizontal dowel. A piece of opaque plastic was placed at the mid-point of the dowel, thus blocking each nymph's view of the other. After a short rest period the plastic barrier was removed, and the behaviour of nymphs was recorded. The most blatant aggressive behaviour noted was designated "labial strike" (Fig. 7.5 A). This entailed a rapid and complete extension of the labium toward the opposing nymph which, in turn, would often swim away (usually the smaller of the pair). Baker cautioned that this behaviour might be either an attempt to displace a potential competitor or an act of predation. A second behaviour used to deflect aggressors was "slash", defined as a lateral bending of the abdomen, resulting in the caudal lamellae being displaced 90º or more (Fig. 7.5 B). Nymphs used slash

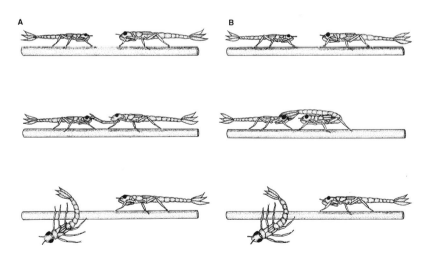

Fig. 7.5   Aggressive acts between damselfly nymphs: (A) labial strike; (B) slash. Larger nymphs usually displace smaller nymphs, which may retreat by swimming off the perch.

if another nymph approached within 1 or 2 cm, or as a defensive behaviour following a labial strike. Baker combined the above analysis with trials that tested the propensity of nymphs to remain near feeding sites, and concluded that the use of space was the result of a general dominance system whereby nymphs tend to remain near areas of food concentration.

In a stream in southern England, predaceous larvae of the net-spinning caddisfly *Plectrocnemia conspersa* aggregate in patches for most of the year (Hildrew and Townsend, 1980). Patch selection (establishing a net) is influenced by the initial success of larvae in capturing prey, and continued residency is affected by subsequent foraging conquests. Hildrew and Townsend considered whether intraspecific encounters between residents and intruders at the net-spinning site might also influence the microdistribution of larvae. In an artificial stream channel that contained small gravel and one centrally-located large stone, one larva was released (the resident), followed shortly by the release of a second larva (the intruder), and the behaviour of each was monitored. Upon detecting intruders (probably via vibrations on the net) residents would stretch their prothoracic legs beyond their head capsules and move towards the disturbance. Larvae would meet "face to face", whereby they often made contact using their front legs or labra. This behaviour was usually followed by "rearing-up" (Fig. 7.6), which entailed a number of striking movements in which a rapid forward and downward movement of the head was combined with biting. If the resident lost the encounter it would withdraw into its net and escape through the opposite end. If the intruder lost it would withdraw on the substrate or enter the drift. In either case, the winner of the conflict, usually the larger of the two, assumed ownership of the net. Hemphill (1988) recorded comparable agonistic behaviours between the caddisfly *Hydropsyche oslari* and the black fly *Simulium virgatum*. Of 140 interactions observed, 43 (31%) involved *H. oslari* "nipping" at *S. virgatum*. Of black flies bitten, 7% were captured, and of these 33% were killed.

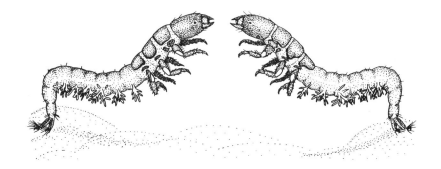

Fig. 7.6   Caddisfly larvae 'rearing up' in an agonistic interaction. Anterior abdominal segments are raised and jaws are opened (redrawn after Hildrew and Townsend, 1980).

Consequently, simuliids avoided *H. oslari*. Similar intraspecific conflicts between simuliids have been observed in the field (e.g. Harding and Colbo, 1981). Hart (1986) has demonstrated that this behaviour may be adaptive as, by displacing conspecifics, simuliids can enhance their capacity to filter particles from the water column.

To assess quantitatively the types of behaviour expressed between potentially competing stoneflies, Peckarsky and Penton (1985) ran a series of experiments using flow-through observation boxes placed at the edge of a small stream. The influence of sympatrically-occurring pairs of stoneflies (*Acroneuria carolinensis/Agnetina capitata* and *Megarcys signata/Kogotus modestus*) on one another's behaviour was assessed by quantifying behavioural patterns of nymphs when released singly into observation boxes (primed with prey), and later when pairs were released together. Contrary to the dramatic displays of antagonism observed in damselflies, caddisflies, and black flies, stoneflies behaved passively towards each other, despite the fact that the presence of a competitor generally reduced their feeding rate. *Acroneuria carolinensis* (the larger of the pair) was perhaps slightly more aggressive than *A. capitata*, but generally both reacted to encounters by using evasive or submissive responses. *Megarcys signata* and *K. modestus* were similarly evasive/submissive in response to competitors, again with the larger, *M. signata*, being slightly more aggressive.

Additional studies support the contention that stoneflies interfere with one another in a relatively benign fashion. For example, following a series of observation experiments, Walde and Davies (1984) suggested that the stonefly *Kogotus nonus* acts "aggressively" towards conspecifics by interrupting them during assaults on prey or by misdirecting attacks (due to distraction). When assessing contests for shelters and hunting sites in *Dinocras cephalotes*, Sjostrom (1985) found that resident nymphs almost always prevailed in retaining the resource. Size differences between nymphs only had a minor effect on the outcome of disputes. When nymphs did fight, they frequently only attacked the opponents cerci or antennae, seldom causing harm.

Agonistic encounters between aquatic insects generally do not result in death or serious injury to combatants. This empirical evidence supports Maynard Smith's (1974, 1976) suggestion that selection should favour conflict-solving processes that minimize costs to aggressors. Potential combatants may "assess" the costs and benefits of an interaction by "considering" asymmetries in both fighting abilities and the value of the disputed resource. Thus, despite the overt nature of some agonistic interactions and the benign nature of others, the utility of these displays may be in creating the "illusion" of threat.

Evidence for exploitative competition   Recent field studies demonstrate that despite blatant expressions of "hostility" between aquatic insects (e.g. damselfly nymphs), exploitative competition may be a more prevalent force influencing population dynamics than interference competition.

Field data collected by Baker (1987, 1989) tend to negate interference (as shown by spacing behaviour) as a dominant factor affecting the population dynamics of the damselflies *Enallagma ebrium* and *Ischnura verticalis*. The evidence for this is that: (1) under laboratory conditions larger nymphs of both species excluded smaller individuals from perches, yet field tests did not yield results consistent with size-related dispersal (Baker, 1987). (2) Nymphs that occurred on macrophytes (preferred resources) were not behaviourally dominant over those found on artificial substrates. This was determined by placing nymphs from both substrates on perches (a horizontal dowel), and the individual that swam from the dowel was considered to be the subordinate of the pair. (3) Nymphal dispersal in the field was not linked with food limitation (based on analysis of the faecal mass of nymphs). (4) There were few data to suggest seasonal food limitations. In the laboratory, Baker (1989) determined the effects of food shortage on condition (mass per unit head width) of larval *I. verticalis*, and then used these findings to estimate the degree of food limitation of *I. verticalis* in the field (based on a 6 month survey). During periods when food conditions were poor, there was no measurable change in prey densities, and the density of nymphal odonates was low. Baker concluded that competition was not an important factor affecting feeding rates and development of *I. verticalis*. He cautioned, however, that discrepancies in his findings, relative to those that support the competition hypothesis (e.g. Crowley *et al.*, 1987; Van Buskirk, 1987), may have been influenced by the presence of fishes in the ponds he surveyed, and/or by the slow-moving nature of *I. verticalis*, both of which would have tended to diminish competition.

An experimental approach was used by Anholt (1990) to separate the effects of interference and exploitative competition within a population of nymphs of the damselfly *Enallagma boreale* by altering three factors: food availability, nymphal density, and habitat complexity. Since damselfly nymphs are "sit-and-wait" predators, interference competition was examined by altering perch density within enclosures while simultaneously holding the availability of prey (zooplankton) constant. Anholt predicted that if interference affected damselfly nymphs, then increasing perch density should lower mortality rates, increase growth rates, increase development rates, and reduce variation in development and growth rates among nymphs. Alternatively, exploitative competition was manipulated by altering the availability of predators and prey, while simultaneously holding the density of perches constant. A field design based on the above postulates demonstrated that as density per perch increased nymphs responded by spacing themselves more evenly. Manipulating prey and damselfly densities (i.e. exploitative factors) significantly affected nymphal survival, the timing of emergence, and mass at emergence, whereas no significant effect of habitat complexity (interference factor) was detected. Apparently the utility of "aggressive" acts between damselflies is to space them evenly, with relatively little associated cost. The degree to which damselflies are food limited, and whether this limitation is density dependent and affects population regulation, remains unknown (Anholt, 1990).

Resource depression by a lotic herbivore, the caddisfly larva *Glossosoma* sp., was examined using artificial substrates placed in a stream (McAuliffe, 1984b). Initial trials demonstrated that reducing *Glossosoma* density resulted in significant increases in both periphyton density and colonization by grazing mayflies (*Ephemerella doddsi* and *Baetis* sp.) on artificial substrates (bricks). Subsequent experiments focused on the short-term colonization behaviour of mayflies in response to: (1) high periphyton density with *Glossosoma* removed from the substrate (i.e. negating the potential for competitive interference); (2) low periphyton density with *Glossosoma* removed; and (3) low periphyton density with high *Glossosoma* density. Comparative analysis indicated that *Glossosoma* only influenced colonization by mayflies through reducing periphyton density (i.e. exploitation). When periphyton density was held constant, altering the number of caddisflies had no direct effect on mayfly behaviour. Thus, McAuliffe concluded that *Glossosoma* influenced mayflies via exploitative competition, and that interference competition was negligible.

A study of intraspecific competition for food, and the relative effects of exploitative and interference competition between *Glossosoma nigrior* larvae (Hart, 1987), yielded results similar to those of McAuliffe (1984b). After establishing that the mass of pupae declined significantly as a function of increasing larval density, Hart used path analysis (Sokal and Rohlf, 1981) to determine the relative contribution of exploitative and interference mechanisms to competitive interactions. A model based purely on exploitation explained 44% of the variance in final pupal mass, whereas a combined exploitative/interference model explained 48%. The additional 4% explained by interference was not significant, strongly suggesting that exploitation was the predominant competitive mechanism.

It would be premature to make general conclusions regarding the relative contributions of exploitative and interference competition to the population dynamics of aquatic insects. However, despite a paucity of studies, contemporary evidence suggests that exploitation may be of greater significance as a competitive force than originally thought, particularly as studies reveal that food is more limiting as a resource (Richardson, 1991) than was previously indicated. As suggested by Anholt (1990), future studies aimed at separating feeding- and mortality-related interference competition from exploitation may be done most effectively through direct manipulations of the behaviour of individual competitors.

7.2.3   Interphyletic competition

An increasing body of research demonstrates significant competition between phylogenetically distant species (e.g. Brown and Davidson, 1977; Hurlbert *et al.*, 1986; Schluter, 1986). Despite this, few studies have focused on competition between aquatic insects and distantly related taxa (e.g. Morin *et al.*, 1988), suggesting that calculations of connectivity (May, 1975) and the frequency of interspecific competition (Schoener, 1983) within aquatic food webs are underestimated.

In a series of artificial ponds, Morin *et al.* (1988) showed interphyletic competition for periphyton between herbivorous aquatic insects and anuran tadpoles. It was demonstrated that naturally occurring densities of aquatic insects reduced the size of tadpoles at metamorphosis which, in turn, was estimated to reduce anuran fitness (fecundity in frogs is size dependent; Smith, 1987). A comparison of the competitive abilities of aquatic insects and tadpoles demonstrated that interspecific and intraspecific interactions were of comparable strengths, indicating that interphyletic competition may be a prominent factor influencing anuran populations. Although this study focused on the negative impact of aquatic insects on anuran tadpoles, the phenomenon demonstrated probably has wide applicability. A holistic perspective on the potential for competition to affect the population dynamics of aquatic insects will suffer significantly until emphasis is placed on discerning interphyletic relationships.

8 TROPHIC RELATIONSHIPS

The first part of this chapter describes the types of food found in discrete sections of lakes and streams, followed by a discussion of functional feeding groups, food conditioning processes, assimilation efficiency, insect growth, physiological mechanisms that enable insects to absorb nutrients, and the role of insect faeces and nutrient spiralling in aquatic systems. The second part covers aspects of food web form and function, by following a time series of pertinent studies from past to present, and outlines current problems in the area of food web research.

8.1 *Potential food in aquatic systems*

The efficient exploitation of available food is a vital requirement of all animals (Emlen, 1966), and as such many studies of aquatic insects have focused on this process (e.g. Cummins and Klug, 1979; Anderson and Cargill, 1987). As used here, "available food" refers to that to which aquatic insects can gain access and exploit, which is usually a subcomponent of the potential food base present in a system (Tavares-Cromar, 1990).

The origin and type of food supply differs significantly both within and between lotic and lentic systems. For example, the food store in many lotic systems is primarily allochthonous in origin (i.e. from outside the system; Fisher and Likens, 1973), whereas lentic systems are largely characterized by autochthonous supplies (i.e. from within the system; Odum and Prentki, 1978). Despite this generality, the proportions of internal and external food supply will differ, either along the course of a stream or river (e.g. Vannote *et al.*, 1980), or as a function of the size and morphometry of a lake or pond (Wetzel, 1983).

Lakes may be divided into four zones (see Chapter 4), each of which is characterized by distinct food types and varieties of insect and plant life. The *littoral zone*, the nutrient-rich inshore area, is typically dominated by macrophytes and supports the greatest diversity of aquatic insects. Here, coarse particulate organic matter (CPOM: Table 8.1) is derived from the leaves of shoreline trees and from streams as they transport larger detrital particles (Anderson and Cargill, 1987). In shallow, near-shore areas, large quantities of particulate organic matter (POM) may be produced from emergent and submerged macrophytes. For example, in an English reedswamp, Mason and Bryant (1975) calculated an annual littoral production of > 2,500 g/m^2. The *sublittoral zone* is a transitional region which extends from the lower edge of rooted macrophytes in the littoral zone, to the upper level of the hypolimnetic water. The sublittoral tends to contain smaller detrital particles than those found in the

Table 8.1 General characteristics of detritus in streams (particles < 0.5 μm are considered to be dissolved organic material) (after Cummins and Klug, 1979).

Category	Size	Description	Associated microbes	Caloric content (kcal/g AFDW*)	%Ash	Carbon:nitrogen ratio
CPOM (Coarse Particulate Organic Matter) Non-woody	> 1 mm	Leaves, needles, other non-woody plant parts	Fungi, bacteria; surface and matrix colonized. Lesser importance: protozoans, rotifers, nematodes, microarthropods	4.8	10-30	20-80:1
Woody	> 1 mm	All wood and woody parts	As above but lower densities. Primarily surface-colonized until late in decomposition	4.5	40-50	220-1340:1
FPOM (Fine Particulate Organic Matter)	< 1 mm > 50 μm	Fragments of CPOM; faeces; flocculated material from DOM; microbial cells; sloughed algae; organic films on mineral particles	Primarily bacteria colonizing surfaces; protozoans on particles >240μm	4.5	--	7.4-37.8:1
UPOM (Ultra Fine Particulate Organic Matter)	< 50 μm > 0.5 μm	Fragments of CPOM and FPOM; flocculated material from DOM; organic films on clay particles	Bacteria sparsely colonizing surfaces or associated with flocculated organics	--	--	--

*AFDW = Ash Free Dry Weight

littoral zone. The open-water *pelagic zone* receives sunlight which fuels photosynthesis, thus in this layer phytoplankton production is high and there are numerous species of herbivorous/carnivorous zooplankton. Despite a large contribution from the macrophytes of inshore areas, the bulk of detritus in lakes originates from moribund phytoplankton that settles out as fine-particle "ooze" (Anderson and Cargill, 1987). The *profundal zone* is not penetrated by sunlight, and coldwater species of fish, such as trout, and dipteran larvae are the typical inhabitants. The contribution of each zone to the food-base of the system can differ if, for example, the littoral zone is shallow and extensive (thus increasing primary productivity: characteristic of eutrophic lakes) vs. narrow and precipitous (characteristic of oligotrophic lakes)(Wetzel, 1983). Riparian vegetation, basin substrate composition, turnover time, and lake surface area also will affect productivity in each zone.

In the headwater reaches of streams, riparian vegetation (deciduous leaves and coniferous needles) contributes greatly to the food supply because of the high ratio of shoreline to stream bottom area. Primary productivity in these reaches is generally low, as canopy trees tend to attenuate sunlight (Anderson and Cargill, 1987). Insects which feed on non-living POM are most abundant in these headwater sections. In higher order streams (4 - 6), the wider channels negate the influence of canopy trees on primary productivity, while also lowering direct inputs of CPOM. Detritus in these sections consists of fine particulate organic matter (FPOM) transported from upstream, and high densities of herbivores occur. In very large rivers (orders >7) primary productivity is limited by depth and turbidity, and detrital particle size is generally reduced.

The general characteristics described for food supply and zonation in lakes have global applicability. However (as discussed in section 4.1.1) the patterns of change portrayed for streams apply primarily to North America, and Winterbourn *et al.* (1981) suggest that streams in New Zealand and Australia, for example, do not conform to these patterns. This is thought to be due to differences in the gradient, deciduous riparian cover, degree of stochasticity in rainfall events and temperature, which, cumulatively, affect many Southern Hemisphere streams differently. Many Australian and New Zealand forests are of the "warm" evergreen type and should thus be compared with those found in Mediterranean climates where, unfortunately, little work has yet been done. Thus, as outlined here, the pattern of food input and growth, and changes in aquatic insect feeding groups along the reaches of streams, is probably most applicable to Northern Hemisphere systems.

8.2   *Role of aquatic insects in processing food*

8.2.1   *Functional feeding groups*

To delineate the means by which food resources may be exploited, Cummins

Table 8.2   A general categorization of the trophic mechanisms and food types of aquatic insects (after Cummins, 1973).

Category based on feeding mechanism	General particle size range of food (μm)	Subdivision based on feeding mechanisms	Subdivision based on dominant food	Aquatic insect taxa containing predominant examples
Shredders	$>10^3$	Chewers and miners	Herbivores: living vascular plant tissue	Trichoptera (Phryganeidae, Leptoceridae) Lepidoptera Coleoptera (Chrysomelidae) Diptera (Chironomidae, Ephydridae)
		Chewers and miners	Detritivores (large particle detritivores): decomposing vascular plant tissue	Plecoptera (Filipalpia) Trichoptera (Limnephilidae, Lepidostomatidae) Diptera (Tipulidae, Chironomidae)
Collectors	$<10^3$	Filter or suspension feeders	Herbivore-detritivores: living algal cells, decomposing particulate organic matter	Ephemeroptera (Siphlonuridae) Trichoptera (Philopotamidae, Psychomyiidae, Hydropsychidae, Brachycentridae) Lepidoptera Diptera (Simuliidae, Chironomidae, Culicidae)
		Sediment or deposit (surface) feeders	Detritivores (fine particle detritivores): decomposing particulate organic matter	Ephemeroptera (Caenidae, Ephemerellidae, Leptophlebiidae, Baetidae, Ephemeridae, Heptageniidae) Hemiptera (Gerridae) Coleoptera (Hydrophilidae) Diptera (Chironomidae, Ceratopogonidae)
Scrapers	$<10^3$	Mineral scrapers	Herbivores: algae and associated microflora attached to living and nonliving substrates	Ephemeroptera (Heptageniidae, Baetidae, Ephemerellidae) Trichoptera (Glossosomatidae, Helicopsychidae, Goeridae Molannidae, Odontoceridae) Lepidoptera Coleoptera (Elmidae, Psephenidae) Diptera (Chironomidae, Tabanidae)
		Organic scrapers	Herbivores: algae and associated attached microflora	Ephemeroptera (Caenidae, Leptophlebiidae, Heptageniidae, Baetidae) Hemiptera (Corixidae) Trichoptera (Leptoceridae) Diptera (Chironomidae)
Predators	$>10^3$	Swallowers	Carnivores: whole animals (or parts)	Odonata Plecoptera (Setipalpia) Megaloptera Trichoptera (Rhyacophilidae, Polycentropodidae, Hydropsychidae) Coleoptera (Dytiscidae, Gyrinidae) Diptera (Chironomidae)
		Piercers	Carnivores: cell and tissue fluids	Hemiptera (Belastomatidae, Nepidae, Notonectidae, Naucoridae) Diptera (Rhagionidae)

(1973) categorized aquatic insects into functional feeding groups based on the feeding mechanisms and range of particle sizes that they ingest (Table 8.2). Primary distinctions were made between detritivory (consumption of non-living organic matter), carnivory (consumption of living animal tissue), and herbivory (consumption of vascular plant or algal tissue), as indicated by the dominant foods ingested. Observations suggest that facultative feeders, which eat a wide array of food types, occupy a broader niche space in lakes and streams than do more specialized, obligate feeders (Cummins and Klug, 1979).

At the family level, larval caddisflies (Trichoptera) are the most diversely represented insects among the functional feeding groups. Analysis by Wiggins and Mackay (1978) indicated that caddisfly shredders tend to dominate the headwaters of streams whereas, in mid-sized reaches, collector-scraper genera are more abundant. Beetles (Coleoptera) are the second most widely represented among the functional groups. Mayfly nymphs (Ephemeroptera) are primarily collector-gatherers (consuming FPOM) or scrapers. Stoneflies (Plecoptera) are either predators or CPOM shredders. All dragonflies/damselflies (Odonata), dobsonflies (Megaloptera), and most bugs (Hemiptera) are predators. Craneflies (Diptera) may be shredders, collector gatherers, or predators. Mosquitoes and black flies are collector-filterers (consuming FPOM). Aquatic moth larvae (Lepidoptera) may be either shredders, collectors or scrapers.

8.2.2   Conditioning food

Although detritivores represent the largest numerical component of the functional feeding categories, few aquatic ecologists considered the fate of organic material, primarily terrestrial, autumn-shed leaves, until Kaushik and Hynes (1971) examined the process in detail. They discovered that in the early stages of decay the nitrogen (protein) content of leaves frequently increased and that this increase only occurred if leaves were exposed to microbes. Further, more protein was present when fungi, rather than bacteria, were the principal microorganisms. Kaushik and Hynes determined that fungi actually add protein to decomposing leaves, and this action has since been described as "conditioning" (e.g. Barlocher, 1985). Kaushik and Hynes also determined that detritivorous insects, when given a choice, selected conditioned leaves over those that were fresh, because of their enhanced nutritional value. Since the work of Kaushik and Hynes many studies have: (1) detailed the action of microbes on leaves; (2) demonstrated selection for conditioned food resources; and (3) shown that growth in insects can vary significantly as a function of the quality of the food ingested.

The mechanisms used by fungi to degrade leaves were examined by Suberkropp and Klug (1980), by initially isolating five species of aquatic hyphomycete (*Alatospora acuminata*, *Clavariopsis aquatica*, *Flagellospora curvula*, *Lemonniera aquatica*, and *Tetracladium marchalianum*) and growing them in pure culture with hickory leaves. Enzymatic activity of each fungal species resulted in the skeletonization of leaves via the maceration of the leaf matrix and the subsequent release of leaf cells as FPOM. After incubation, further fractionation and analysis of leaf material indicated that all fungal species metabolized (degraded) cellulose, and that two species (*T. marchalianum* and *F. curvula*) metabolized hemicelluloses. In cultures treated with *T. marchalianum*, the release of fine particulates was coincident with increases in fungal biomass (as measured by ATP) and enzymatic activity in the supernatant, which degraded carboxymethycellulose, xylan and polygalacturonic acid. Macerating activity increased with pH, thus indicating that pectin transeliminase was involved in the softening of leaf tissue by *T. marchalianum*. This finding prompted Suberkropp and Klug to suggest that transeliminase is more important in the release of leaf cells than hydrolytic enzymes with lower pH optima.

Aquatic insects show a positive selection for conditioned food. For example, Iversen (1973) found that the caddisfly larva *Sericostoma personatum* would not feed on autumn-shed leaves until after they were enriched with microbially-derived nitrogen. Using artificial streams, Petersen and Cummins (1974) demonstrated that invertebrate colonization of leaf packs varied as a function of microbial colonization and conditioning. Using [14C] glucose labelling experiments, Winterbourn and Davis (1976) demonstrated that much of the microflora present on dead beech leaves was assimilated by the trichopteran *Zelandopsyche ingens*. Ontogenetic changes in the feeding preferences of the caddisflies *Clistoronia magnifica*, *Hydatophylax hesperus*, and *Lepidostoma quercina*, were demonstrated when they were presented with leaves in various stages of decomposition (Cargill *et al.*, 1985). In contrast to early instar larvae, older larvae preferred lipid-coated leaves, presumably because of their different nutritional needs (i.e. triglycerides derived from lipids) for metamorphosis and reproduction.

Bacteria also can condition detrital material, but evidence suggests that fungi may be more effective in enhancing nutritional value. For example, when Mackay and Kalff (1973) presented leaf discs from fungal and bacterial cultures to the caddisflies *Pycnopsyche gentilis* and *P. luculenta*, fungal discs were preferred. However, bacteria have been demonstrated to contribute significantly to the nutritional requirements of *Simulium* sp. and *Chironomus* sp. (e.g. Baker and Bradnam, 1976; Wotton, 1980). Apparently conditioning affects the desirability of food in a species specific manner.

8.2.3   Assimilation efficiency and insect growth

Assimilation efficiency (i.e. the degree to which food is used for respiration and net biomass production) varies significantly among species of aquatic insect (Wetzel, 1983). For example, assimilation efficiencies for caddisflies vary from 8-24% (*Zelandopsyche ingens*; Winterbourn and Davis, 1976), 15-40% (*Potamophylax cingulatus*; Otto, 1974), and 10-25% (*Pseudostenophylax edwardsi*; Anderson and Grafius, 1975). Assimilation efficiency in a cranefly (*Tipula abdominalis*; Vannote, 1969) has been measured at 33%. Assimilation efficiency may be affected by the amount and type(s) of organic coatings on the ingested materials as, for example, Lawson *et al.* (1984) found that the presence of microbes on pignut hickory leaves significantly affected assimilation efficiency in the tipulid *T. abdominalis*. The presence of inorganic particles may break up cells in the guts of insects, which may enhance assimilation efficiency (Wetzel, 1983). In general, assimilation efficiency in aquatic insects seems to range from 10 to 40%.

Many studies demonstrate that the growth and development of aquatic insects vary as a function of nutritional input and assimilation efficiency (Waters, 1979). Anderson and Cummins (1979) considered representative shredders, collectors, scrapers and predators, and determined that growth and development varied according to the nutritional value of food - which they ranked along an ascending gradient from: (1) wood; (2) terrestrial leaf litter; (3) fine particulate organic matter (FPOM); (4) decomposing vascular hydrophytes and filamentous algae; (5) living algae (primarily diatoms); to (6) animal tissues. A marked effect of food on insect growth was noted by Cummins *et al.* (1973), who reported higher growth rates in larvae of *Tipula* sp. when they were fed rapidly-processed vs. slowly-processed leaves. The availability of rapidly-decomposing leaf species in a Swedish stream resulted in an increase in lipid reserves in the caddisfly *Potamophylax cingulatus* (Otto, 1974). Fuller and Mackay (1981) found that the collector caddisflies, *Hydropsyche sparna*, *H. betteni*, and *H. slossonae*, grew 7 times faster when fed animal material diets, and 4.6 times faster when fed diatom diets, than when they were maintained on diets of either leaves or faecal detritus. Scraper caddisflies, *Glossosoma nigrior*, from first-order streams were much smaller than those found in third-order streams. Gut content analysis revealed that the larvae in headwater reaches consumed CPOM, whereas those downstream ingested large proportions of FPOM (Cummins, 1973, 1975).

Predaceous odonate nymphs have been used extensively to demonstrate the relationship between insect diet and growth. Lawton *et al.* (1980) found that increases in the size of nymphs reared on "poor" quality diets were much less than those of nymphs fed on "high" quality diets. Johnson *et al.* (1984) postulated that a food shortage at high odonate densities caused a reduction in nymphal condition (weight/head-width). Baker and Feltmate (1987) found that a higher availability of food to newly-moulted *Ischnura verticalis* nymphs increased not only growth rate, but also the relative size between moults, and condition. In the presence of rainbow trout, nymphs of the predaceous stonefly *Paragnetina media* feed less which results in reductions in the size and condition of immatures and the condition and fecundity of adults (Feltmate and Williams, 1991b). In addition to the effects of food on the growth of insects, thermal conditions are also important (Sweeney and Vannote, 1978). Thus, in any study of growth, both food quality and temperature effects should be considered (see sections 5.3 and 11.2 for further discussion of temperature effects).

8.2.4   Physiological adaptations

The physiology of digestion in aquatic insects has received little attention (Martin *et al.*, 1980), moreover the few existing studies have tended to focus on detritivores.

Correlations between the abundance of functional feeding groups and food quantity led Kesler (1982) to determine whether the digestive processes of various groups were specifically adapted to the locally available food supply. Logically, insects found in areas that receive allochthonous input high in cellulose should possess digestive enzymes capable of decomposing cellulose, collectively known as cellulase. Kesler examined cellulase activity in the guts of a shredder (*Oecetis* sp.), two collector-filterers (*Hydropsyche* sp. and *Simulium* sp.), and a predator (*Sialis* sp.), and found the greatest activity in the shredders (Table 8.3). He suggested that such systematic variation in enzyme activity amongst the functional feeding groups may be both widespread and influential in maintaining a separation of the resource base among species.

Table 8.3   Cellulase activity (reported as μg glucose) and functional feeding group of four species of aquatic insect (after Kesler, 1982).

Taxon	Functional group	μg glucose/min/ mg protein
Oecetis sp. (Trichoptera)	shredder-herbivore	34
Hydropsyche sp. (Trichoptera)	collector-filterer	10
Simulium sp. (Diptera)	collector-filterer	0
Sialis sp. (Megaloptera)	engulfer (predator)	0

Aquatic insects use various means to facilitate the digestion of amino acids. Larvae of the detritivorous cranefly *Tipula abdominalis* have a gut pH of 11, which helps to release detrital-bound protein from the phenolic complexes of leaves. At pH values >11, certain atypical proteolytic enzymes can facilitate further digestion (Martin *et al.*, 1980). pH values of >10 have also been recorded in the mid-guts of detritus-feeding larval black flies (Lacey and Federici, 1979) and mosquitoes (Dadd, 1975).

An exception to the typical situation of "mid-gut alkalinity" occurs in the nearly neutral gut of the detritivorous caddisfly *Pycnopsyche guttifer* (Martin *et al.*, 1981*a*). In this case, larvae may meet their nutritional requirements for nitrogen by digesting the protein of microorganisms, diatoms, other algae, and invertebrate prey that are inadvertently consumed with plant litter.

The digestive systems of most detritivores are poorly designed to decompose dietary polysaccharides (Martin *et al.*, 1981a), thus suggesting that dietary lipids may be more important in the nutrition of detritivores than has been generally recognized. Cargill *et al.* (1985) reared the caddisfly *Clistoronia magnifica*, from third instar, to adult, to hatching of the next generation, on a series of diets. Larvae fed on detritus alone did not complete development, those fed fungi plus high-nitrogen leaf detritus matured but did not reproduce, and larvae fed detritus plus wheat grains (high fat content) completed development and reproduced successfully. Larvae fed on all diets grew identically until the final instar, at which time their metabolism shifted to the synthesis and storage of triglycerides (derived from wheat starch). Apparently the high energy reserves of triglycerides are necessary for successful reproduction. This study is one of very few that have addressed ontogenetic changes in the nutritional requirements of aquatic insects. Virtually no studies have examined, in detail, physiological and morphological changes in the digestive system that accompany this nutritional shift (Martin *et al.*, 1981b).

8.2.5 Role of insect faeces and nutrient spiralling

As aquatic insects comminute organic matter, their faeces provide an important food supply for fine-particle collector-gatherers (Short and Maslin, 1977; Grafius and Anderson, 1979; Shepard and Minshall, 1981). Faeces may occur as pellets encased in a peritrophic membrane (Reeve *et al.*, 1975), or they may be bound by mucus and gelatinous substances which are added to undigested food as it passes through the gut (Bandel, 1974). After being voided from the body, pellets encased by membranes often absorb water, causing them to swell and sometimes burst, thus "fractionating" their contents into FPOM (Lautenschlager *et al.*, 1978).

Coprophagy (the ingestion of faecal material) is commonly used by aquatic insects to access nutrient rich material. Wotton (1980) suggested that black flies derive many compounds from faeces because of preconditioning by digestive enzymes and prolonged exposure to bacteria. In areas where the amount and quality of primary food sources are already high, such as lake outlets where many black flies congregate (Carlsson *et al.*, 1977), coprophagy may convey an additional advantage by delaying and offsetting nutrient loss to downstream areas.

Several authors have demonstrated that faecal mass varies as a function of the amount (Folsom and Collins, 1982; Pierce, 1988) and quality (Baker, 1986) of food material ingested. In these studies, changes in the mass of insect faeces were used to detect reductions in food consumption because of the presence of predators or competitors.

Faecal material does not persist *ad infinitum*. In lotic systems it is ultimately entrained in the *"nutrient spiralling"* process, a term used to describe the reprocessing (i.e. ingestion, egestion, reingestion, entrainment and oxidation) of particulate organic matter as it is transported downstream (Newbold *et al.*, 1981). To measure the efficiency of this process, the average distance associated with one complete cycle of a nutrient atom (usually phosphorus or nitrogen) may be measured. This distance tends to be shorter (i.e. the spirals are "tighter") in nutrient poor (oligotrophic) systems, but longer in eutrophic sytems.

8.3   Food webs

The movement of food energy from the stage of primary producers (plants), to herbivores and ultimately to carnivores, is referred to as a food chain. Early researchers who recognized this process were Hardy (1924), who focused primarily on marine systems, and Elton (1927), who examined terrestrial systems. Both of these pioneers realized that many species belong to several food chains and that, cumulatively, these chains form food webs. Although, since their conception, a large body of theoretical work has been developed around food webs, there is a marked paucity of field data against which theories and models can be tested. This lack of an adequate data base has significantly retarded our understanding of food web dynamics in aquatic ecosystems.

Hardy's (1924) characterization of the food web of the North Sea herring, *Clupea harengus*, not only gives historical perspective to the field of food web research, but also demonstrates: (1) basic static and dynamic principles of food web structure and function that apply directly to aquatic insects and their role in freshwater ecosystems; and (2) that much of what is being advocated today as a direction for more holistic and realistic food web research, particularly the inclusion of ontogenetic changes in trophic feeding status (Paine, 1988; Pimm and Kitching, 1988), was deemed imperative by Hardy back in 1924. It is testimony to the rigour of Hardy's work that his study still remains one of the most detailed accounts of the feeding relationships of an aquatic animal (Fig. 8.1). The illustration summarizes data for 10 months in each of two consecutive years for fishes collected from a wide range of localities in the North Sea. Although herring do feed on several species of prey at any one location, the range of links presented would not occur simultaneously. Nevertheless, the changes depicted in herring feeding as they pass through increasing size classes from young to adult are real. This view of an animal passing through multiple series of linear food chains during its life cycle is probably more the rule than the exception. Another general phenomenon of food webs, depicted in Fig. 8.1, is that as organisms grow they capture larger food organisms and are consequently elevated to higher trophic levels, and as a result smaller prey organisms are released from predation by the predator as larger prey become available to it. For example, as herring pass from the smallest size class, less

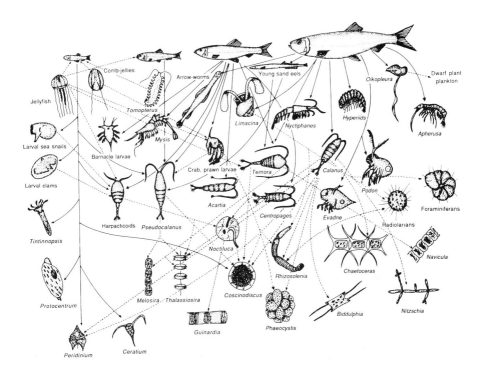

Fig. 8.1 Changes in the feeding habits of herring as they grow from a length of 1 cm on the left to greater than 13 cm on the right. Major food chains are indicated by dark arrows; other organisms important in the food chains supporting herring are indicated by light arrows (redrawn after Hardy, 1924).

pressure is exerted on larval sea snails and clams, while more is immediately placed upon copepods. Such periods of alleviated pressure, if predictable, may coincide with critical life history stages of prey development. A final, generally applicable point to extract from Hardy's work is that individuals near the top trophic level tend to exploit two or more preceding trophic levels (primary carnivores and herbivores), while those near the bottom feed mainly on herbivores.

To assemble a data base that includes both spatial and temporal changes in feeding trajectories of organisms is, as the above study indicates, a daunting task. It is not surprising that rather few researchers have attempted such projects on other systems. It is fortuitous, however, that of the few existing holistic analyses of food webs some have focused on freshwater systems and the role of aquatic insects.

One of the early, more comprehensive, studies of food webs in a stream was that of Jones (1950) on the River Rheidol in Wales, which involved analysis of the gut contents of approximately 1,300 insects. The Rheidol is a fast-flowing stream with a largely cobble substrate and a pH of 5.8-6.4. The dominant non-insect predators, at the time of study, were brown trout (*Salmo trutta*) and Hydracarina (water mites). Newts, sticklebacks (*Gasterosteus aculeatus*), and leeches were rare, and there were no large crustaceans. The available supply of what Jones termed "vegetable" food was classified as: (1) macroflora - bryophytes and flowering plants; (2) algae or microflora; (3) detritus; and (4) plant material originating outside the river (allochthonous material). The microflora of the river was highly variable in density as a result of frequent and violent flooding, with the most significant reductions occurring in diatoms (which attach poorly to substrates). Jones suggested that the most reliable source of food in the stream was detritus. It accumulated amongst the stones and gravel of the stream bed and was strained out of the flowing water by phanerogams (flowering plants). Detritus was thought to be formed through disintegration of the vegetation growing in the water and through deposition by rain and wind. Detritus was always abundant.

When plants and animals were collected from the river they were transported to the laboratory and analysis of gut contents began immediately. Such speed is important when identifying plant material as the chloroplasts of algae are quickly digested (macroinvertebrate prey are processed much more slowly). For mayflies and stoneflies, care was taken to avoid pre-emergent nymphs as they invariably had empty guts; otherwise, specimens for study were chosen at random.

Analyses indicated that detritus and dead leaves were the primary foods of the common phytophagous and omnivorous insects, and that green algae and diatoms were comparatively unimportant during autumn, winter and early spring. During late spring and summer, however, green algae became important as food, while some species of mayfly (e.g. *Ephemerella ?notata*) and caddisfly (e.g. *Hydropsyche instabilis*) fed extensively on diatoms. No molluscs were found in any of the guts, probably because the only common species (*Ancylus fluviatilis*) was protected by its shell and its ability to cling to stones. Although the heavy case of the caddisfly *Glossosoma boltoni* appeared to limit predation by other insects, it did not deter trout. Some species, such as water striders, were thought to avoid capture by predatory stoneflies and caddisflies by living on the water surface.

Jones found that for the majority of species studied, the food "chains" tended to be short and uncomplicated. For example, the scheme outlined for the stonefly *Perlodes mortoni* was more complicated than most, yet it was still simple (Fig. 8.2). Predators and prey were generally polyphagous and phytophagous, respectively. Larger predators, such as *P. mortoni*, probably received some nutritional input from the partially digested food of their prey. Species like *P. mortoni*, which have rapid growth, tended to have full guts, whereas slower growing insects, like the stonefly *P. carlukiana* (which has a 3 year life cycle), fed relatively irregularly and often had empty guts. Jones also observed that

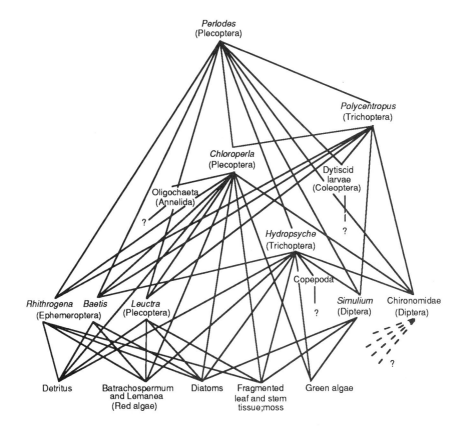

Fig. 8.2   The main features of the food chain of *Perlodes mortoni* in the River
Rheidol, Wales (redrawn after Jones, 1950).

predatory insects were most often associated with fast-flowing shallow water,
whereas phytophagous species occurred in slower-flowing deeper sections.

A more highly resolved study than Jones' was that of Koslucher and
Minshall (1973). The latter focused on seasonal changes in the feeding
relationships of several numerically dominant stream species, to determine
whether food consumption changed as a function of the increasing size of the
consumer, or whether feeding was simply an opportunistic process. They
sampled Deep Creek (on the Idaho-Utah border) seasonally and during each
sampling period representative insects were collected from as many size classes
as possible. Each time that samples of algae and macrophytes were taken, their
relative abundance in the stream was estimated. In the laboratory, insect guts
were removed and their contents analysed. Potential food included detritus,
diatoms, filamentous green algae, aquatic vascular plants, and aquatic

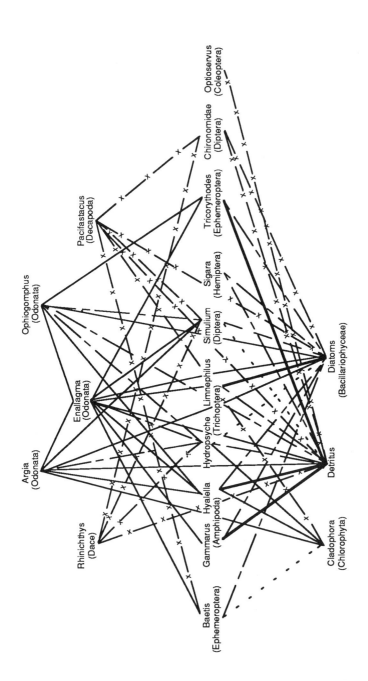

Fig. 8.3   Trophic relationships of some of the more important insect groups in Deep Creek based on a composite of data from all stations and seasons. The relative importance of each pathway is indicated by different lines, as follows: ——— = 0-20%; – – – = 21-35%; ——— = 36-50%; ‒ ‒ ‒ = 51-65%; ——— = 66-80%; x———x = amount unknown (redrawn after Koslucher and Minshall 1973).

invertebrates. All of the insects in Deep Creek proved to be general feeders, consuming a wide variety of food (Fig. 8.3). There was no evidence to suggest concomitant changes between increasing insect and food particle size, although Koslucher and Minshall suggested that limitations in the types of food present in the stream may have forced consumers to be non-discriminatory in their feeding habits. Comparisons of seasonal feeding in a few species indicated no significant patterns for either detritivores or carnivores, and only slight seasonality was observed for those that fed on algae. During winter, diatoms formed a disproportionate component of the diets of the mayfly *Baetis tricaudatus* and the caddisfly *Hydropsyche occidentalis*, suggesting selection for this food type in that season.

In sum, Koslucher and Minshall's study indicated that the immature stages of insects in Deep Creek were more opportunistic in their feeding habits than those in the River Rheidol. Surprisingly, both studies made only passing reference to ontogenetic feeding shifts despite the importance of these factors as demonstrated by Hardy.

8.3.1 Problems and perspectives in food web research

The previous section introduced some of the dynamical aspects of species interactions in food webs as revealed by early studies. Contemporary research attempts to discern the primary mechanisms that govern food web structure, thus creating a framework whereby the impact of perturbations on these systems can be predicted. Before this can be achieved certain problems regarding data collection must be addressed (Paine, 1988).

(1) Because all organisms are not easily observed, most food webs tend to be deficient in the inclusion of "ecologically subtle" species. (2) Species that are observable are often highly mobile, and as such they may be underrepresented in food webs. These species may affect trophic relationships significantly and thus every effort should be made to include them in analyses. (3) The influence of species that are transient and not normally part of any local assemblage is often overlooked. It is difficult to assess the impact, if any, that these nomads might have on an assemblage of interacting species, although Paine was inclined to think that the omission of transients in food webs is probably not serious. (4) Ontogenetic changes in the diets of species are common. However, food web researchers often deny this reality by depicting trophic relationships as static, fixed entities. More accurate characterizations of food webs would include dietary shifts. An additional problem (5), not addressed by Paine, is a general lack of examination of the role of microorganisms and the microbial food loop in affecting the cycling of nutrients at primary levels in food webs.

Despite problems in data collection, research on food webs demonstrates reasonably predictable patterns in structure, suggesting that food webs are not simply stochastic aggregates. Pimm (1982) has recognized the following general characteristics of food webs.

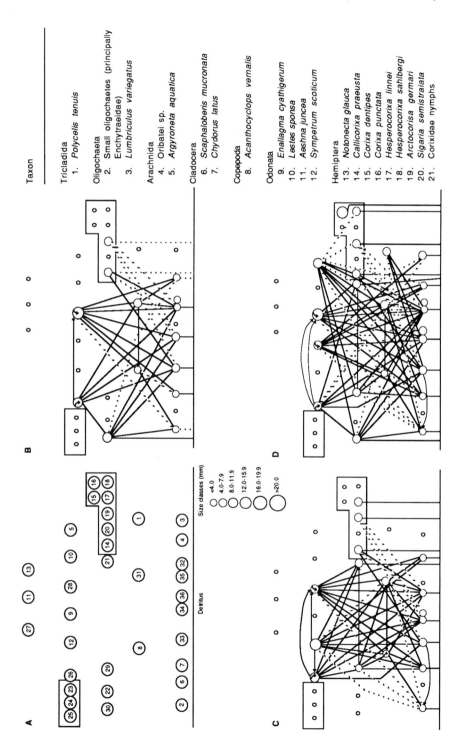

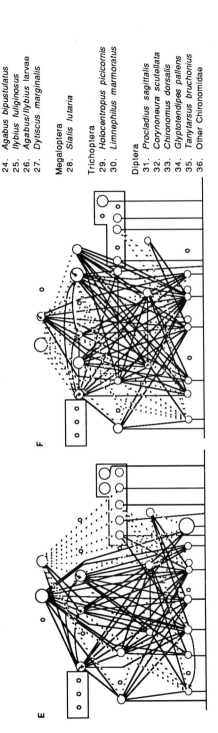

Coleoptera
22. *Hydroporus erythrocephalus*
23. *Agabus sturmii*
24. *Agabus bipustulatus*
25. *Ilybius fuliginosus*
26. *Agabus/Ilybius* larvae
27. *Dytiscus marginalis*

Megaloptera
28. *Sialis lutaria*

Trichoptera
29. *Holocentropus picicornis*
30. *Limnephilus marmoratus*

Diptera
31. *Procladius sagittalis*
32. *Corynoneura scutellata*
33. *Chironomus dorsalis*
34. *Glyptotendipes pallens*
35. *Tanytarsus bruchonius*
36. Other Chironomidae

Fig. 8.4   Food web diagrams for selected date/habitat combinations in a small pond in north Yorkshire. Species are represented by circles and links are directed downwards from consumer to resource unless otherwise indicated. Links within circles denote cannibalism. Very small circles (with no links) represent the positions (on the web templet (A)) of species which were not found in that particular sample. The six sizes of larger circle indicate the mean size of species. The line at the base of the diagram represents a basal resource of detritus, fungi and microorganisms. Broken lines show links derived from literature evidence or inference from other species in the web, solid lines indicate links shown in feeding trials or by gut contents analysis. Boxes enclosing species indicate that the adult and larval stages of those species occurred in the web, but that larval stages were not identified to species. A) Web templet; B) Open water, March (date 1); C) Open water, June (date 3); D) (Open water, October (date 5); E) Margin, March (date 1); F) margin, June (date 3). (redrawn after Warren 1989).

1)   Food chains, which make up food webs, tend to be short.
2)   Connectance (the actual number of links as a proportion of all those possible) decreases hyperbolically with the number of species in the web.
3a)   Omnivory (feeding on more than one trophic level) is rare (there is, however, considerable controversy regarding this point).
  b)   Omnivory is more common in food webs dominated by insect-predator-parasitoid interactions.
4)   Donor-controlled dynamics (the donor [prey] controls the density of the recipient [predator], but not the reverse) are expected in some systems such as detritivore webs, and these webs have complex patterns of omnivory.
5a)   Webs are only occasionally compartmentalized, and when they are they correspond to major habitat divisions.
  b)   Webs are not compartmentalized within habitats.
6)   Loops (such as A eats B, B eats C, C eats A) and oddities, such as predators without prey and two top predators feeding on the same prey, are rare.

8.3.2   Spatial and temporal variation in the structure of freshwater food webs

Two holistic studies of freshwater food webs have recently addressed several of the aforementioned problems related to limited data bases. In both studies the role of aquatic insects as agents of energy transfer was emphasized.

Warren (1989) recognized that spatial and temporal variance in trophic relationships could be significant, as would probably be indicated by a more highly resolved data base. Therefore, over a one year period, Warren partitioned the food web of a pond into subwebs that depicted spatially and temporally distinct habitats.

The study site was a small, 1 m deep permanent pond in North Yorkshire, England. The margins of the pond were heavily vegetated (Habitat 1) while the centre of the pond was mostly open (Habitat 2). Animals were sampled by coring and sweep netting in both habitats during March, May, June, August and October. Trophic links were determined in the laboratory by feeding trials and gut content analysis of collected animals.

Across all sites and seasons the food web included 36 taxa. No fishes or amphibians were found. Most species occurred regularly around the pond margin but were rarely found in the open water. No species occurred exclusively in the open water. Warren suggested that the open water community was, in effect, a subset of that found at the pond margin.

The common food resource was loosely classified as detritus, which Warren, for pragmatic reasons, did not partition into components of living and dead organic material. The macroinvertebrate food webs classified by habitat and sample date showed considerable variation in structure (Fig. 8.4). Specifically, food webs from heavily vegetated pond margins were consistently more complex than those from open water due to greater species richness at the margins. The occurrence of some species in the vegetated areas may, for example, have reflected needs for net-building sites (e.g. for the caddisfly *Holocentropus* sp., and the aquatic spider *Argyroneta aquatica*) or perches (in the case of the odonates). During later winter and early spring, the immature stages of many species were absent or nearing the end of development, thus webs generally had fewer species early in the season but became more complex as the season progressed. Additional complexity was observed in open water following the spread of moss. The average connectance of webs in margin and open water areas was similar, yet the length of food chains in the margins was longer since additional inshore species were predominantly large predators. Omnivory in all locations was more common than would have been predicted based on previous studies.

Warren's holistic perspective on food webs would not have been possible had he not: (1) partitioned the pond spatially and temporally; and (2) conducted his analyses with rigour. This approach enabled him to counter some commonly held beliefs, such as the supposed rarity of omnivory.

In stream ecosystems, data on the trophic relations of aquatic insects are extensive, however most studies have focused on one or two species and no complete lotic community food webs have been constructed. To address this void, Tavares-Cromar (1990) examined the aquatic stages of insects from two riffles in a temperate stream over a period of one year.

Duffin Creek (Ontario, Canada) is approximately 7 m wide, 15-30 cm deep, with riffle sections characterized by substrates which range in size from small pebbles to boulders. The biota of the riffle sections was determined via rock scrapings, water, moss and substrate core samples that were taken at monthly intervals. The distribution of organic and inorganic material in the water column, on the substrate surface, and in the bed interstices was as is indicated in Fig. 8.5.

The food web of Duffin Creek, at the time of study, contained 44 species exhibiting an array of complex relationships (summarized in Fig. 8.6). All insects used detritus as their primary food source, supplemented in some species with biofilm-coated microinorganic particles, diatoms, fungal hyphae and plant material. Omnivorous species included the stonefly *Paracapnia angulata*, and the caddisflies *Hydropsyche sparna*, *H. morosa* and *Cheumatopsyche* sp., which complemented their detritus-dominated diets with animal material in the form of chironomids and mayflies (particularly *Baetis tricaudatus* and *Ephemerella subvaria*). Insects previously considered to be obligate predators, such as the stoneflies *Paragnetina media* and *Agnetina capitata*, appeared to be omnivorous. As they grew, stonefly nymphs consumed less detritus and assumed a more predaceous role, reflected by increased proportions of animal material in their diets. This type of omnivory has been referred to as "life history omnivory" (as opposed to conventional omnivory), which is defined as "different life history stages of a species feeding in trophically different positions in the food web" (Pimm and Rice, 1987).

Pimm's (1982) plea regarding the need for more comprehensive food web data bases was endorsed by Tavares-Cromar (1990). Additionally, Tavares-

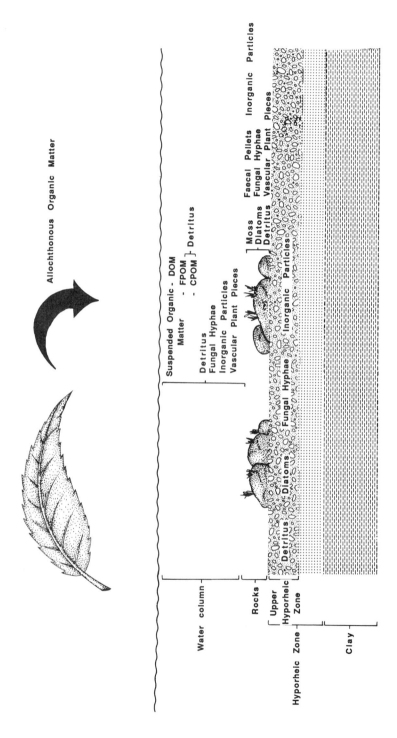

Fig. 8.5   Potential food in the water column and on and beneath the surface of the bed of Duffin Creek (redrawn after Tavares-Cromar, 1990).

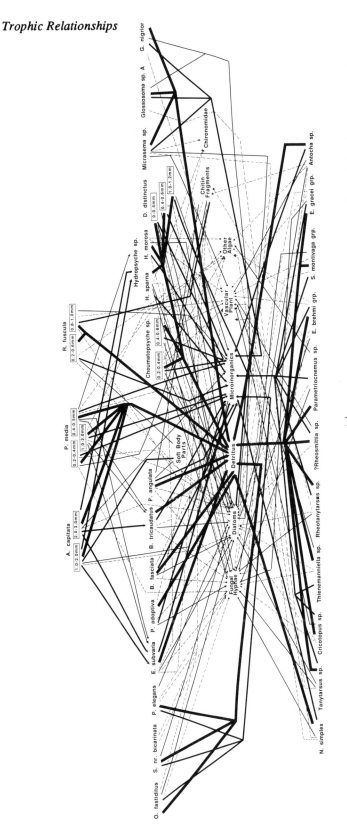

Fig. 8.6 Summary food web of the Duffin Creek riffle insect fauna (the thickness of lines represents the relative importance of the resource; different size classes of some of the more prominent species are indicated in boxes) (after Tavares-Cromar, 1990).

Cromar suggested that although lumping aquatic insects into functional feeding groups may be useful for many descriptive applications, this system should be avoided in food web analyses, as it can mask ontogenetically-related changes in the diets of insects and possible intraspecific differences in feeding among habitats.

9   PREDATION AND ITS CONSEQUENCES

9.1   Brief historical perspective: a case history

During the 1980s, research on predation in freshwater systems became increasingly experimental, with most studies focusing primarily on one or a few prey species, and with little mention made as to how the impact of predators on populations might cascade through communities. However, the need for more holistic examinations was explicitly indicated by Macan (1977), following his 21 year study on predation in an aquatic habitat (one of the first large spatial/temporal-scale manipulations of predators).

Macan studied the influence of fish predators (brown trout, *Salmo trutta*) on the composition of the animal community in Hodson's Tarn (a 4,000 m^2 artificial fishpond), by sampling invertebrates during times of high and low fish density over the period 1955-1975. Most attention was focused on large macro-invertebrates, particularly the Odonata, Trichoptera, Hemiptera, Coleoptera, and one species of amphipod, *Gammarus pulex*. Macan observed that the trout fed mostly in weed beds and that the organisms that appeared to suffer greatest losses were tadpoles, which swim conspicuously, backswimmers, which station themselves at the water surface waiting for prey, and certain Coleoptera whose larvae hunt in the open water. These animals did not seem well adapted to co-exist with trout and Macan suggested that predators limited the distribution of these prey types.

The hemipteran *Hesperocorixa castanea* was found throughout the littoral zone, during times of low trout density but, when predator density was high, nymphs were found only near inshore areas, in sedge beds. Similarly, during periods of high prey density damselfly nymphs (*Lestes sponsa*) restricted their movements to near-shore *Myriophyllum* beds, rather than to those offshore that nymphs could only reach by swimming.

Other than the species mentioned above, Macan found little evidence to suggest that the densities of other prey were significantly impacted by trout, which led him to suggest that some species must have a self-regulatory mechanism which limits predator impact. For example, in the odonate *Pyrrhosoma* sp., larger nymphs excluded small nymphs from perches where food was abundant. As a result of being forced into suboptimal feeding areas, the subordinates either took an extra year to reach adulthood or, more often, they starved to death. However, in the presence of fishes large nymphs were often consumed, thus creating space for smaller nymphs. Thus, the impact of fish predators on the odonate prey population was largely negligible since smaller nymphs that would have died in the absence of fishes now survived. Macan's findings have been borne out by other studies, including those on mammals (e.g. Errington 1963). Thus, to measure the impact of predation requires more than simply counting the number of prey killed; it is necessary to factor out the proportion of the prey population that would have died independently of the influence of predators. This concept of predators consuming primarily "surplus" prey has been revived by Fretwell (1987) in his work on food chain dynamics. He terms the process *"Donor-controlled Predation"*, and suggests that removing donor-controlled predators from localities has negligible effects on prey populations.

That Macan developed such insights regarding the process of predation is impressive, particularly considering that his work was performed on large spatial and temporal scales, and that it was only semi-quantitative and correlative. The process of donor-controlled predation among insects could stand careful examination and evaluation by modern aquatic ecologists.

9.2   Anti-predator defences

In this section the behavioural and morphological adaptations that insects have evolved to circumvent the advances of predators will be considered. Rather than making passing reference to innumerable studies on particular anti-predator defence mechanisms, we shall examine a few studies in greater detail. The examples chosen best illustrate particular mechanisms and, consequently, help develop an appreciation for the concept.

Substrate

Several studies of fishes and benthic insects demonstrate the effectiveness of substrate as a refugium from predators. In artificial streams, Brusven and Rose (1981) created substrates that were composed primarily of either cobbles, pebbles, sand, or combinations of all three. Sculpins (*Cottus rhotheous*) were released into the streams, followed shortly by the introduction of several dozen insects, either *Ephemerella subvaria* (mayfly), *Hesperoperla pacifica* (stonefly), or *Rhyacophila vaccua* (caddisfly). After one day, the remaining density of prey was recorded. Predation on all three insect prey species was generally low when cobbles were present, slightly higher with pebbles, and very high when only sand was present. When combinations of substrate were used, predation still remained least when cobbles were the dominant landscape.

Macrophytes are the lentic anologues of the rock substrates of streams. Pierce (1988) tested the affects of bluegills (*Lepomis macrochirus*) on microhabitat use and vulnerability of nymphs of three species of odonate, *Tetragoneuria cynosura*, *Ladona deplanata* and *Sympetrum semicinctum*. In laboratory aquaria, small wooden dowels were placed vertically and dead leaves covered most of the bottom area. Pierce examined the distribution of nymphs on dowels and leaves under conditions of day/night vs. fish/no-fish. Prior to running trials with fishes, their mouths were sewn shut, thus preventing them from consuming nymphs. Microhabitat use by all three prey species varied significantly with diel period and presence or absence of fishes. Bluegills induced increased use of cover in all three prey species, especially during the day. This behaviour was adaptive, as predation on nymphs was found to be positively correlated with the degree of exposure.

Colour

Colour can have a multi-functional purpose: (1) to help camouflage prey from predators; (2) to help camouflage predators from prey; or (3) to help make prey strikingly visible. The latter refers to the condition of *aposematism* - i.e. advertising unpalatability, often as a result of secondary compounds derived from food plants (Hutchinson, 1981). A clear example of aposematic coloration is seen in scarlet-coloured water mites that are largely ignored by predaceous insects and fishes. Similarly, Ellis and Borden (1970) found that certain chironomids which are deep red in colour, because of haemoglobin, were the least preferred items of a wide choice of prey offered to notonectids (possibly because of distastefulness). It was suggested that such avoidance may have evolved as a secondary mechanism to limit predation by fishes that forage in the sandy/muddy substrate where these chironomids occur.

The adaptive significance of head coloration in case-making caddisfly larvae has been shown to be habitat specific. By suspending the head capsules of caddis larvae on pins in front of brown trout, Otto (1984) found that striped heads were attacked least frequently under conditions of low illumination, but most frequently under conditions of high illumination. The results for larvae with solidly-coloured head capsules were reversed. In a survey of caddisfly larvae, Otto found that larvae with striped heads tended to be found in deeper, more poorly-illuminated lentic habitats, whereas species with solidly-coloured heads tended to be found in shallower, more highly-illuminated lotic habitats.

In combined field and laboratory studies, Feltmate and Williams (1989a) found that substrate colour selection by the dark brown stonefly *Paragnetina media* functioned as a pre-detection (Type 1; Edmunds,1972) defence mechanism. In predation experiments with rainbow trout, more stoneflies were consumed in aquaria containing light vs. dark coloured substrate. When both light and dark coloured substrates were placed in aquaria simultaneously, nymphs selected dark substrate independent of the presence or absence of trout. On a 12:12 h light:dark lighting schedule, selection for dark substrate occurred during daylight and for approximately 2 hours after dark, and ceased completely during the night. Within a few minutes of the lights being turned on, selection for dark substrate by nymphs was restored. When the stoneflies crawled over the surface of substrate patterned with both alternating light and dark coloured patches, the nymphs remained almost exclusively on the dark coloured patches.

Foraging efficiency in the damselfly nymph *Ischnura verticalis* is influenced by the colour of substrate on which it rests. In clear glass beakers in the laboratory, Moum and Baker (1990) placed two dowels, both of which were either green or brown. They then released mayfly nymphs (*Callibaetis* sp.), and a few minutes later a green *I. verticalis* nymph, into each glass. Shortly afterwards they recorded the number of live mayflies remaining. Contrary to expectations, more mayflies were consumed in glasses containing brown rather that green substrate. This differential predation was not related to prey behaviour, as the positioning of *Callibaetis* did not differ between substrate conditions. Moum and Baker suggested that because these particular damselflies are partially transparent, they may be more difficult for prey to discern on a dark- rather than light-coloured background.

Drift

Stream insects are known to drift downstream with the current, especially at night (see also section 7.1). In a small Minnesota stream that contained brook trout (*Salvelinus fontinalis*), Waters (1962) took hourly samples of insects over periods of one day during each of the four seasons. He observed marked increases in drift about 1 hour after sunset, a steady decrease through the night, and a return to daytime lows near sunrise. For all species, drift rates tended to be higher in summer than in winter, perhaps related to the increased activity of invertebrates during summer. Waters suggested, but did not test, the hypothesis that the adaptive significance of nocturnal drift might be to avoid predaceous fishes. Similarly, the depressant effect of moonlight on drift of large stream invertebrates might lower predation by visually feeding predators. Between full moon and dark nights, Anderson (1966) detected a marked difference in the numbers of drifting Ephemeroptera, Plecoptera and Simuliidae (Fig. 9.1).

For nearly two decades after Water's (1962) study, there was still no research to rigorously test whether disproportionate nocturnal drift in insects reduced predation by fishes. Several studies have shown that when given a choice between small and large prey, trout prefer large items over smaller ones and that trout are particularly effective daytime predators (e.g. Ringler, 1979). Allan (1978) postulated that relative to smaller conspecifics, fishes should disproportionately constrain larger invertebrate taxa to nocturnal activity. He also suggested that smaller taxa or stages may be aperiodic or day active. Allan tested his hypotheses by sampling the drift in a Colorado mountain stream, at several locations and dates, throughout the summer and autumn. Samples were collected over 24 hour periods (3 night samples, 5 day samples), and the head widths of up to 30 individuals of each species collected were measured. The night:day drift ratio observed in the mayflies *Cinygmula* sp. and *Baetis bicaudatus* showed a positive correlation with increasing body size. The size composition of drift showed that, for many species, larger individuals were more common by night than by day, thus lending general support to Allan's theory. Allan cautioned,

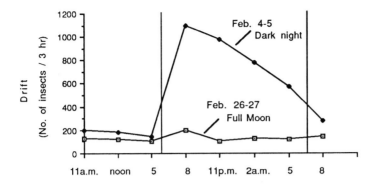

Fig. 9.1   Numbers of insects (Ephemeroptera, Plecoptera, Simuliidae) collected per 3 hour sampling period on a dark night and a moonlit night. The vertical lines indicate approximate times of sunset and sunrise (redrawn after Anderson, 1966).

however, that drift of many species could not be explained by fish predation alone. Other factors such as escape from invertebrate predators and/or aggressive conspecifics, respiratory movements, or some "accidental" factor might be influential also.

The responses of mayfly prey to stonefly predators were examined by Peckarsky (1980) in two small streams (one in Wisconsin and one in Colorado). Plexiglass observation boxes with screen ends and a painted silicon and sand substrate were placed longitudinally near the stream edge. Observations indicated that slow-moving mayflies such as *Ephemerella* avoided stonefly predators by using long-distance chemical cues, whereas faster-moving mayflies like *Baetis* avoided capture by drift-swimming (i.e. initially drifting and then swimming), drifting, and/or crawling away from the predator. Particularly interesting was the observation that once a *Baetis* nymph had been attacked by a stonefly the remaining mayflies aggregated in the corners and ends of the observation boxes. No mechanism to explain such behaviour was suggested but possibly an "alarm substance" was released.

Reductions in movement and feeding

A plethora of studies has recently (post 1980) demonstrated that aquatic insects may avoid being detected, by fish and invertebrate predators, by moving infrequently and feeding less.

The influence of a predator, *Notonecta glauca*, on patch choice and feeding rate of nymphs of the damselfly *Ischnura elegans*, was examined by Heads (1986). In opposite corners of a laboratory aquarium, patches of either high prey density and low cover from predators, or low prey density and high cover from predators, were created. High and low cover were created by increasing or decreasing the number of strips of mesh cloth per corner. The experiment was run under dim light and a high density of prey, *Daphnia magna*, was maintained by shining a bright beam of light on a corner to which the *Daphnia* were attracted. A notonectid was held in a screen enclosure adjacent to the low cover/high prey patch area. When damselflies were released and free to choose between patches they selected high cover areas despite the low availability of prey. Nymphs only selected the high food patch in the absence of notonectids. Even in cases where damselfly nymphs had been starved for 12 days, nymphs were still risk adverse to predators. As a result of reduced feeding in the presence of notonectids, potential losses in *Ischnura's* rate of development (i.e. length of time between instars) were calculated to lie between 4 and 18%. Reductions in growth could negatively affect nymphal survival and adult reproductive fitness.

In an experiment similar in concept to that of Heads, Sih (1980) examined whether juvenile notonectids could balance conflicting demands between the need to feed, which he related to fitness, while simultaneously avoiding predators. The model system involved using adult *Notonecta hoffmanni* as the predator and juvenile *N. hoffmanni* (which are prey for adults) as the forager, both of which ate wingless fruit fly prey. In laboratory aquaria, distinct regions of high and low prey density were created, and within each of these regions adult notonectids were released, three in the high prey density region and none in the low prey density region. Barriers were used to prohibit movement of insects between regions. To determine whether ontogenetic changes in the feeding rate of foragers varied as a function of predator stress and food availability, $1st$ to $5th$ instar foragers were released into both regions of the aquaria and their feeding rate on fruitflies was recorded. Surprisingly, the adult notonectids did not consume the foragers. However, feeding rate of early instar notonectids was less in high predator/high prey regions than in low predator/low prey regions, because of young foragers spending a disproportionate amount of time avoiding predators in the predator sector. Mid-instar forager feeding rate did not differ between regions and late instar nymphs fed at higher rates in the high prey density sector, as predicted, since adults represented less of a threat to this group. Having established the feeding rate of foragers under each condition of predator stress and prey availability, Sih then tested whether each instar of forager would distribute itself optimally between regions when barriers were removed. Fortunately, adult notonectids, which also fed on the fruit flies, still remained in the high prey regions in the absence of barriers. As predicted, early instar foragers avoided high predator/high prey regions, whereas mid- to late instar nymphs showed the opposite behaviour. These results clearly indicate that juvenile *N. hoffmanni* are capable of adaptively balancing the conflicting demands of avoiding predators and foraging efficiently. Sih stressed that in past formulations of optimal foraging models, usually only behaviour that maximized feeding rate was considered. He suggested that such models would be more rigorous if innate tendencies to avoid predators were also included.

Several studies on lotic systems have demonstrated the influence of fishes in affecting the feeding rate of aquatic insects. In artificial streams, Kohler and McPeek (1989) used the benthic feeding, mottled sculpin, *Cottus bairdi*, and the grazers *Baetis tricaudatus* (mayfly) and *Glossosoma nigrior* (caddisfly), to examine prey behaviour when predation risk varied temporally and spatially among sections of stream that also differed in foraging profitability. In this system, it was necessary for *Baetis* and *Glossosoma* to expose themselves to sculpins while feeding on epilithic algae. However, *Glossosoma* is largely protected from predators by a heavy stone case that encompasses most of its body whereas *Baetis* is unprotected. Thus *Baetis* and *Glossosoma* were considered reasonable models of stream insects that were, respectively, vulnerable and invulnerable to predaceous fishes. Using such a system the behavioural response of grazers to the presence of sculpins, and whether grazers make adaptive compromises between feeding and minimizing predation risk, was assessed. To understand the adaptive significance of the differential responses of grazers to sculpins, the relative vulnerability of both species was also considered. A multi-factorial design was used to examine grazer behaviour under conditions of sculpin presence/absence, food abundance on the substrate surface (high/low), grazer hunger level (fed/starved), and time of day (night/day). The results of these feeding trials indicated that *Baetis* was much more vulnerable to sculpin predation than was *Glossosoma*. Only the foraging behaviour of *Baetis* was significantly affected by predators. In the presence of sculpins, the mayflies tended to avoid the substrate surface (where food was present), and those that did venture on to the surface remained there for less time and moved less frequently than was observed in the absence of sculpins. The rate at which mayflies drifted (when, of course, they cannot feed) was also greater in the presence of sculpins. It would be interesting to examine whether the presence of drift-feeding fishes, such as trout, might counter the tendency to drift, thus suggesting that mayflies have evolved the adaptive capacity to distinguish between benthivorous and drift-feeding predators. Lack of influence of sculpins on the feeding behaviour of *Glossosoma* larvae, compared with the significant impact on *Baetis*, suggests that armoured species are not as vulnerable to predaceous fishes, whereas soft-bodied species counter the threat of predation by behavioural avoidance.

Whether the feeding rate of aquatic insects may be differentially affected by predators that present differing degrees of risk was considered by Feltmate and Williams (1989b), using rainbow trout and the predaceous stonefly, *Paragnetina media*. In laboratory aquaria, the feeding rate of stoneflies on enchytraeid worms was examined under conditions of large, small, or no rainbow trout present. Small trout would approach and sometimes bite stoneflies, but they could not consume them. To prevent large trout from eating nymphs, hatchery-reared trout, that did not feed for approximately 1 week after arriving in the laboratory, were used. Previous experiments established that, when these dark-coloured stonefly nymphs rested on dark brown substrate, they were under a low threat of predation (because of crypsis). Thus, a dark substrate (in the form of a tile cube, 5 cm on a side) was placed at one end of each aquarium, and a feeding cup containing enchytraeids was placed at the other. Nymphs had to leave the safe refuge of the cube in order to feed on the worms. To determine whether the degree of hunger of nymphs affected their behaviour, the experiment was run with nymphs that had been either well-fed or starved prior to trials. When either small or large trout were present in aquaria, nymphal feeding rate was reduced by approximately 50% for both starved and well-fed nymphs.

Reductions in the feeding rate of prey in the presence of predators would inevitably result in reductions in prey growth and development. Growth refers to an increase in organism size, and development is the morphological and physiological progression toward reproductive maturity. The body weight of most female aquatic insects increases substantially prior to emergence, and this increase is related to reproduction as maturation of ovarioles occurs in the final nymphal instars. Thus, reductions in nymphal feeding in the presence of predators would, predictably, result in reductions in nymphal and adult growth and development, and adult fecundity.

Thanatosis, freezing, reflex bleeding, deimatic behaviour, and cyclomorphosis

Aquatic insects possess several putatively rare behavioural and morphological defence mechanisms. The supposed rarity of defences discussed in this section may reflect reality, or it may simply be an artifact of limited investigation.

Nymphs of the large detritivorous stonefly *Pteronarcys dorsata* respond to different predators by behaving thanatotically (feigning death in a curled-up position), freezing in position, or reflex bleeding. These mechanisms were demonstrated by dividing aquaria into two halves using glass dividers. In one half were released two *P. dorsata* nymphs, and in the other half either a sculpin (*Cottus bairdi*), a rainbow trout (*Oncorhynchus mykiss*), or a crayfish (*Cambarus robustus*) (Moore and Williams, 1990). Sculpins and rainbow trout are benthic and pelagic predators, respectively. Following removal of the glass divider, if an active stonefly touched a sculpin with its antennae, the nymph would freeze and cling to a rock surface. However, if an inactive stonefly was attacked by a sculpin, the nymph exhibited thanatosis. If, during an attack, a trout lifted a stonefly off the substrate the nymph generally feigned death, whereas if the trout was not successful in loosening the nymph's grip on the substrate the nymph generally would freeze. When the bodies of dead *P. dorsata* nymphs were straightened (from an otherwise curled position) and both curled and straightened nymphs were presented to trout, curled nymphs were almost always rejected whereas straightened nymphs were consumed, thus demonstrating the utility of thanatotic behaviour.

When nymphs of *P. dorsata* were attacked by crayfishes, they responded by reflex bleeding (autohaemorrhaging). Fluid was forcibly expelled into the water as a milky cloud that coated the crayfish's antennae with a viscous film; this

Fig. 9.2   Crayfish attacking nymph (A), and the nymph responding by autohaemorrhaging. Arrows indicate the edge of the cloud of secretion (B). The crayfish then immediately releases the stonefly (C) and retreats (D) (from Moore and Williams, 1990).

effectively masked mechanoreceptors, spines and hairs. The crayfish would immediately release the stonefly and retreat (Fig. 9.2) to clean its mouth parts, chelae and antennae. Scanning electron micrographs (SEM) showed that secretions were expelled through pores located on the trochanteral segments of the stonefly's metathoracic legs. Bread that been soaked in water or secreted haemolymph was eaten readily by both crayfishes and fishes, suggesting that there were no toxic substances in the secretion, and that the repellent action, in the case of the crayfish, was due to the cloaking of the antennal sense organs.

Deimatic behaviour is characterized by a posture or display that prey may use to startle, intimidate or frighten a predator. Peckarsky (1980) documented a dramatic example of deimaticism in the 'scorpion' display of nymphs of the mayfly *Ephemerella subvaria*, practised in the presence of the predaceous stonefly *Acroneuria lycorias* (Fig. 9.3). Presumably, from the largely tactile perspective of the predator the scorpion posture increases the apparent size of the prey which is then rejected by the predator.

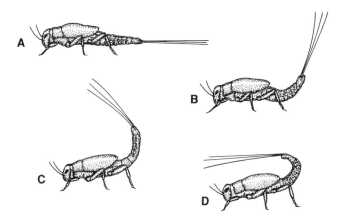

Fig. 9.3   *Ephemerella subvaria* (mayfly): (A) normal resting posture; (B) low-intensity scorpion posture; (C) moderate-intensity scorpion posture; (D) high-intensity scorpion posture (redrawn after Peckarsky, 1980).

Cyclomorphosis is cyclic morphological change in a species which is usually displayed by successive generations that reproduce several times a year (Dodson, 1989). This phenomenon is supposedly rare in aquatic insects, yet it is seen in a variety of other organisms such as freshwater ciliates, algae, rotifers, cladocerans, intertidal snails, barnacles and bryozoans. The morphological shift is usually towards the production or elongation of spines or teeth, the utility of which is presumed to limit predator success. Most examples of cyclomorphosis are seen in zooplankton, a notable example being the response of *Daphnia pulex*

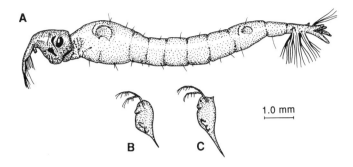

A

1.0 mm

B   C

Fig. 9.4   A model system for the study of predator-induced cyclomorphosis.
            (A) *Chaoborus americanus* larva (the predator). Neonates of *Daphnia*
            pulex (the prey) grown in the absence (B) and presence (C) of
            Chaoborus or water from a *Chaoborus* culture (redrawn from Dodson,
            1989).

to the phantom midge *Chaoborus*. Phantom midges generally consume young
Daphnia, prefering individuals of certain sizes and shapes. In response to
Chaoborus feeding, *D. pulex* produces neckteeth and elongated tailspines as a
result of detecting the chemical signal of the predator (Fig. 9.4). An example of
the energetic cost of producing cyclomorphic structures (thus explaining why
such armour is not permanent) has been demonstrated in rotifers, which show
reductions in growth rate coincident with cyclomorphic change (Stemberger,
1988).

The only species of aquatic insect known to show cyclomorphosis is the
aquatic midge *Cricotopus sylvestris* (Hershey and Dodson, 1987). The abdomen
of this midge supports lateral hairs that tend to be longer in spring generations
which co-occur with predaceous *Hydra*. Laboratory work has shown that the long
hairs strip nematocysts from *Hydra* tentacles, thereby temporarily disarming the
predator and allowing the prey to escape.

Chemical cues

The relationship between the presence of predators and their chemically-mediated
detection by prey has only recently received due recognition in the literature.
The limited number of studies suggest a strong adaptive utility for being able to
detect predators at a distance.

Cues used by prey to gauge predation risk were examined by Sih (1986), in
experiments with predaceous *Notonecta undulata* and larval mosquitoes, *Aedes*
aegypti and *Culex pipiens*. *Culex pipiens* often co-occurs with *N. undulata*,
whereas *A. aegypti* has had virtually no contact with notonectids over
evolutionary time. *Aedes aegypti* typically lives in treeholes and small, discrete

habitats which notonectids do not occupy. Thus Sih reasoned that *C. pipiens* should be more adept at detecting and responding to notonectids than the evolutionarily naive *A. aegypti*. He focused specifically on differences in chemical detection. Sih released mosquito larvae, of either *A. aegypti* or *C. pipiens*, in laboratory aquaria. Around the periphery of the aquaria were multiple strands of cord that represented areas of safe refuge for the larvae. The behaviour of each species was recorded before and after the addition of a container of either filtered tap water (control) or a container with tap water and either: (1) notonectid adults; (2) mosquito larvae ground by a homogenizer; (3) larvae pierced with a dissecting pin; or (4) larvae and notonectids that had been allowed to kill prey. Only this final treatment, predators which had actually killed prey, could have produced a chemical signal that exactly modelled the actual act of predation. Ground, pierced or predator-killed larvae were always the same species as those being tested. The last trial was the one that elicited the greatest response from *C. pipiens*, both in terms of the percentage time that the mosquito larvae spent moving and in the time that they spent in the centre of the aquaria. Notonectid water, and water containing ground-up mosquito larvae, affected *C. pipiens* behaviour marginally, and water in which larvae had only been pierced produced no significant effects. In contrast, no chemically-mediated effects on the behaviour of *A. aegypti* larvae were detected. The obvious conclusion is that natural selection has resulted in the evolution of a chemically-mediated defence mechanism in *C. pipiens*, whereas no such evidence exists for the non-evolutionarily related *A. aegypti*.

Mayflies use non-contact chemical stimuli to detect the presence of predaceous stoneflies, as demonstrated by Peckarsky (1980) in Wisconsin and Colorado streams. A screened tube, used to restrain insects, was inserted through a hole in the clear cover of a plexiglass flow-through box, and the box was placed in shallow water. Predaceous stoneflies were released into the screened tube and changes in the distribution of mayflies downstream of the tube were monitored. In test trials, dye was released from the tube to determine the pattern of dispersion that stonefly odour would take downstream. Following placement of stoneflies in tubes, densities of several species of *Ephemerella* and *Stenonema* declined significantly in the downstream odour region. A subsequent increase in mayfly density was observed following stonefly removal. However, not all mayfly species (e.g. *Baetis phoebus*) showed non-contact responses to predators. Peckarsky suggested that the fast escape response (swimming) in some species is sufficient to avoid being caught.

In lotic systems there must literally be a plume of dilute odour constantly flowing downstream, and one would think it impossible for prey to distinguish between predator and non-predator odours. However, the odour of predators immediately upstream of prey would be concentrated, thus facilitating detection and allowing sufficient time for prey to leave areas where the risk of predation is high. From an evolutionary perspective it is curious that predators produce "warning" odours at all. Perhaps these chemicals convey some hitherto undiscovered advantage, or are simply the released by-products of predator metabolism.

Learning to avoid predators

Learning in aquatic insects has been largely ignored, although numerous studies have demonstrated learning in terrestrial insects. It is reasonable to postulate that aquatic insects, having evolved under environmental conditions of comparable complexity to their terrestrial analogues, also would have evolved a capacity for learning. As used here learning refers to a change in the behaviour of an animal because of experience.

An example of learned predator avoidance in a terrestrial insect was recently observed in the butterfly *Heliconius erato petiverana*. Mallet *et al.* (1987) showed that adults, once handled, decreased their rate of return to the site where they were originally captured, and that this decrease was due to behavioural avoidance of the capture site, rather than dispersal or increased mortality. Although these authors were interested in how avoidance behaviour might influence mark-recapture population estimates, they speculated that the natural equivalent of handling effects could enable *Heliconius* to avoid areas where predators might be likely to attack or sites with poor resources.

Research, in both the laboratory and field, has shown that nymphs of the stonefly *Paragnetina media* select substrate beneath which they can crawl and thereby avoid being detected by fishes. Feltmate and Williams (1991a) reasoned that nymphs, having located suitably sized substrate, might be able to leave and later rapidly relocate it using learned environmental cues. This ability could be adaptive in decreasing the time needed to locate refugia and consequent exposure to predators. The ability of stoneflies to learn the location of a group of stones that provide shelter from predators was tested in laboratory aquaria. In the first experiment, the stoneflies displayed a capacity for *path learning*. During successive trials, stoneflies crawled between starting points and substrate assemblages (stone groupings) more quickly when both positions were fixed, than did stoneflies in aquaria in which the starting points were fixed but substrate positions were not. In a second experiment *P. media* demonstrated an ability to learn to locate stone assemblages when released at variable points in aquaria (i.e. *spatial learning*). A group of stoneflies was held for 5 days in aquaria in which the spatial arrangement of cues and rewards was changed daily. On the sixth day, stoneflies that had been held in aquaria with environmentally constant conditions were able to locate substrate assemblages more quickly than did stoneflies held under variable conditions. Stoneflies that had learned the location of shelters experienced significantly fewer attacks from rainbow trout than did naive individuals. A capacity for learning probably exists in many aquatic insects, as fitness would be enhanced, for example, by the ability to relocate refugia or food patches quickly.

9.3   *Influence of predators on aquatic insect population dynamics and community structure*

Stress is defined as an environmental condition that impairs Darwinian fitness, as demonstrated by reductions in the size or condition of immature or adult organisms, their survivorship, or fecundity (Sibly and Calow, 1989). Whether predaceous fishes and insects exert a significant stress on insects found in ponds and lakes, as indicated by these or a broader array of fitness components, is uncertain. This is due to a general lack of large spatial/temporal scale manipulations of predators in lentic systems. To date, most studies of predation in ponds and lakes have been able to demonstrate only that fishes and predaceous invertebrates affect the movement, feeding, density and morphology of insects (as outlined earlier in this chapter). It is generally assumed, however, that predator-induced costs to prey fitness are significant.

During the early to mid-1980s, studies on predation in streams gradually evolved into complementary works which indicated that fishes had little impact on invertebrate prey populations and community structure (e.g. Allan, 1982; Reice and Edwards, 1986; Culp, 1986). However, such conclusions may be erroneous because of limitations in the design of each of these studies, specifically: (1) manipulations of the independent variate were conservative (fish predators were of limited size and density); (2) ontogenetic changes in prey growth and development were not monitored (missing potentially sensitive life cycle stages); (3) spatial (cages) and temporal (1-2 weeks) scales were small (except Allan, 1982); and (4) only prey density and rate of drift, among many potential response variables, were considered in detail. However, these studies did indicate that macroinvertebrates in streams were not sensitive to minor perturbations in predation pressure. The response of prey to "substantial and sustained" (Carpenter, 1989) levels of predation was still unknown.

Current research has addressed some of the criticisms outlined above and suggests that fishes can affect the density and size distribution of stream insects. For example, following the release of "large" (11.6 g) and "small" (1.3 g) brown trout into 1 m diameter enclosures, Schofield *et al.* (1988) observed a significant reduction in density of the trichopteran *Plectronemia conspersa*, in less than 8 days in enclosures with large trout. Gilliam *et al.* (1989) found that six creek chub (*Semotilus atromaculatus*) per m^2 in artificial streams reduced total invertebrate volume and density over 3 months, and that the mean size distribution of oligochaetes and isopods was also reduced. Schlosser and Ebel (1989) showed that 4 weeks of cyprinid predation in an artificial stream altered insect abundance, and that the effects were variable between habitats. For example, they found that the density of pool-dwelling chironomids and crustaceans decreased more in the presence of cyprinids than did riffle-dwelling hydropsychids and simuliids. A large enclosure/exclosure field experiment, used to examine the effects of rainbow trout on stonefly prey populations, showed that the activity of nymphs was sufficiently depressed (i.e. nymphs developed a "slow life-style"; Sih, 1987), that density, size and condition (weight/head-width ratio) of immature stoneflies were negatively impacted by trout prior to emergence (Feltmate and Williams, 1991b). Furthermore, adult stoneflies emerging from predator-inhabited sections of stream suffered losses in the total number that emerged, adult condition, and fecundity. These findings demonstrated that direct and indirect effects of a predaceous fish can cascade from immature to adult prey and from aquatic to terrestrial ecosystems. In a survey of streams in southern Ontario, Canada, Bowlby and Roff (1986) found that predators (top-down control) explained more variance in invertebrate assemblages than did food limitation (bottom-up effects). However, Bowlby and Roff made no claims of cause and effect since their study was purely observational.

As the majority of aquatic insects spend their adult and reproductive stages on land, more studies should consider the impact of predators on these terrestrial components of the life cycle.

To avoid erroneous results in future studies on predator/prey dynamics in aquatic systems, as exemplified in studies on streams during the 1980s, caution should be taken to ensure that: (1) predator manipulations are performed; (2) spatial and temporal scales are sufficiently large; and (3) a wide range of potential response variables is considered. Surveying streams, ponds and lakes, or sections of stream with and without predaceous fishes (e.g. above and below waterfalls), would help aquatic ecologists discern life history characteristics that prey have evolved as a defence against predators. For example, in tropical streams, Reznick and Endler (1982) have used this approach to study successfully the life history adaptations that guppies, *Poecilia reticulata*, have evolved to limit the impact of predaceous fishes. Large-scale, comparative approaches, particularly involving insects, have yet to receive due consideration in the study of predator/prey population and community dynamics in aquatic systems.

10 EXPERIMENTAL DESIGN AND SAMPLING TECHNIQUES

10.1 Designing field and laboratory studies

Most early studies on aquatic insects were primarily descriptive, basing inferences on direct sampling of the benthos or on data gathered from placing artificial substrates into streams, ponds or lakes. Although research has subsequently become more sophisticated as greater emphasis is placed on field manipulations, experiments on aquatic insects often do not meet the criteria for modern experimental design and analysis. For example, subsampling is frequently mistaken for true replication, and sample sizes are often inadequate for tests of significance (Eberhardt and Thomas, 1991).

A classification and description of eight methods for designing field studies, emphasizing some techniques as yet little used in the study of aquatic insects, is outlined in Fig. 10.1. This system is defined in terms of the type of control imposed by the observer, by the presence or absence of a perturbation, and by the domain of the study. Although presented with emphasis on field studies, the principles outlined in the sections on replicated and unreplicated experiments are equally applicable to laboratory investigations.

10.1.1 Replicated experiments

The strongest inferences about cause and effect relationships can be determined using randomized, replicated experimental approaches. However, because there is generally an inverse relationship between replicate number and the size of sampling units, *a priori* power analyses should be used to determine the number of replicates necessary to detect treatment effects between experimental and control units (Peterman, 1990a). Failure to do so increases the probability of committing *Type II* errors (i.e. accepting false "null" hypotheses), a reality largely ignored in aquatic studies. Peterman (1990b) found that 98% of recently surveyed papers in fisheries and aquatic sciences that did not reject the null hypothesis (Ho), failed to report b (the probability of making a Type II error), and of those papers 52% drew conclusions as if Ho were true. Power calculations (b) should be used in all impact assessment studies, particularly in settling disputes between industry and regulatory agencies (Peterman, 1990a). Failure to do so may result in irreversible damage to an ecosystem.

A problem commonly observed in replicated experimental studies is the lack of distinction between "fixed effects" and "random effects" models in analysis of variance (i.e. ANOVA). The fixed effects model is the most commonly used, yet

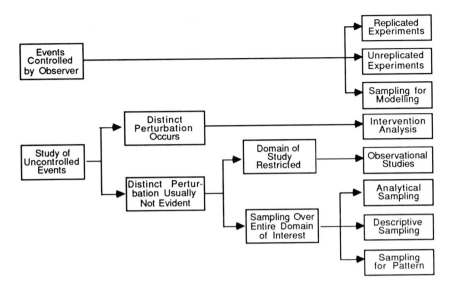

Fig. 10.1   Methods for designing field studies, along a gradient of high degree of experimenter control (top), to none (redrawn after Eberhardt and Thomas, 1991).

it is the least commonly justified, for ecological data sets. It is generally applicable to a fixed and finite set of entities and thus supports only limited and specific inferences. Most ecological studies require analyses based on a conceptually-indefinitely-large population and therefore require a random effects model. Eberhardt and Thomas suggest that a review and analysis of existing data sets are required to help ecologists understand the inferential differences between these models.

Stricter adherence to the underlying assumptions of parametric vs. non-parametric analyses also would decrease the probability of falsely accepting or rejecting hypotheses. There are three primary assumptions which apply to most parametric statistical tests (Zar, 1984): (1) The sampling of individuals from the population should be at random. For example, in a study of the effects of current speed on the distribution of mayflies on substrate, nymphs allocated to each treatment should be selected at random. If they are not randomly selected, those subjected to a particular treatment could be consistently "different" from those applied to another treatment, and therefore the results would not reflect a true or unbiased estimate of the treatment (in this case current) effect. Non-randomness of sample selection may be reflected in a non-normal distribution, lack of homogeneity of variances, or non-independence of items (Sokal and Rohlf, 1981). Measures to ensure random sampling during experiments, when sampling in the field or from a stock pool in the laboratory, are essential. It is, in some instances, possible to address a lack of independence in data using tests specifically designed for such circumstances (e.g. Repeated Measures Analysis of Variance, RMANOVA). (2) Samples should be drawn from populations that are normally distributed. Either a graphical analysis or a Goodness of Fit test (e.g. Kolmogorov-Smirnov) may be used to test for normality. To correct for or to "normalize" data, one of several transformations (e.g. square root, logarithmic [x+1], arcsin $\sqrt{x}$) may be applied. (3) Sample variance between groups should be homogeneous within the bounds of random variation. The equality of variances is referred to as the condition of *homoscedasticity*, and a quick check for this can be made using an F-max test. This test relies on the cumulative probability distribution of a statistic that is the ratio of the largest to the smallest of several sample variances.

When making multiple comparisons of means using pair-wise comparisons, an experiment-wise or Bonferroni error adjustment to the alpha value should be made. This adjustment essentially splits the level of significance evenly among all comparisons, thus lowering the probability of committing a *Type I* error (i.e. when performing multiple comparisons the probability of finding a significant difference, due to chance, increases with the number of comparisons).

It is often the case that parametric criteria cannot be met, hence necessitating the use of non-parametric tests, which are distribution free and relatively quick and easy to perform (Lehner, 1979). However, because there are fewer constraints on using nonparametric tests they tend to be less powerful in detecting treatment effects. So researchers may choose to transform data to meet the assumptions of normality and homoscedasticity, and then apply parametric analyses. Some argue (e.g. Kirk, 1968) that parametric tests are sufficiently robust that assumptions can sometimes be violated without affecting the validity of tests. However, such violations, without clearly defined limits, enter into a nebulous realm of statistics that can make parametric analyses questionable (i.e. when does violating test criteria become unacceptable?). Thus it is probably prudent to adhere to the axioms of parametric analysis.

The degree to which experimental criteria are violated, because of ignorance or apathy, was clearly indicated by Hurlbert (1984) in his focus on pseudo-replication, defined as, "the use of inferential statistics to test treatment effects with data from experiments where either treatments are not replicated (though samples may be) or replicates are not statistically independent". He found that 48% of experimental studies which used inferential statistics, published between 1960-1984, involved pseudoreplication. Although the incidence of pseudo-replication is greatest in studies of marine benthos and small mammals, it is certainly common in studies on aquatic insects.

A diagrammatic representation of various experimental designs is shown in Fig. 10.2. Each scheme may, for example, depict the placement of experimental cages across a stream to measure a treatment effect. Arrangements A-1, A-2 and A-3 would be acceptable for testing such an effect, whereas the lower five could only indicate spatial differences. The top three arrangements are sufficiently

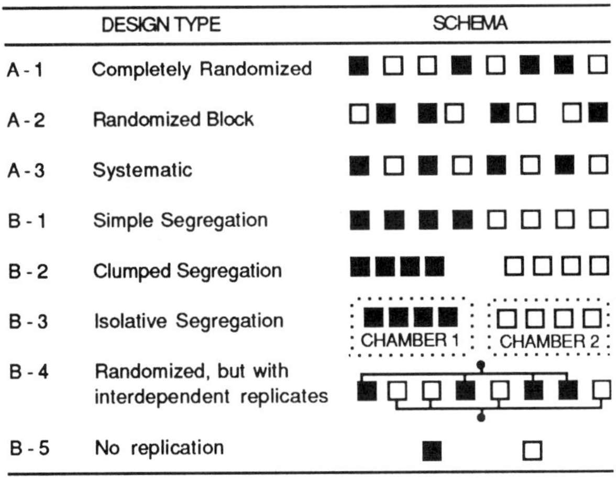

	DESIGN TYPE	SCHEMA
A - 1	Completely Randomized	
A - 2	Randomized Block	
A - 3	Systematic	
B - 1	Simple Segregation	
B - 2	Clumped Segregation	
B - 3	Isolative Segregation	CHAMBER 1 CHAMBER 2
B - 4	Randomized, but with interdependent replicates	
B - 5	No replication	

Fig. 10.2   Representation of acceptable methods of interspersing replicate units (A 1-3) of two treatments (light, dark), and various unacceptable (B 1-5) arrangements (redrawn after Hurlbert, 1984).

interspersed that any significant differences in findings between the shaded and open squares could, reasonably, be ascribed only to the treatment, whereas schemes B-1, B-2 and B-3 might reflect inherent side-to-side differences in the stream. Although scheme B-4 initially looks satisfactory, it depicts a situation whereby all treatments are maintained by one system and controls by another. This segregation would preclude attributing significant differences in response variables to a treatment, as they could also be artifacts of the different maintenance systems. The unreplicated B-5 design could only suggest treatment effects if precautionary steps, described below, were taken.

10.1.2   Unreplicated experiments

The number of unreplicated experiments in the ecological literature is surprising, considering Fisher and Wishart's (1930) comment that, "No one would now dream of testing the response to a treatment by comparing two plots, one treated and the other untreated". However, if one considers Fisher and Wishart's supposition carefully, and similar sentiments made by Hurlbert (1984), both accept the reality that it is often impossible to replicate experiments (because of physical or monetary restrictions). They simply emphasize that caution should be taken when assessing cause and effect relationships, and that significant differences without replication indicate only spatial or temporal effects, which may or may not be attributable to a treatment.

An example of a pseudoreplicated field experiment, with precautionary checks, was performed by Feltmate and Williams (1991b) when examining the influence of predaceous fishes (rainbow trout) on stonefly prey populations in streams. A reach of stream was divided into two sections using a fence divider. Control (No Predators) and experimental (Predators) sections were upstream and downstream of one-another (analogous to Hurlbert's B-5 design, Fig. 10.2). Trout were released in the lower section of stream, rather than the upper, to preclude the possibility that concentrated fish odours might have affected behaviour of the stoneflies (e.g. frequency of drift, movement, feeding rate). To measure the impact of trout, multiple Hess sample collections were taken in both the experimental and control sections of stream (i.e. samples were treated as replicates), and characteristics of the prey populations were compared using *t*-tests. However, the pseudoreplicative design of the experiment precluded attributing differences in the response of prey to predators, and statistical analyses could *only* indicate spatial differences between the two sections of stream. Nevertheless, the authors made the argument that spatial differences were attributable to the presence of predators. As precautionary steps, they established that prior to introducing trout there were no differences in physical (current speed, depth, width, substrate composition) or chemical (oxygen concentration, pH) characteristics of the stream, nor were there initial differences in prey response variables, between the predator and non-predator sections. Thus it was logical to assume that spatial differences in characteristics of the prey populations were attributable to predators.

An additional precautionary step to be used in unreplicated designs is to perform a post-manipulation comparison of response variables. Presumably if the independent variate is reduced, response variables will approach equality.

Since response variables may change concomitantly with other variables such as, for example, weather, Eberhardt (1976) proposed the use of a "station-pairs" or "ratio-method" approach. In this method, baseline or premanipulation data are obtained for paired control and manipulated stations, and the effects of the independent variate are determined by comparing the change of the ratio of the station pairs following intervention. A similar technique was described by Stewart-Oaten *et al.* (1986) to measure the effects of effluent discharge into a river. He suggested taking samples, replicated in time, *Before* a manipulation begins and *After* it has begun, at both the *Control* and *Impact* sites (i.e. the BACI design). Both control and impact areas should be sufficiently close that the same range of long-term natural phenomena affects both areas equally. If local events cause populations at control and impact sites to have different long-term responses, then the design cannot be used.

10.1.3   Sampling for modelling

The focus here is on efficient experimentation for parameter estimation in specified non-linear models (i.e. models where the relationship between independent and dependent factors is not constant). The method was originally developed for research in industrial experimentation and has been placed under the guise of events controlled by the observer.

Two problems arise when considering the use of modelling for studies on aquatic insects. The first is that methods for estimating single parameters are, at best, in their infancy (i.e. as discussed at the beginning of this chapter, experimental work on aquatic insect ecology is a recent phenomenon). Indeed, for ecological and environmental studies generally, there are few detailed sources on sampling for modelling (Eberhardt and Thomas, 1991). Second, the cumulative probability of committing an error in estimating two or more parameters increases multiplicatively, necessarily forcing modellers to keep models simple relative to the complexity of nature. As Krebs (1988) asserts, most modelling is directed towards explaining events *a posteriori* and that, given enough time and simplification, a model can be built to explain anything. He suggests that, presently, this is not the most effective way to advance ecological knowledge, and that more work to test predictions using field experiments is necessary. Although Krebs was directing his comments towards research on the population regulation of small mammals the message is equally applicable to studies on aquatic insects.

10.1.4   Intervention analysis

This method may be used to study the effects of a known perturbation (intervention) using a retrospective analysis of time series data. This approach could fall under "events controlled by the observer", yet it is placed here because the pattern of time trends in "experimental" and "control" units is properly a semi-qualitative/quantitative evaluative measure (Eberhardt and Thomas, 1991).

The limiting factor in time series or intervention analysis is the requirement for long sequences of pre- and post-manipulation data. Often such data are not available, thus diminishing the robustness or statistical power of the test. However, if a reasonable data base is available this analysis can be very useful in determining the impact of a perturbation on a system. For example, the analysis could be used to assess whether the release of an effluent into a stream or lake affected resident insects, by testing whether a significant difference in the mean level of the response variable existed before versus after the intervention.

10.1.5   Observational studies

The concern here is to compare the effects of a factor(s) by contrasting selected groups of individuals, or other units of the population, subjected to different levels (including zero) of the factor. Observational studies therefore initially resemble experiments, with the exception that the investigator does not control which individual gets what treatment. For example, a researcher may, after a broad perusal of ponds, ask whether insects which occur in weed beds are larger than those of the same species found in open areas. This could be done observationally by sampling groups of individuals in weed and weedless inshore areas and subsequently testing for significant differences. In contrast, if this were to be done experimentally, individuals would be assigned to levels of "weediness", under a controlled regime (with multiple independent replicates) and, after a period of time, data would be collected and tested for differences.

To limit extraneous effects in observational studies, Cochran (1983) makes three suggestions: (1) To apply the greatest degree of rigour that can be practically achieved in collecting data. Minimal effort spent working with different collecting devices or sampling methods will often significantly reduce sampling error. (2) Individuals exposed and not-exposed to a factor may be "matched", so that as many extraneous factors which could affect the response are eliminated. This necessitates a large sample size. (3) The influence of extraneous factors may be removed using a statistical or mathematical approach. This can be done, for example, using analysis of covariance (ANCOVA) or partial correlations.

10.1.6   Analytical sampling

This approach involves making inferences or comparisons by means of sampling an entire population (i.e. it is more broadly based than observational studies). Thus, caution should be taken to ensure that the sampling regime is inclusive of the entire population, and that it is not affected by disproportionate representation from isolated, potentially idiosyncratic, groups.

10.1.7   Descriptive sampling

A common goal in studies of aquatic insects is to estimate the abundance of one or more species in the water column, on substrate, etc. Descriptive or survey sampling is used to estimate parameters such as means, standard errors, or the total number of species per unit area, using a variety of techniques.

The most popular technique used in descriptive sampling is the simple random sample. For example, if one wished to sample the benthos of a section of stream, an artificial grid can be placed over the stream (a physical grid is not absolutely necessary, it can simply be estimated), and sampling locations may be determined by consulting a random numbers table. Any one of a number of schemes can be used to ensure that there is a one-to-one correspondence between the randomly-selected number and grid location. A specific sampling unit is usually not surveyed twice, as many sampling methods are destructive and subsequent sampling would yield biased estimates.

When dealing with large populations it is generally not important, when calculating confidence limits, to distinguish between sampling with and without replacement (i.e. returning the test "unit", with minimal disturbance, to its place of origin). However, as Eberhardt and Thomas (1991) suggest, two aspects of survey sampling merit more attention. Firstly, when dealing with small populations a finite population correction factor may be necessary in calculating variances. Secondly, sampling with replacement should be considered if measurements may change with time.

For convenience, researchers often sample using a systematic protocol. However, there is inherent danger in doing so if, by chance, periodicity in the sampling regime coincides with some natural repetition (thus biasing estimates). To limit potential error, the investigator should be sufficiently familiar with the system to ensure that the sampling regime does not coincide with any multiple periodicity in the population (Hansen *et al.*, 1953).

10.1.8   Sampling for pattern

Sampling for pattern originally emerged from the geological literature to assess the value of a body of ore or the expanse of an oil field. Often the data base in geological studies was haphazard (i.e. oil well test holes were drilled sporadically) and large gaps in data sets existed.

Sampling for pattern methods are useful in reducing biases resulting from haphazard sampling, thus enabling researchers to formulate data into easily interpretable summary maps. These methods may be applied to studies of aquatic insects to produce, for example, maps of species richness or distributional range using data sets collected over broad geographical regions. Allowing for "gaps" in data sets makes the technique particularly useful as, in many surveys, it is often difficult to sample in isolated regions.

Multivariate analyses

Multivariate analyses provide several primary functions, including the initial perusal of field data for hypothesis seeking, classifying or grouping data according to similarities, and hypothesis testing (Lehner, 1979). Aspey and Blakenship (1977) suggest that the utility of multivariate analyses is as diagnostic tools to: "(1) uncover homogeneous subgroups from naturally-selected heterogeneous samples; and (2) identify relationships among multiple variables when the underlying source, or biological basis, or individual variation is unknown." They caution that researchers should be careful to match the specific analysis and study groups to the question under investigation, and not the reverse. Additionally, although the assumption of a multivariate normal distribution in data applies to many multivariate analyses, researchers seldom report whether their data meet this criterion. If this assumption is not met transformation procedures should be used (Manly, 1986; Digby and Kempton, 1987).

Advances in both the power and "user-friendliness" of microcomputers and software have facilitated access by researchers to a variety of powerful multivariate techniques. Examples of some of the most widely used analyses include: (1) **R-Factor Analysis**, a grouping procedure which can be used to organize a large number of variables into a smaller number of factors based on their underlying similarities; (2) **Principal-Components Analysis**, which is used to separate units (e.g., species) of a sample into a few composite components. Principal Component 1 encompasses the greatest individual differences, while Principal Component 2 delineates the combination of variables, orthogonal to (not correlated with) Principal Component 1, that account for the largest proportion of remaining individual differences; (3) **Discriminant Function Analysis**, a procedure used to assess relationships, or differences, among pre-defined groups, or to place single characters into appropriate groups; and (4) **CANOCO** (CANOnical Community Ordination), a useful method for relating species composition to naturally occurring environmental variables, pollution gradients or management regimes. Data analysis with CANOCO may be either confirmatory or exploratory. In the exploratory mode it displays an ordination diagram of samples, species and environmental variables which show how community composition varies within a system. In the confirmatory mode it may be used to test the effects of various environmental factors on community composition. The utility of CANOCO was, for example, demonstrated by Williams, N.E. (1991), to reveal general trends and some regional and habitat-related differences in the assemblages of caddisfly larvae found in 25 springs across Canada.

10.2   Holism and reductionism

Definitions of "holism" and "reductionism" are somewhat nebulous. In a survey of 27 ecologists, Redfield (1988) found that most respondents viewed reductionism as a perspective in which the components of research are examined as the basis for explanation which, as Wiegert (1988) affirms, is the traditional view of reductionism. Holistic research was deemed to be that which explains phenomena in the context of the appropriate ecosystem (i.e. traditional, large scale explanations). Alternatively, Lidicker (1988) defined holism as, "looking out of the context of the system", and reductionism as, "looking into the system of interest". For purposes of discussing aquatic insects we will customize definitions so that small-scale manipulations (e.g. insects in field cages or aquaria) will be considered reductionistic, and large-scale manipulations (e.g. involving the bays of lakes or large reaches of streams) will be considered holistic.

There are costs and benefits associated with using both reductionistic and holistic approaches. Small scale manipulations in aquaria or cages usually yield unequivocal answers to specific questions since: (a) extraneous factors are significantly reduced; and (b) high replication is possible. However, by reducing systems an artificiality is imposed that can place doubt on the applicability of findings to the natural world. Although responses observed under controlled conditions probably do have a natural origin (i.e. no evolutionary pressure selects for animals to behave in a specified manner when surrounded by "laboratory glassware"; Hynes, 1970), it would be wrong to assume that the degree of expression would be equivalent in the field. For example, Baker (1980) found that initially, in relatively featureless aquaria, large nymphs of *Coenagrion resolutum* (Zygoptera) consistently displaced smaller conspecifics in contests for perches adjacent to feeding sites, yet subsequent field examinations yielded no evidence of size-dependent displacement (Baker, 1987, 1989). Aggressive interactions probably did function in the field yet the behaviour was masked in this more complex system. Alternatively, although animals behave more "typically" in natural settings, a multiplicity of 3rd and 4th order interactions often render interpretations impossible. Detailed knowledge of an insect's biology, by complementary holistic and reductionistic approaches, is the only way to comprehend such interactions effectively.

The tendency in the literature on aquatic insects is to discuss findings in the context of the species under examination. However, greater emphasis on taking the specific findings of a study and placing them in a broader conceptual framework would facilitate our understanding of aquatic insect ecology. This could be accomplished by conducting investigations using analogues which mimic types of systems, relationships or organisms, and discerning generally applicable principles. This requires that questions be: (1) holistically conceived; (2) reduced to manageable units for investigation; and (3) later transposed back into the holistic framework.

10.3 Perturbation experiments

A perturbation is the selective alteration of one or more members of a community which, as Bender *et al.* (1984) assert, may be achieved in two distinctly different ways. Firstly, there is a *pulse* perturbation, which is an instantaneous change in the density of insects, followed by a period of "relaxation", whereby a system may revert back to its previous state of equilibrium. Secondly, there is a *press* perturbation, which is a sustained alteration of the density of insects (often involving the elimination of species) that is maintained until a new level of equilibrium is reached. Since 1984, these definitions have been liberalized so that pulse often refers to any ephemeral manipulation of a system and press denotes sustained pressure.

An important difference between these two approaches pertains to the mechanism(s) by which change occurs. Pulse experiments generally yield information only on direct interactions, whereas press experiments can indicate both direct and indirect effects. Thus, following a pulse experiment, insect species B may increase in density immediately following, and as a direct result of, the removal of insect species A. In contrast, in a press experiment, the decline or removal of species A may, over time, affect the density of species B_1, which affects species B_2, ..., B_n, which affects species C (Yodzis, 1988). Thus, in the press experiment, species A never directly affected species C. When designing experiments it is important to consider these differences and, particularly, to be cautious when interpreting the results of press experiments.

Decisions regarding the degree to which systems should be pressed or pulsed are often difficult. If one is simply interested in whether or not the system is capable of responding to a particular perturbation, then a "sledgehammer" type of approach (*sensu* Schindler, 1987) may be used. However, it is generally not surprising, nor productive, to find that a species or system will respond to impacts that far exceed natural limits. Even if effects were measured using a sledgehammer approach, the procedure is usually repeated at a level within natural bounds. For example, to determine whether predaceous fishes could affect species richness of aquatic insects in lakes, it would not be surprising to find that, at 10 times ambient density, effects were significant. Even with this information the subsequent (and more relevant) question would be whether species richness is affected at high, yet natural, densities of predaceous fishes. Alternatively, a logical argument in favour of "sledgehammers" is that when replication is not possible (usually with large-scale perturbations), substantial and sustained manipulations may cause sufficiently large changes that they are convincing in the absence of statistical analyses (Hurlbert, 1984: Carpenter, 1990). In contrast to sledgehammer approaches are sensitivity studies which are used to determine whether systems will respond to relatively benign levels of perturbation. Sensitivity studies are most often used in toxicological studies to determine whether trace toxic substances might affect insects.

10.4   Spatial and temporal scale

The results of any study may vary as an idiosyncratic function of spatial and/or temporal scale. Yet, in studies with aquatic insects, the duration and scale of experiments are often not rationalized.

If an experiment is of "short" duration, there may not be time for a response to occur, thus increasing the probability of committing a Type II error. Alternatively, if an organism or system is perturbed disproportionately longer than would occur naturally, there is an increased risk of making a Type I error. The potential for error may be minimized by adjusting the time frame of a manipulation to a specific natural event. For example, the response of an insect to a perturbation may be directly related to its size, which is correlated with ontogenetic changes in growth. If, for pragmatic reasons, this cannot be done, precautionary statements regarding the interpretation of results should be made.

Most field experiments on aquatic insects are performed on small spatial scales (e.g. in cages). This was evident from a survey of more than 250 journal papers on predation experiments, in which lentic and lotic systems, and cage vs. large-scale (whole lake, pond, pool, divided lakes and ponds, and sections of stream) manipulations, were considered separately (Cooper *et al.*, 1990). The number of cage vs. large-scale manipulations was, respectively, 14 and 5 for streams, and 32 and 13 for lakes and ponds. Although the survey focused on predation experiments, cage-type manipulations are used in many other types of study (e.g. competition experiments). Despite their popularity, cage experiments dependent on immigration and emigration of insects are subject to a potential flaw: the movement of insects into and out of cages will vary as a function of the perimeter:area (P:A) ratio. For example, if a cage is 10 x 10 cm square, the P:A ratio would be 2:5, whereas if the perimeter is doubled to 20 x 20 cm, the P:A ratio would be 1:5. Since perimeter and area increase as linear and squared functions respectively, the potential to affect a change in the total number of insects in a cage decreases as perimeter increases, thus imposing an artefact on emigration and immigration studies (e.g. Culp, 1986; Reice and Edwards, 1986; Hart, 1987).

Inspection or premanipulation studies should be used to establish that the range of natural conditions that may affect an insect's behaviour can be encompassed within the space of an experimental unit. An effective way to help ensure that the spatial environment is sufficiently realistic to "satisfy" a test insect is to construct ethograms to check for behaviour indicative of an "unstressed" animal. For example, if an insect was observed feeding shortly after release into a cage or aquarium, this would suggest that the scale and "naturalness" of the design was sufficient to elicit typical responses. Precautionary checks such as this add an air of confidence when extrapolating findings to nature.

10.5   Sampling techniques

To ensure that the goals of an experiment are realized it is imperative to select, and use properly, appropriate sampling equipment, of which an arsenal has been designed to sample aquatic insects (see Elliott and Tullett, 1978). Indeed, as Cummins (1962) commented, the number of aquatic insect sampling devices roughly reflects the number of researchers. In this section we will focus on common types of equipment used to sample immature (mostly aquatic) and adult (mostly aerial) insects in a variety of aquatic habitats.

Sampling immature insects in small streams and ponds Equipment to sample aquatic insects typically can be used in both standing and flowing water. Since many sampling devices must be hand-held, water depth is usually the factor limiting use.

The most commonly used device for sampling aquatic insects is the D-frame net (Fig. 10.3A). Insects can be sampled in lakes or ponds by forcing the net through vegetation or surface layers of substrate. In streams, insects may be sampled by holding the D-net or hand screen collector (Fig. 10.3B) downstream while the substrate immediately upstream is agitated by foot. Both of these

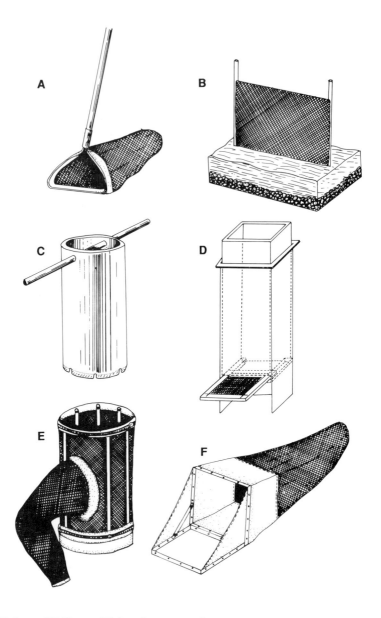

Fig. 10.3   (A) D-net; (B) hand screen collector; (C) Wilding sampler (Wilding, 1940); (D) Gerking sampler (Gerking, 1957); (E) Hess sampler (Hess, 1941); (F) Surber sampler (Surber, 1937).

techniques are typically used for qualitative, descriptive purposes but, when applied carefully, may be used to obtain semi-quantitative data.

To sample littoral insects (including the macrophytes upon which they may sit) either a Wilding (Fig. 10.3C) or a Gerking (Fig. 10.3D) sampler may be used. The Gerking sampler is more effective in sampling deeply rooted vegetation, as the edge of the frame encasing the bottom screen can be sharpened to cut stems. The difficulty with using either of these devices is that the action of placing them over vegetation causes insects to disperse. Thus, density estimates based on these methods are usually conservative.

The Surber sampler (Fig. 10.3E) is used to sample stream benthos quantitatively. The device is placed with the opening facing upstream, while substrate immediately in front of the opening is agitated, thus dislodging materials which float downstream into the net. Larger substrate, coated with algae, calcium carbonate, etc., should be held in the net opening and scrubbed with a brush to dislodge attached insects. A limitation of the Surber sampler can result from back-pressure in the net, caused by clogging of the mesh openings, which sometimes causes material to escape. The Hess sampler (Fig. 10.3F), which can be used anywhere a Surber sampler can, addresses this problem because mesh completely surrounds the sample unit. Hess samplers can be used also in standing water by disturbing the benthos and fanning suspended material into the collecting sock.

Drift nets (Fig. 10.4A) can be placed at various depths in streams (using stakes or floats) to measure the downstream movement of aquatic insects. As a precaution, dye may be released upstream of the nets, both before and after trials, to ensure that water flow through the nets is adequate. For more rigorous, quantitative measures of drift, a current meter may be used to calculate the discharge of water through the net so that an appropriate correction factor can be applied. Drift nets usually require greater night-time maintenance as animals drift more at night than during the day (Anderson, 1966).

The term *hyporheos* refers to the fauna which occurs within the substrate of streams in a zone bordered by the epigean water of the stream above and by the true groundwater below (Orghidan, 1959; Williams and Hynes, 1974). Although it has long been established that hyporheic animals can occur at high densities (e.g. Racovitza, 1907), quantitative studies are still comparatively rare. Paucity of information on the hyporheos is undoubtedly the result of the physical difficulty in sampling this zone, despite many recent sampling innovations. One device which may be used to sample the hyporheos is the freeze corer (Fig. 10.4B). Basically this apparatus functions by placing a protective cap (A) on top of the standpipe (B) and then pounding it into the substrate using the weighted sleeve (C). The sleeve is then removed, and liquid nitrogen is passed down the inside of the standpipe by means of a flexible tube. Moving the tip of the tube, periodically, prevents the coolant from building up in the bottom of the standpipe which results in a more even core (Pugsley and Hynes, 1983). After approximately 20 minutes, the surrounding substrate will have frozen to

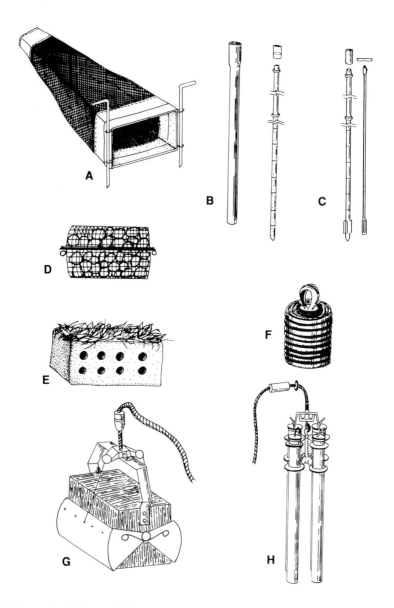

Fig. 10.4   (A) Drift net (Elliott, 1970); (B) Freeze core sampler (Stocker and Williams, 1972); (C) Standpipe corer (Williams and Hynes, 1974); (D) Substrate sampler; (E) Leaf pack; (F) Multi-plate sampler; (G) Ekman Grab (Ekman, 1911); (H) Multi-core sampler (Flannagan, 1970).

the standpipe. The standpipe plus the substrate sample can then be removed using a winch. The core is then placed horizontally on plastic sheeting, wrapped in ice, and moved to the laboratory for analysis. Initial problems of interstitial insects moving away from the standpipe ahead of the cold-front can be overcome by applying a pre-freezing electrical shock to the sediment to kill insects *in situ* (Bretschko, 1985). Chironomids are the most common component of the hyporheos, with most occurring < 1 metre below the stream bed surface. A less cumbersome technique for sampling sediments is the standpipe corer (Fig. 10.4C), although this is limited in the amount of sediment that can be removed at any one time. Repeated sampling, however, can reveal accurate distribution patterns of interstitial organisms (Williams, 1989).

Rather than sampling insects directly, substrate-filled trays, leaf packs, or multi-plate samplers (Fig. 10.4 D,E,F) may be used to make collections. However, seemingly irrelevant aspects of design can significantly affect the results obtained using any of these methods. For example, if a substrate-filled tray is devoid of food, this will usually affect colonization. Similarly, leaf packs should be sufficiently conditioned by microorganisms to be attractive to insects. Again, if the interstitial space between plates of a multiplate sampler is too narrow, this may prevent larger species from colonizing.

Sampling immature insects in large rivers and lakes Many of the methods, described above, to sample aquatic insects in small lakes and streams can also be used to sample along the shorelines of large streams and lakes. However, to sample in open water, beyond the limits of waders, different equipment is required.

An Ekman grab (Fig. 10.4G) is a semi-quantitative device used to sample the benthos of soft-substrate regions of lakes or river pools. Once it is lowered onto the substrate surface, a spring-loaded trigger is released thus forcing two jaws to instantly close and "grab" a sample of substrate. The device can be operated from a boat using a rope attachment or, in shallow water, extension rods may be used. Using the Ekman grab in inshore areas can be frustrating as logs, sticks or rocks often prevent the jaws from closing.

Another device used to sample the benthos of lakes with soft sediments is a multi-core sampler (Fig. 10.4H), which can be lowered over the side of a boat and, under its own weight, will sink into the sediment. A weighted messenger, which travels down the rope, is used to close caps over the ends of the tubes before the unit is retrieved.

An air-lift sampler (Fig. 10.5A) may be used to sample insects in lakes, ponds, streams, or shifting sand rivers (Hogg and Norris, 1991). The system involves placing the large cylindrical end onto the substrate and releasing a flow of air within it; this creates an upward flow of animals and detritus which collect in a mesh cup. The sample generally requires little sorting as most of the fine detrital material escapes through the mesh. A disadvantage is that the system requires a source of compressed air, such as a Scuba tank.

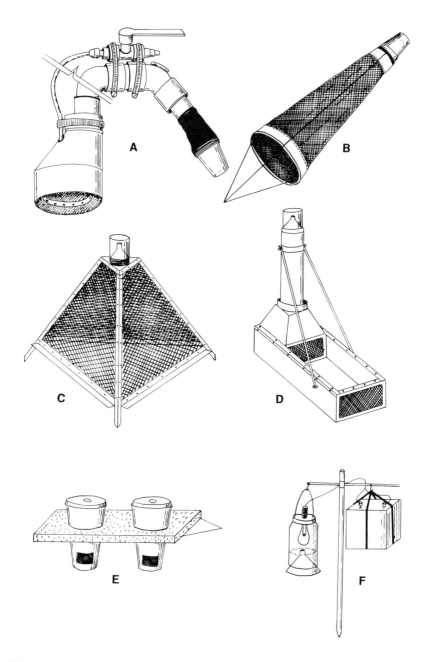

Fig. 10.5   (A) Air-lift sampler (Hogg and Norris, 1991); (B) Plankton net (Elliott, 1970); (C) Mundie pyramid sampler (Mundie, 1956); (D) Hamilton sampler (Hamilton, 1969); (E) Edmunds sampler (Edmunds *et al.*, 1976); (F) Subaquatic light trap.

Species found in the water column (nekton) of lakes (e.g. *Chaoborus*) can be collected by towing a plankton net (Fig. 10.5B) behind a boat. As with drift nets, if quantitative estimates of density are required a calculation of net efficiency must be made. Since many nektonic species exhibit diel vertical migrations (Beeton, 1960: Wells, 1960), time of day will affect density estimates. Quantitative estimates of nektonic insects also may be made using large versions of standard plankton samplers, such as the Schindler trap (Schindler, 1969).

Sampling adult insects Two types of trap are most often used to collect adult insects in streams: the Mundie pyramid trap (Fig. 10.5C) and the Hamilton box trap (Fig. 10.5D). When aquatic insects emerge into their aerial form they instinctively crawl or fly upwards. Thus, when either of these cages is placed over a section of stream emerging insects travel upwards into collecting jars.

Many types of emergence trap can be floated on the surface of lakes or ponds. A simple design by Edmunds *et al.* (1976) is based on suspending cups from a styrofoam board (Fig. 10.5E). Each cup is open on the bottom, allowing emerging insects access, while the upper half of the cup rises above the surface of the water and is enclosed with mesh. This system is attractive as it can be affordably made.

Since many insects emerge at night and are positively phototropic (Chapman, 1982), researchers have designed traps that use light as an attractant. The subaquatic light trap (Fig. 10.5F) is one of several designs in which a battery is used to power a low voltage light bulb placed within a collecting vessel. Although these systems generally work efficiently the battery makes them bulky and expensive to replicate.

11 RELATIONSHIPS WITH MAN

11.1 Pollution

The effects of anthropogenic stress on aquatic and terrestrial ecosystems drew greater world attention during the 1980s, largely because of the tangible manifestation of these effects on human sensibilities (e.g. acid deposition diminishing sport fishing in northeastern North America and western and central Europe; Falk and Dunson, 1977; Okland and Okland, 1980). International recognition of environmental problems was also enhanced by the United Nations report on sustainable development - i.e. *Our Common Future* (1987). Early accounts of the effects of pollution on aquatic systems and insects, aimed at both the public and scientific communities, were made formal by Carson (1962) and Hynes (1960), respectively.

Pollution is a semi-nebulous term used to describe changes in the physical, chemical or biological characteristics of water, air or soil, that can affect the health, survival, or activities of living entities (Miller, 1988). Organisms respond to pollution usually in one of two ways, acutely or chronically. Acute effects result in serious injury to, or death of, the organism shortly after exposure to high concentrations of a pollutant. Chronic effects are realized following exposure to low concentrations of a pollutant, the results of which appear over time, often as serious diseases (e.g. cancers). In this section we will examine major types of pollutants that affect aquatic insect population and community dynamics.

11.1.1 Physical disturbances

Urban land development Two major effects of urbanization on aquatic systems and insects are (a) increases in sediment loads during construction phases (Guy and Ferguson, 1962), and (b) post-storm increases in the discharge of streams and rivers downstream from developments (Cordery, 1976). During construction, urban developers routinely remove native vegetation, disrupting the biological balance within the soil and thereby accelerating natural erosion processes. Natural drainage channels are frequently eliminated during construction and this, combined with the relentless movement of machinery over bare sub-soil, also accelerates erosion. As machinery compacts lower layers of soil the infiltration capacity of water is restricted. Thus, during each rainfall, an upper layer of soil, which is loosely aggregated because of the movement of machinery, is easily eroded by a higher volume of "runoff" into streams or the inshore areas of nearby lakes. Once an urban centre has been developed, flood

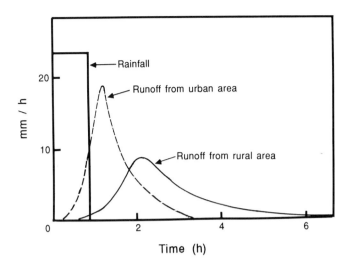

Fig. 11.1   Comparison of runoff from similar urban and rural catchments (redrawn from Cordery, 1976).

peaks in streams and rivers may increase 2-4 times in comparison with pre-urbanization discharges (Fig. 11.1). These increases are due to the paving and roofing-over of large areas (usually 15-60% in suburban areas and up to 100% in business or industrial centres; James, 1965), which reduces the amount of rain that infiltrates the soil while also increasing the amount of overland flow. As streams enlarge to accommodate higher discharges, sediment loads are further enhanced by bankside erosion.

Higher levels of sedimentation can affect aquatic insects by altering biochemical conditions, food resources, respiratory diffusion gradients, and habitable space (Scott, 1966; Brusven and Prather, 1974; McClelland and Brusven, 1980). Indeed, as Cummins (1966) suggested, there is perhaps no other physical factor that influences aquatic insects more than substrate composition. McClelland and Brusven (1980) found that, in a laboratory stream, several species of mayfly, stonefly and caddisfly, when given a choice between sedimented and unsedimented regions, all selected unsedimented substrate. Avoidance of sedimented regions was due to the loss of interstitial space between stones, and behavioural observations revealed that the insects would not excavate fine particles.

Several field studies have considered the effects of urban development and increased levels of sedimentation on aquatic insect community structure. For example, in a 3rd order stream which flows through the city of Edmonton (population, ca. 500,000), Alberta, Whiting and Clifford (1983) found that many aquatic insects which were common upstream of the city were rare or absent within Edmonton, whereas, for other species (e.g. chironomids and tubificid worms), the reverse occurred. Although the density of insects was higher within the city, diversity and richness (number of species and individuals) of the fauna was much lower. Changes in the urban insect community were attributed to higher silt loads and organic runoff through sewers. Routine sampling of the stream did not show significant differences in chemical characteristics between urban and non-urban sites, probably because of the sporadic nature of storm sewer runoff. Aquatic insects, therefore, exhibited a historical response to discharges within the city and served as better indicators of water quality than could be detected by routine chemical monitoring. A similar study of insect communities above and below an urban centre was conducted by Garie and McIntosh (1986) in a New Jersey stream. They found that the number of taxa above the urban area was significantly greater than occurred downstream. Concentrations of heavy metals (lead, chromium, nickel, copper, zinc, cadmium) at downstream sites were significantly higher than upstream, again attributable to the effects of urban runoff. In the Murrumbidgee River, Australia, Hogg and Norris (1991) studied aquatic insect assemblages above and below the point of confluence with Tuggeranong Creek, an urban stream. Monitoring indicated high concentrations of suspended solids in the water column of Tuggeranong Creek during storms (i.e. a maximum of 560 mg/l) which ultimately resulted in high concentrations of fine organic material (< 250 μm) in downstream reaches of the Murrumbidgee River. As predicted, the number of taxa and density of aquatic insects decreased significantly downstream, catalysed by the cascading effects of urban development and subsequent changes in substrate composition. In each of the above studies, reductions in densities of aquatic insects in areas of stream exposed to heavy siltation were related to increases in catastrophic and behavioural drift (Rosenberg and Wiens, 1978).

Motorvehicles impose an additional urban-related stress on aquatic insects (Cordery, 1976). Angino *et al.* (1972) demonstrated higher values of COD (chemical oxygen demand) in urban surface waters, which they suggested were due to hydrocarbons leaked from automobiles (i.e. oil and petrol). As runoff enters streams and lakes, higher COD levels will deplete the available oxygen supply, thus stressing many species of insect. Urban runoff also contains significant quantities of lead, which Bryan (1972) related to the proportion of catchment area allotted to motor vehicles and to the density of traffic.

Forestry practices The impact of logging on aquatic insects is related primarily to two activities, road construction and tree felling. Studies which have focused on road construction have demonstrated effects that are very similar to those seen during urbanization (i.e. increasing sedimentation rates reduce insect richness; e.g. Richardson, 1985). Extence (1978) suggested that these effects are sufficiently damaging that a concerted effort should be made to develop a code of environmental ethics for logging and erosion control.

In logged areas, where roads do not run close to streams (i.e. where sediment loading is not excessive), increases in both the number of taxa and the density of aquatic insects have been observed (Silsbee and Larson, 1983; Wallace and Gurtz, 1986). This occurs because after trees are felled, higher levels of sunlight, temperature and nutrient supply result in increases in stream primary productivity (Likens *et al.*, 1970). In response, insects with short generation times and rapid colonization rates (Newbold *et al.*, 1980) are quick to use this rich food supply and their numbers therefore increase. Densities of less opportunistic species (those with long generation times and/or slow colonization rates) tend to decrease.

To curtail the impact of logging on aquatic insects Newbold *et al.* (1980) suggested that a 30 m wide buffer strip be left along shorelines. Using several comparisons, these authors detected no significant differences in macro-invertebrate communities in streams with wide buffers (> 30 m), as compared with controlled (unlogged) areas. However, in logged areas, the diversity of aquatic insects in wide-buffered sections of stream was significantly higher than in sections where no buffer strips were present.

Dams and reservoirs   Anthropogenic control over the flow of running water, usually by means of dams and reservoirs, has influenced nearly all of the world's major river systems (Ward and Stanford, 1979). Among the very first large reservoirs to be constructed were the Gorkee and Kuybyshev on the Volga River, U.S.S.R. (now the Commonwealth of Independent States) (Baxter, 1977). These reservoirs were filled in late autumn and by spring most resident insect species had disappeared, until July, when *Chironomus plumosus* appeared. This opportunistic species became so abundant that during its emergence period it interfered with navigation (Zhadin and Gerd, 1963). Increases in other motile aquatic arthropods and the freshwater mussel, *Dreissena* sp., followed.

Chironomids are ideally suited to colonize newly-formed reservoirs as they are aerially dispersed in both the immature (on the feet of waterfowl) and winged adult stages (Davies, 1976), many species have a high reproductive capacity (Baxter, 1977), and the larvae can endure the low oxygen concentrations that characterize these systems. Numerous studies demonstrate the adaptive capacity of chironomids to survive in newly-formed reservoirs. For example, following the building of a beaver dam on a stream in Algonquin Park, Ontario, Sprules (1940) found that large numbers of mayflies, caddisflies and stoneflies disappeared almost immediately but were replaced by high densities of midges. Paterson and Fernando (1969) similarly found that, following the formation of an impoundment, chironomid and oligochaete numbers increased significantly, while other species showed little or no response. Within two years of construction of a reservoir on the Kananaskis River, Alberta, chironomids formed the vast majority of the invertebrate biomass (Nursall, 1952).

The physical characteristics of streams below dams differ according to whether the dam is of a surface-release or deep-release type (Table 11.1). However, the response of insects to either set of characteristics is generally a

Table 11.1   Some generalized qualitative responses of downstream reaches
              to stream regulation (relative to comparable unregulated streams)
              (after Ward, 1982).

	Receiving stream	
Variable	Surface-release	Deep-release
Turbidity	clarification	clarification
Nutrients	reduced	reduced less
Plankton	enhanced	present
Detritus	reduced	reduced
Discharge	variable	variable
Temperature range	increased	decreased
Dissolved oxygen	saturated	anoxic to supersaturated

reduction in species diversity and a significant alteration in taxonomic composition (Ward, 1982). Entire groups of insects (e.g. stoneflies, heptageniid mayflies) are often absent immediately below dams.

Changes in the thermal regime below deep-release dams (which release cold hypolimnetic water) can significantly alter aquatic insect community structure under conditions that are otherwise favourable (Ward, 1976; Gore, 1977). A notable example of such a thermal effect on insects was demonstrated by Lehmkuhl (1972) while working on the South Saskatchewan River. He concluded that 15 species of aquatic insect (mostly mayflies) were absent below a dam, and that this was due to either an absence of the required temperature sequence for breaking diapause and stimulating egg hatching and growth, or insufficient degree days (especially in spring) needed to promote adult emergence. A suboptimal thermal regime may affect reductions in density by: (1) reducing growth efficiency and/or fecundity; (2) shifting predator - prey relationships; or (3) creating an environment for competitive exclusion (Ward, 1976). In contrast to deep-release dams, shallow or surface-release reservoirs may elevate downstream temperature profiles to the point that indigenous cold water species are threatened (Fraley, 1979; Kondratieff and Voshell, 1981).

Where the flow of water below deep-release (Ward, 1976) and surface-release (Ward and Short, 1978) reservoirs is fairly constant, extremely high densities of aquatic insects have been recorded. In contrast, severe fluctuations in flow can deplete the insect fauna, directly or indirectly, by altering the food base or by reducing habitat diversity (Ward, 1982; Gray and Ward, 1982). A fluctuating flow regime can also facilitate drift in insects (Brooker and Hemsworth, 1978), the effects of which can be particularly severe in streams with few tributaries to supply replacement animals.

A common phenomenon below surface-release impoundments and natural lakes is ultra-high densities of filter feeders such as net-spinning caddisflies and black flies. Cushing (1963) attributed these differences to richer supplies of epilimnetic food particles.

11.1.2 Chemical disturbances

Pesticides Immediately following the publication *Silent Spring* (Carson, 1962), the public outcry over the use of pesticides was sufficiently great that even the Kennedy White House responded. From that moment on, a controversy was launched between environmentalists and pesticide industry officials that still rages (Miller, 1988).

The ideal pest control agent would: (1) kill only target species; (2) have no long- or short-term effects on non-target species; (3) break down into harmless chemicals in a short period of time; (4) not select for genetic resistance in the target organisms; (5) not affect predator/prey relationships or competitive interactions; and (6) be more economical than not using pest control (Miller, 1988). Unfortunately, no such pesticide exists. A well known example of a control agent initially thought to be ideal, and later discovered not to be so, is DDT (dichlorodiphenyltrichloroethane). By means of a circuitous path, aquatic insects can link DDT to the death of higher vertebrates.

DDT was discovered in 1874. When its potency as a pesticide was shown in 1939 it was used extensively; DDT was cheap to produce, it killed many types of pest, and its effects persisted over time because it is highly stable. The World Health Organization (WHO) advocated using DDT to control many insect-transmitted diseases, such as malaria carried by *Anopheles* mosquitoes (Miller, 1988). Yet, by 1985, WHO reported that 51 of 60 malaria-carrying mosquito species had developed genetic resistance to DDT. Furthermore, it was discovered that declines in populations of predatory birds (e.g. osprey, bald eagle, peregrine falcon) during the 1950s and 1960s were directly attributable to DDT (it was concentrated, in an upward path through the food web, via "biomagnification"). Basically, DDT was incorporated into the bodies of aquatic insects and zooplankton as they fed and absorbed elements indirectly through their cuticle. As larger insects consumed smaller ones, they effectively concentrated DDT in their tissues. Many fishes preyed on the larger insects (further concentrating the DDT) and, in turn, birds preyed on the fishes. Concentrated levels of DDT in birds has been shown to break down to form DDE (1, 1-dichloro-2, 2-bis p-chlorophenyl ethylene), which reduces calcium deposition in the shells of eggs, which are then inadvertently crushed by the parent birds. The behaviour of DDT clearly illustrates how a chemical can transmit its effects across the aquatic/terrestrial boundary. DDT was banned in the United States in 1972, yet it is still used extensively in Less Developed Countries (LDCs) around the world - undoubtedly because its persistent nature effectively controls adult mosquitoes in houses.

Extensive aerial spraying has been used to control forest insects such as the spruce budworm (*Choristoneura fumiferana*), eastern tent caterpillar (*Malacosoma americanum*), and gypsy moth (*Porthetria dispar*). However, these spraying programmes can significantly impact non-target species of aquatic insect, as is often indicated by significant increases in the rate of stream drift following an aerial spray. For example, in a series of small woodland streams in Maine, Courtemanch and Gibbs (1980) found that in streams sprayed with carbaryl (Sevin-4-oil; a spruce budworm and gypsy moth control agent) for the first time, drift rates increased markedly (170 times) relative to untreated streams. Subsequent decline in the benthic densities of stoneflies, mayflies and caddisflies were recorded. Similarly, Eidt (1975) found that fenitrothion (another spruce budworm spray) applied aerially to woodland streams resulted in significant increases in the drift of stoneflies and mayflies. Friesen *et al.* (1983) determined that permethrin (an agricultural/horticultural spray), when applied aerially, contaminated sediment and was toxic to the burrowing mayfly *Hexagenia rigida*. Following abatement, the effects of sprays or their derivatives can be detected within communities for one or more years (Courtemanch and Gibbs, 1980).

Attempts are often made to control aerial "pests" which have an aquatic stage while these pests are still water-bound (e.g. black flies). The chemical most often used to kill larval black flies is methoxychlor (Fredeen *et al.*, 1975; Wallace *et al.*, 1976). Despite its popularity, methoxychlor also kills many non-target aquatic insects and, even in very small doses, its effects can be observed over incredible distances downstream. For example, a 15 minute injection of 0.3 ppm methoxychlor into the Athabasca River, Alberta, caused catastrophic drift in insects over a distance of 400 km. Four weeks after treatment, recolonization by non-target invertebrates was minimal.

An interesting paper by Muirhead-Thomson (1973) focused on the differential effects of pesticides on a variety of aquatic insects, and considered the potential effects such responses might have on predator/prey interactions. Tests were performed in the laboratory as field tests were thought to be too difficult to interpret. Releasing concentrations of 20 ppm of DDT for 1 hour had virtually no effect on dragonfly nymphs and *Nepa* sp., whereas *Baetis* nymphs and *Simulium* larvae suffered 100% mortality. In contrast, Lebaycid (fenthion) proved lethal to dragonfly nymphs, *Baetis* and *Simulium*. Indeed, dragonfly nymphs were even more susceptible to Lebaycid than were *Hydropsyche* sp. larvae. Muirhead-Thomson hypothesized that these differential responses may facilitate the release of prey by predators (e.g. as a result of dragonfly nymphs being killed by Lebaycid - thus facilitating increases in prey density), or that facultative predators, such as *Hydropsyche* sp., may assume more prominent roles. There is a marked paucity of studies that have examined how changes in biotic interactions (predator/prey, competition, etc.) might affect aquatic insect assemblages following exposure to insecticidal sprays.

Industrial pollutants Most types of pollutant discussed thus far have been from non-point sources. In contrast, industrial pollutants are generally point source in origin as they are usually discharged through pipes, ditches and sewers into bodies of water, at specific locations. Upon entering water, the chemical nature and concentration of pollutants will usually change as a result of four natural processes: dilution, biodegradation, biological amplification, and sedimentation. The rates at which these processes occur (particularly dilution and the oxygen-consuming process of biodegradation) vary directly as a function of the turn-over time of water in a system. When streams or lakes are overloaded with contaminants, and sediments become anaerobic and/or laden with heavy metals, the impact on aquatic insect communities can be severe.

Water bodies that suffer from industrial pollution are generally characterized by high densities of chironomids and an absence of mayflies and stoneflies. Winner *et al.* (1980) found that aquatic insects, in a small stream in Ohio, responded drastically to heavy metals (Cu, Cr, Zn) downstream of a metal-plating plant which had discharged waste into the stream for 8 years. In the most heavily contaminated sections of stream, chironomids and tubificid worms were virtually the only taxa to survive. At collecting stations near the source of the pollution, midges constituted 86% of all insects collected, whereas, at less polluted downstream sites, they made up only 10% of the insects. Caddisflies increased in density at downstream sites where the pollution was of intermediate severity, whereas mayflies occurred only where pollutants were immeasurable. As there appeared to be a direct relationship between the fraction of the community composed of chironomids and the degree of pollution, Winner *et al.* suggested that the percentage of chironomids in samples may be a useful index of heavy-metal pollution (see section 11.1.3). In the River Cynon, a trout stream in south-east Wales, Learner *et al.* (1971) examined aquatic insect assemblages above and below a point source of sewage and industrial effluent. Again, the upstream faunal assemblage was found to be quite diverse, whereas chironomids and oligochaetes dominated downstream reaches. Saether (1980) similarly recorded significant increases in the number of midge larvae while examining the effects of eutrophication on the benthic invertebrates of deep lakes.

The time required for insect assemblages to return to their natural state, following disturbances such as those described above, can be on the order of many years for streams, and decades for lakes (Miller, 1988). For example, constant monitoring of an aquatic community following the closure of a sulphite pulp mill showed that only after 8 years had passed did the community approximate that which had occurred 40 years earlier (Rosenberg, 1976).

Mechanisms that aid chironomids in withstanding the effects of pollution include a higher oxygen storage capacity because of the presence of haemoglobin (e.g. in *Chironomus anthracinus* and *C. plumosus*; Nagell and Landahl, 1978), and an ability to avoid heavy metals by burrowing into the sediment (Wentsel *et al.*, 1977; Rosenberg and Wiens, 1980). Other means of limiting the impact of pollutants include: (1) body and gill movements to enhance oxygen uptake (as in perlid stoneflies and ephemerellid mayflies; Gaufin *et al.*, 1974); (2) breathing at the water surface by means of tracheal tubes (various Hemiptera; Gaufin, 1973); (3) adjustment of life-cycle to avoid periods of pollution stress (e.g. as in *Ephemerella ignata*; Sodergren, 1976); and (4) having generation times short enough to avoid stressful periods (e.g. as in *Baetis* and *Nemoura*; Newbold *et al.*, 1980).

Chronic responses to pollutants are often recorded. Donald (1980) recorded deformities in the antennae, maxillae, labium and cerci of capniid stoneflies downstream from industrial and domestic sewage outfalls. In streams receiving chlorinated wastes, Simpson (1980) reported deformities in the tracheal gill development of both stoneflies and caddisflies. Anomalies in the structure of hydropsychid capture nets have been recorded in areas of streams receiving heavy metal and toxic chemical wastes (Fig. 11.2). The frequency of net anomalies is negatively correlated with distance from the source of pollution, suggesting that anomalous structures may be reasonable indicators of pollution (particularly considering the ubiquitous distribution of hydropsychids: see section 11.1.3).

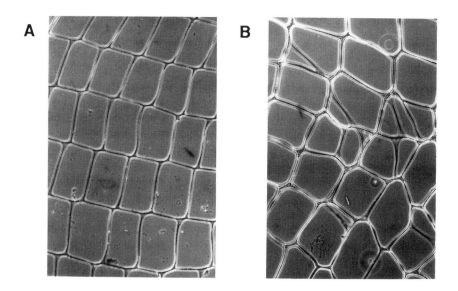

Fig. 11.2   Photomicrographs of *Hydropsyche* nets at 100X showing: (A) normal mesh structure and (B) an abnormal net with structural anomalies (from Petersen and Petersen, 1983).

Oil spills   Although concerted efforts have been focused on examining the ecological consequences of oil spills in marine systems, only limited consideration has been given to the accidental release of oil in freshwater habitats

(Rosenberg and Wiens, 1976). Further exacerbating the threat of accidental oil spills to the ecosystem is the deliberate release of massive amounts of oil as a means of environmental terrorism or as an act of aggression during war (e.g. Persian Gulf War, 1991). Thus, the need to understand the manner in which oil spills affect freshwater biota is ever greater.

Barton and Wallace (1979a) found that populations of various stoneflies and mayflies were significantly depleted in a section of river that flowed through a natural deposit of tar sand. The density of insects on the smooth, featureless tar substratum was about half of that observed on more heterogeneous, natural substrates. In a subsequent study, Barton and Wallace (1979b) poured tailings sludge into a river and observed a rapid decline (60%) in the standing stock of aquatic insects in the immediate vicinity of the spill. Particularly sensitive species (*Ephemerella* spp., *Baetis* spp., *Hastaperla brevis*, *Taeniopteryx nivalis*, *Hydropsyche* sp., *Corynoneura* sp.) remained suppressed for at least 4 weeks. The factors which affected species were a combination of fine silt, sticky oils, and the direct action of toxic compounds (Rosenberg, 1986).

One group of insects, chironomids of the subfamily Orthocladiinae, has been shown to respond positively to oiled vs. non-oiled artificial substrates (Rosenberg and Wiens, 1976). Apparently the oil stimulates algal growth which attracts these largely algivorous larvae.

Mine waste The harmful effects of mine waste on aquatic insects vary according to the type of mineral extracted, the size of the operation, the type of mine (surface or subsurface, hard or soft rock), and the local topography and climate. Generally, subsurface mining is less damaging to aquatic systems as, for each unit of mineral extracted, only one-tenth as much land is disturbed as would occur by extracting the equivalent unit from a surface mine (Miller, 1988). The factors that affect aquatic insects most severely are the release of toxic (mostly heavy metals) substances, increased silt loads, and higher levels of acidity.

Most methods to assess the impact of mine effluents on aquatic biota have been based on upstream/downstream comparisons. This approach was used by Norris (1986) to examine the impact of particulate matter and mine workings (sources of metal contamination) on aquatic insects in the Molonglo River, Australia. He found that the variety and numbers of insect taxa at downstream sites were significantly lower than upstream, with one exception, leptocerid caddisflies. This result reflected previous findings of Weatherley *et al.* (1967), suggesting that leptocerids are somehow resistant to trace metal contaminations. Norris *et al.* (1982) performed a similar study on a 170 km section of the South Esk River, Tasmania, where the effects of waste on aquatic insects were detected over a distance of 80 km. Hierarchical and non-hierarchical classification systems were then used to define three groups of aquatic insect based on their responses to mining effluents: (1) taxa that were abundant at both contaminated and uncontaminated sites (a leptocerid caddisfly and a baetid mayfly); (2) taxa that were most abundant at sites upstream from the contaminant (four species of leptophlebiid mayfly and five species of caddisfly); and (3) taxa whose numbers were highest at sites downstream from the contaminant (six dipteran species and four species of caddisfly). These findings are not surprising, as dipterans (mostly chironomids), and many species of caddisfly, are known to withstand a wide array of environmental stresses.

The effects of pollutants from coal mines on the aquatic fauna of a small stream in South Wales revealed that higher levels of acid discharge in the upper reaches of the stream eliminated most insects (Scullion and Edwards, 1980). The insects most vulnerable were mayflies, stoneflies, beetles and some caddisflies, while chironomids responded positively and dominated the community.

A comparison of lentic insects between ponds affected by coal strip-mining and a non-affected control pond revealed that the diversity of insects was least in spoil ponds, probably due to higher nitrate and sulphate levels (Canton and Ward, 1981). In contrast, the density of insects was highest in the contaminated ponds, which was attributable, again, to high numbers of chironomids.

The recovery of insect populations following cessation of mining activities, even when combined with terrestrial restoration programmes (e.g. planting vegetation), is very slow. Matter and Ney (1981) compared the benthic invertebrate and fish communities, between streams which had been undisturbed by mining actions for approximately seven years, and streams which had never been affected. They found that even several years later, alkalinity, hardness, sulphate, conductivity and fine-particle suspended solids in the mining (coal) streams were significantly elevated, and that this was reflected by depauperate insect and fish assemblages. These authors emphasized that terrestrial reclamation programmes do not assure aquatic restoration, and they recommended that water quality criteria be given greater consideration in reclamation procedures for mined lands.

Road salt   In many cold weather countries, road salts ($NaCl$, $CaCl_2$, and KCl) are regularly applied in an attempt to de-ice motorways. The fate of these salts is to enter surface water and groundwater supplies, where they have both a proximal impact on stream- and lake-dwelling insects and a distal impact on those species found in groundwater outflows.

The lethal effects of sodium and potassium chloride solutions on two benthic insects common in Lake Michigan (the caddisfly, *Hydroptila angusta*, and the midge, *Cricotopus trifascia*) were clearly demonstrated by Hamilton *et al.* (1975). They found that $NaCl$, ranging from 3,735-10,136 mg/l, and KCl, ranging from 204-6,713 mg/l, resulted in 100% mortality of both species within 48 hours. Despite the fact that these concentrations greatly exceeded levels in Lake Michigan (affirmed by Hoffman *et al.*, 1981), the physiological impact of lower concentrations was still thought to diminish species fitness.

The influence of sodium chloride on the rate of drift of the caddisflies *Hydropsyche betteni* and *Cheumatopsyche analis* was examined by Crowther and

Hynes (1977). In the field, salt concentrations that exceeded 1000 mg/l induced significant increases in drift and, as maximum Cl⁻ concentrations of 1770 mg/l had been recorded in the field during the summer, the authors concluded that road salt applications do have a debilitating effect on stream insects.

Several studies have demonstrated effects of elevated salt concentrations on the physiology of aquatic insects. For example, Shaw (1955) found that the megalopteran *Sialis lutaria* could not correct osmotic water loss or produce a hypertonic excretory fluid when held in hypertonic solutions. Sutcliffe (1961) demonstrated that the brackish water trichopteran *Limnephilus affinis* was able to concentrate rectal fluid hypertonic to external media, but suggested that obligate freshwater trichopterans could not do so. Kamento and Goodnight (1956) found that increasing dissolved salts affected oligochaete reproductive processes, and they suggested that insects may respond similarly.

Acid deposition In western and central Europe, Scandinavia, the north-eastern United States, southeastern Canada, and southeastern China, acid deposition is a serious problem affecting aquatic and terrestrial systems. Although acid deposition is more commonly referred to as *acid rain*, this is a misnomer, as acids and acid forming substances are also deposited in snow, sleet, fog, dew and as dry particles and gas.

Precipitation has a natural pH value of approximately 5.1 (5.0-5.6, depending on location), which forms when carbon dioxide (CO_2), trace amounts of sulphur and nitrogen compounds, and atmospheric organic acids dissolve in atmospheric water. However, elevated levels of acidity result when the primary air pollutants, sulphur dioxide (SO_2) and nitrous oxide (NO), enter the atmosphere at disproportionately high rates (due primarily to the burning of fossil fuels[1]), and react to form secondary air pollutants. For example, sulphur dioxide combines with oxygen to form the secondary pollutant sulphur trioxide ($2SO_2 + O_2 = 2SO_3$), which then reacts with water vapour to form droplets of sulphuric acid ($SO_3 + H_2O = H_2SO_4$), another secondary air pollutant. If the acid deposition lands in regions that contain limestone ($CaCO_3$) or other alkaline (basic) substances, then the effects on the ecosystem are neutralized (although this buffering capacity does not last *ad infinitum*; Dillon *et al.*, 1987; Schindler, 1988). If, however, the acid deposition lands in regions with little buffering capacity, such as on granite or some types of sandstone, damage to the ecosystem may be significant. If the region contains bedrock with a high aluminium content the effects are further exacerbated, as the aluminium will dissolve and, in ionic form, cause asphyxiation in fishes through the impairment of gill function (Schofield, 1976). The role of aluminium in affecting the

---

[1] Almost half of all SO_2 emissions and one-fourth of NO emissions in the U.S.A. come from coal and oil burning plants in seven central and upper mid-west states.

physiology of aquatic insects is not as clearly defined (e.g. Hall *et al.*, 1980; Havas and Hutchinson, 1982). Despite the problems in determining the specific pathways by which stress is manifested, many studies clearly demonstrate the impact of acid fallout on aquatic insects, particularly those which occur in non-buffered streams and lakes (e.g. Bell, 1971; Herricks and Cairns, 1974; Zischke *et al.*, 1983).

To determine the effects of elevated acidity on the ecology of aquatic systems, Hall *et al.* (1980) added dilute concentrations of sulphuric acid to Norris Brook, a stream in the Hubbard Brook Experimental Forest, New Hampshire. They compared biotic and abiotic characteristics of the system between upstream (control) and downstream (acidified) locations. They found a 37% reduction in the mean number of emerging adult insects, and a maximum increase in drift of 13 times in the acidified section of stream, as compared with the control. The insects which showed the largest drift response were the mayflies *Epeorus* and *Ephemerella*, the black fly *Prosimulium*, and the stoneflies *Nemoura*, *Malirekus*, and *Isoperla*. Overall, benthic samples from the acidified zone of Norris Brook contained significantly fewer (75%) individuals. The addition of sulphuric acid resulted in significant increases in aluminium, calcium, magnesium and potassium ions in the stream water. Hall *et al.* proposed three mechanisms which could have influenced the responses of insects to elevated hydrogen ion concentrations: (1) a direct effect on insect physiology; (2) a toxic effect of heavy metals; and (3) indirect effects on primary production and/or bacterial decomposition. Although others (Sutcliffe and Carrick, 1973) have found evidence in support of the third hypothesis, Hall *et al.* (1980) dismissed it, as observed reductions in density occurred before any diminution of the food supply. Instead, they subscribed to the first and second hypotheses and to a possible synergistic interaction between the two. Other researchers (Minshall and Minshall, 1978) have proposed that chemical factors, and not pH *per se*, restrict the activity of aquatic insects. Hall *et al.* (1980) suspected that the ion exchange capacity and salt uptake mechanisms in insects may be negatively affected by aluminium in a manner analogous to that which occurs in freshwater fishes (e.g. Driscoll *et al.*, 1980). In a study of acidic and alkaline tundra ponds at the Smoking Hills, N.W.T., Canada, Havas and Hutchinson (1982) found evidence to implicate aluminium further in restricting the distribution of aquatic insects, specifically the caddisfly *Limnephilus pallens* and the chironomid *Orthocladius riparius*.

A survey of the aquatic insects found in two low alkalinity streams in Algonquin Provincial Park, Ontario, was first made in 1937 (Ide, 1940). Hall and Ide (1987) revisited the collection sites 48 years later and, using identical methods, repeated insect collections and water survey measurements. One of the two streams had remained chemically consistent over the years (as it was mediated by an upstream, non-acidic feeder lake), whereas the other stream had become more acidic (it flowed over granitic bedrock and thin soil). Twenty mayfly species and 17 stoneflies species present in the non-acidified stream still remained 48 years later. Additionally, three acid-tolerant mayflies were found (*Leptophlebia cupida, L. nebulosa* and *Eurylophella funeralis*). In contrast, 10 of 13 species collected previously had disappeared in the acidified stream. The extinct species were those insects known to be acid intolerant. Hall and Ide also found that, in the acid stream, seven species not found in 1937 appeared in the 1985 survey, and of these recent colonizers most were acid tolerant. Thus, the insect assemblage in the acid stream had switched from an acid intolerant to an acid tolerant one.

The effects of acid stress on aquatic life can be extreme in geographical regions characterized by spring snow melts. For example, Schofield (1977) recorded that the episodic release of hydrogen ions caused the pH of 1st to 3rd order streams to drop to 4.0 for periods of 3-7 days in the Adirondack Mountains of New York. Both insects (Hall *et al.*, 1980) and fishes (Leivestad and Muniz, 1976; Schofield, 1977) are known to suffer significant mortality when exposed to such periodic releases of acid.

11.1.3 Biodiversity indices

The use of aquatic insects as indicators of water quality is not a recent idea (Mol, 1982). As early as 1848 it was demonstrated that the absence of caddisfly larvae from a stream was the result of a city upstream (Kolenati, 1848). The first comprehensive system of biological water quality assessment was developed in Europe, by Kolkwitz and Marsson (1908, 1909), and was known as the "Saprobiensystem".

A very comprehensive index developed to monitor organic pollution in freshwater systems was designed by Hilsenhoff (1977, 1982, 1987). This index relies on the differential ability of aquatic invertebrates (primarily insects) to survive under specific levels of dissolved oxygen availability, which directly reflects levels of organic wastes in a system. Every species or genus is assigned a predetermined tolerance value of from 0 to 10 (Hilsenhoff, 1987, has classified hundreds of aquatic invertebrates in this way), with 0 indicating organisms least tolerant of organic pollutants. Following collection of a sample from a water body, all the organisms are picked out, identified, and assigned values, which are then averaged to calculate the Biotic Index (BI). The lower the mean, the more polluted the system. To use the BI requires taxonomic expertise.

A simpler method of assessing the degree of pollution in an aquatic system was developed by Cairns and Dickson (1971), whereby species diversity is calculated based on the ratio of the "number of runs"[1] of specimens to the total number of animals collected within a sample - thus the name, the Sequential Comparison Index (SCI). A small SCI represents a lower species diversity (indicative of stressed conditions), which may be contrasted with a similarly

---

1 "Species", or different types of insect, are removed at random from a sample, and each "run" is defined as a sequential removal of identical "species".

derived calculation from another site, or to the theoretical maximum, unity. The SCI requires little taxonomic expertise and is based on differences in the shape, colour and size of animals.

Indicator organisms may be used, very simply, to detect anthropogenic stress in its early stages. For example, in Europe, the mayflies *Baetis rhodani*, *B. lapponicus* and *B. macani* usually disappear from acid systems before fish stocks show any response (Minshall and Minshall, 1978; Raddum and Fjellheim, 1984).

A widely used method of calculating biotic diversity in aquatic and terrestrial ecosystems is the Shannon-Wiener Index, which is expressed as:

$$H = - \sum_{i}^{s} (p_i)(\log_2 p_i)$$

where H = index of species diversity; S = number of species; p_i = proportion of total sample belonging to the ith species. A large H value indicates greater diversity, as influenced by a greater number and/or a more equitable distribution of species (Lloyd and Ghelardi, 1964). More recently, Wright *et al.* (1989) have developed a predictive model, called RIVPACS, as a more objective means of assessing the impact of pollution at a site. This technique compares the observed fauna with that predicted in the absence of any pollution.

11.1.4   Environmental impact assessment

EIA is defined as, "the process of doing predictive studies on a proposed development, and analysing and evaluating the results of that development" (Lash *et al.*, 1974; Rosenberg *et al.*, 1986). Although these are admirable goals, EIA, as presently practised, does not make the contribution it might to environmental science (Larkin, 1984; Schindler, 1987). This is particularly true of two aspects of impact assessment using aquatic insects. Firstly, impact studies focus almost exclusively on the short-term effects of abiotic impacts (physical or chemical) on population and community dynamics, while only cursory interest is paid to the effects of potential long-term (multi-generational) changes in biotic interactions (e.g. predator/prey dynamics, competitive interactions, immigration/emigration). Secondly, sensitivity studies aimed at assessing anthropogenic stress in its early stages have not received due attention (Schindler, 1987). These problem areas could be addressed by: (1) placing greater emphasis on long-term, large-scale perturbations of various anthropogenic stresses (identified throughout section 11.1) on aquatic insects; and (2) focusing more attention on the development of "early-warning", species response indices to impending stress.

11.2   The potential impact of global warming on the ecology of aquatic insects

Anthropogenic influences have elevated environmental stress to levels that mandate new integrated approaches to the study of aquatic and terrestrial ecosystems. This reality has received formal recognition by the Ecological Society of America as the Sustainable Biosphere Initiative (SBI: Lubchenco *et al.*, 1991), a programme that focuses on the imperative of ecological research to manage the earth's life support systems effectively. Specific milestones of the SBI relate global warming (i.e. the greenhouse effect) directly to freshwater ecology and aquatic insects, stressing the need for complementary field and laboratory experimental approaches. This issue is sufficiently important to merit special consideration.

Increased emphasis on combined field and laboratory investigations will be needed to address the contemporary problem of linking species assemblages and ecosystems with climate, and to predict ecological responses under climates that do not presently exist (Lubchenco *et al.*, 1991). Predicted temperatures for the next century will exceed any experienced by biota over the past several million years (Jaeger, 1988), and projected rates of change may be more than an order of magnitude faster than any global change in the past 2 million years. On a proximal scale, predicted changes in climatic temperatures over the next century are estimated to occur at minimal rates of 0.06°C/decade, to upper limits of 0.8°C/decade, thus giving projected changes in the range indicated on Fig. 11.3 (Jaeger, 1988). More recent predictions by Hengevald (1990) suggest that during the next century the climate may warm as much as 1.5-4.5°C. The Intergovernmental Panel on Climate Change (IPCC, 1990) categorically states that emissions resulting from human activities are substantially increasing atmospheric concentrations of the greenhouse gases (carbon dioxide, methane, chlorofluorocarbons [CFCs], and nitrous oxide) which will ultimately warm the earth's surface. How global warming will affect aquatic insect communities is not well understood.

Aquatic insects may respond behaviourally to conditions of elevated temperature by emigrating from, or changing distribution within, stressed regions, or they may react physiologically by adjusting growth and development in immature stages, and size, condition and fecundity as adults. Behavioural and physiological responses of aquatic insects exposed to manipulated thermal environments will be outlined, before briefly discussing laboratory and field research directions related to global warming and aquatic insect ecology.

Although few studies have examined whether the microdistribution of aquatic insects within habitats is influenced by temperature, the collective evidence thus far indicates that such habitat selection does occur. For example, when placed in a horizontal gradient, 4th instar larvae of the cranefly *Tipula plutonis* displayed a diel rhythm of temperature selection whereby they chose warmer (16-18°C) and cooler (12-14°C) temperatures during night and day

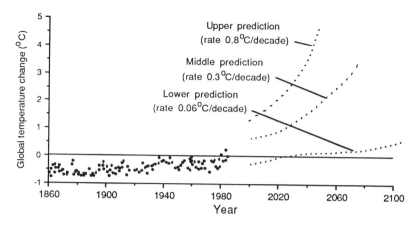

Fig. 11.3   Three predictions for global temperature change to the year 2100, based on projected levels of trace greenhouse gas emissions (redrawn from Jaeger, 1988).

phases, respectively (Kavaliers, 1981). Adults of the water-strider *Gerris pingreensis* have been observed to dip their bodies below the water surface in an attempt to avoid cold temperatures (termed "underwater basking"; Spence *et al.* 1980). Temperature selection was observed in the damselfly *Argia vivida*, as nymphs remained in a temperature range of 23-32°C when placed in a gradient of 9-41°C (Leggot and Pritchard, 1986). Similarly, Baker and Feltmate (1989) found that when nymphs of the damselfly *Ischnura verticalis* were released into either stratified or isothermal water columns, they tended to stay disproportionately near the surface, under both conditions, where waters are typically warmer. Behavioural thermoregulation, as demonstrated in the above examples, may allow individuals to take better advantage of thermal regimes, thus enhancing their fitness (Magnuson *et al.*, 1979).

Several studies at the species level suggest that changing temperature regimes can have significant physiological impacts on growth and development of both immature and adult aquatic insects. For example, Sweeney *et al.* (1986) found that manipulating temperature affected development of nymphs and the size of adults of the leaf-shredding stonefly *Soyedina carolinensis*. Nymphs reared under ambient field conditions, ambient + 3°C, or ambient + 6°C, were differentially affected so that adding 6°C to the normal temperature regime was lethal to 99% of the nymphs, whereas warming by 3°C did not affect rate of development yet reduced adult size. In a similar study, Elliott (1987) examined the European stonefly *Leuctra nigra* to determine whether temperature could influence changes in life cycle duration from 1 year to 2 years. In the laboratory, he examined nymphal growth and development at six discrete temperatures (ranging from 6-20°C). He found that mean growth rate increased directly with increasing temperature; thus the life cycle could change from being semivoltine to univoltine simply because of an increase in temperature. However, Elliott also discovered that at higher temperatures mortality increased exponentially and egg production dropped significantly, and he concluded that the optimum habitat for the species was a summer cool stream where the life cycle was 2 years from egg to adult. Numerous other studies have also shown that egg development, fecundity, size and condition of immature and adult insects are negatively affected by elevated temperatures (see section 5.3).

It is presently impossible to predict how global warming will affect aquatic insect community structure since impacts at the species level vary greatly. However, it is predictable that such specific variability will exacerbate asynchrony in life cycles. How these changes would manifest themselves over time is difficult to estimate. Thus, large spatial/temporal scale field experiments, where the seasonal temperature profiles of streams, ponds and lakes are elevated relative to control (unmanipulated) areas, are needed. Experiments should be preceded by premanipulation trials to establish that response variables do not differ initially between experimental and control sections, and trials should run for a minimum of 2 years (to include generations of predominantly univoltine species). This field approach should be complemented with laboratory experiments which focus on physiological or behavioural changes in aquatic insects (e.g. changes in microdistribution, feeding rate, frequency of movement, and growth rate) under various temperature regimes to discern causal mechanisms that influence the observed responses of insects.

In sum, to determine how aquatic insects may respond to the reality of global warming, increased emphasis on companion field and laboratory studies is necessary. Field models also require longer and larger temporal/spatial scale manipulation of the thermal regimes of aquatic systems than has been considered to date.

11.3   Aquatic insects as vectors of disease

Human disease, along with that of other animals, arises from two causes: (1) collapse of an individual's internal immune system leading to the development of organic disease; and (2) the presence of an introduced pathogen which disrupts normal body function. Aquatic insects cause only the latter forms of disease. Some terrestrial insects, such as lice, are themselves the pathogens but more often, and especially in the case of aquatic insects, the insect acts as a carrier (vector) of the pathogen from one host to another. Insects may act as vectors in one of two ways. *Mechanical transmission* occurs when the insect picks up the pathogen from one source and deposits it elsewhere, where it may infect a new host. This is rare in aquatic insects although, in Australia, the myxomatosis virus of rabbits is known to have been spread, initially, by mosquitoes. *Biological transmission* occurs when the only natural route for the pathogen to take from one host to another is through an insect, a process that often involves a period of pathogen development within the insect's body (Kettle, 1984). The part played by the insect is thus crucial in biological transmission, and effective control of diseases spread in this way is, more often than not arrived at by destruction of the insect species, either as adults or immature stages.

Most pathogens spread by aquatic insects are living organisms, particularly arboviruses, virus- and bacteria-like microorganisms, protozoans and nematode worms, but in some instances the tissue of the insect itself may act as an allergen. Mass emergences of the adults of some aquatic insects, such as mayflies and caddisflies, produce large numbers of exuviae. These cast skins, when fragmented and wind-borne, become inhalent allergens (Frazier, 1969).

It is the aquatic Diptera that are most significant in the biological transmission of disease, and the following five families in particular - although forms of myiasis are initiated by some members of the semi-aquatic Scarcophagidae.

Culicidae - mosquitoes serve as vectors of three main groups of human pathogens: (1) the **Haemosporina**, containing four species of human malaria parasite belonging to the genus *Plasmodium*. Although these produce similar illnesses and are all grouped under the heading "malaria" they are four different diseases, each with its own specific pattern; the vectors are all species of *Anopheles*. Despite extensive control programmes, malaria is still the most widespread and persistent disease of mankind - it was estimated that, in 1981, half of the world's population (mostly in the tropics and subtropics) was still at risk to malaria; mortality among infants and young children is particularly high. (2) the **Filarioidea**, nematode worms belonging to the genera *Wuchereria* and *Brugia*, which live in the lymphatic system and obstruct the flow of lymph, resulting in a chronic debilitating condition that may culminate in elephantiasis. The greatest concentrations of lymphatic filariasis are in Africa and India where populations of the vector *Culex quinquefasciatus* have increased significantly in urban areas (Nelson, 1978). (3) the **arboviruses**, of which nearly 100 types are known to be transmitted by mosquitoes. They manifest themselves as a variety of "fevers", including dengue, encephalitis and yellow fever (Mattingly, 1969).

Only about 60 of the 400 or so species of *Anopheles* mosquito transmit malaria. Whether or not a species carries the disease depends on a number of factors primary among which are: (1) there must be no biochemical deterrent to prevent the parasite from completing its development within the body of the mosquito; and (2) the mosquito must bite people. For example, *Anopheles gambiae* is perhaps the world's most efficient vector species, and especially so in tropical Africa. This efficiency is due to its clear preference for feeding on Man, a long adult life, prolific fecundity, particularly around human habitations, and its high infection rate (estimated to be between 4 and 5%). In contrast, *A. culicifacies*, the main vector in India, often bites cattle in preference to Man, has a lower survival rate, and has a lower infection rate (~ 0.1%); consequently, it is not as good a vector (Service, 1986). Proximity of larval habitats to human settlements is another factor important in the transmission of disease, or the evolution of transmission to humans. Man's activities often create, incidentally, ideal breeding sites (see Chapter 4).

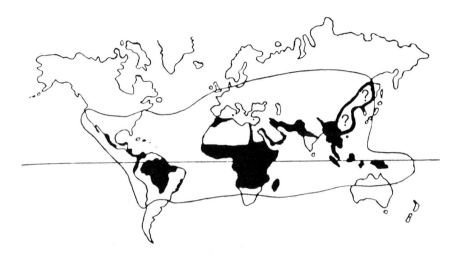

Fig. 11.4   Past (outline) and present (black) distribution of malaria (redrawn after Busvine, 1979).

Although malaria remains the most important tropical disease, its current range is considerably less than in the recent past (Fig. 11.4). This is due to a combination of factors including effective mosquito control, use of anti-malarial drugs, eradication of larval habitats, etc. Interestingly, the present day distribution of malaria does not totally coincide with the distribution of its vector species. This discrepancy applies to some other mosquito-borne diseases as well. Areas of Europe, for example, abound with anopheline mosquitoes, some of which carry malaria - in fact malaria in the form of *Plasmodium vivax* was endemic as far north as Scandinavia until the middle of the 19th Century. Since that time, the disease has retreated from Europe and transmission is now rare. Change in human lifestyle is thought to have been one contributing factor, but there has also been a shift in host selection brought about by changes in the ratio of the number of humans to domestic animals - a consequence of declining human birth rates coupled with more efficient animal husbandry. In a malaria-free region, anopheline mosquitoes are 400 times as likely to feed on animals other than Man, whereas in malarious regions, one in eight feeds from Man. It is possible to calculate from this (see method in Lehane, 1991) that people in malarious regions are 2,500 times as likely to contract malaria than people living in non-malarious areas.

Wuchereria bancrofti and *Brugia malayi* are two of the most important species of mosquito-borne filariases transmitted to Man. Some 400 million people are estimated to be at risk from such diseases, although not all will develop severe symptoms. *Wuchereria bancrofti* is particularly common in both urban and rural areas in the tropics. Microfilariae show marked nocturnal periodicity and are abundant in the peripheral blood of the human host between 22.00 and 01.00 hours. They are picked up, therefore, only by night-biting mosquitoes, especially *Culex quinquefasciatus*, which breeds in polluted urban waters. This species has a variable vectorial capacity being, for example, a poor transmitter in West Africa but a more effective one in East Africa (Service, 1986). In rural areas, transmission is more likely to be through species of *Anopheles* whose larvae tend to favour cleaner water habitats. *Brugia malayi* occurs primarily in rural populations in Asia. It has two forms, the more common of which is nocturnal and is transmitted by night-biting species of *Anopheles* and *Mansonia*. Filariasis control is through the use of drugs, such as diethylcarbamazine which kills the microfilariae while they are resident in people, although this has had limited success, and by eradication of the vectors.

Most mosquito-borne arboviruses occur in the tropics. The viruses, themselves, are a heterogeneous group of rod-shaped or spherical particles containing RNA and, when transmitted into a susceptible host, they multiply rapidly. In yellow fever, for example, this incubation period can be as little as 3 to 6 days after which human hosts show symptoms of fever, pains in the legs and lower abdomen, vomiting and diarrhoea. Jaundice frequently follows and the end results may be fatal. Yellow fever also occurs in monkeys, spread by species of mosquito that inhabit the forest canopy such as those belonging to the genus *Haemagogus* in South America and by *Aedes africanus* in Africa (Busvine, 1979). In urban areas, on both continents, it is spread by the anthropophile *Aedes aegypti* which is also the primary vector of dengue in Africa and Asia. Yellow fever, however, has not yet reached Asia although, clearly, its vector is already resident there.

The encephalitis viruses are spread mainly by *Culex* and *Aedes* mosquitoes. The primary hosts are typically birds, rodents and horses, with Man as an incidental host. Western equine encephalitis occurs mostly in the western U.S.A. although outbreaks have been recorded as far south as Argentina and up into southern Canada. The primary vector is *Culex tarsalis* which feeds mostly on birds. Prior to epidemics, there seems to be a pattern of virus population build-up in house sparrows before it manifests itself in domestic fowls and humans (Mattingly, 1969). Within Man, the virus invades the central nervous system and brain, causing damage to motor nerves and death in about 10% of cases. Eastern equine encephalitis ranges from the eastern U.S.A. down to South America and is transmitted by *Culiseta melanura* among birds but by species of *Aedes* to Man and horses (Service, 1986). The virus of Japanese "B" encephalitis in Japan, builds up in herons and ibises in the spring before being spread by *Culex tritaeniorhynchus* to domestic pigs in the summer and subsequently to

Man. Control of arboviruses is achieved chiefly through destruction of the vector species with insecticides.

Ceratopogonidae   Sandflies, punkies, biting midges or no-see-ums are among the smallest of the blood-sucking flies. Only four of the more than 50 genera affect Man significantly, by transmitting parasites to both humans and domestic animals. Species of *Lasiohelea* and *Leptoconops* produce very persistent skin reactions at the bite site. Species of *Culicoides* are economically important as vectors of arboviruses to livestock and of protozoans to poultry. They also transmit filarial worms, such as those belonging to the genus *Mansonella*, to both Man and livestock, but these particular nematodes do not appear to be highly pathenogenic, and are therefore of little clinical significance (Kettle, 1984). In Queensland, Australia, however, *Culicoides* is the vector of *Onchocerca gibsoni* in cattle. Damage to the host's skin, caused by worm nodules, lowers the market value of the hides. Species of *Culicoides* are also responsible for spreading equine onchocerciasis in both the holarctic and australian regions (Crosskey, 1990). This same genus is responsible for the spread of bluetongue fever, a viral disease of sheep that causes serious economic losses in many parts of the world.

Psychodidae   Also known as sandflies, members of the subfamily Phlebotominae are another group of small biting dipterans. They are the vectors of various pathogens, the most serious of which are species of the flagellate protozoan genus *Leishmania*. Leishmaniasis (or "kala-azar"), infection of the skin and viscera, is patchily but widely distributed throughout warmer parts of the globe. It is carried by species of *Phlebotomus* and *Lutzomyia*. Leishmaniasis cannot be controlled very effectively as it does not respond well to drugs. There have been some attempts to reduce the number of infections by using DDT, fine-meshed sleeping nets and, in China, by destroying dogs which act as major reservoirs for the protozoan (Service, 1986). Other phlebotomine-carried diseases include bartonellosis (restricted to people living in high altitude valleys in the Andes) and papatasi (or sandfly-) fever, a non-fatal, short-term but sharp fever common from the Mediterranean through the desert regions of the Middle East, to Pakistan. The latter is spread by *Phlebotomus papatasi*.

Simuliidae   Black flies, or buffalo gnats, transmit several pathogenic organisms including viruses and protozoan blood parasites, primarily in birds, and filarial nematodes in mammals. It is because of the latter that black flies are of considerable importance in medical entomology. The most serious disease of Man associated with simuliids is onchocerciasis, or river blindness. Common in the tropical regions of Africa and the Americas, this disease is caused by the filarial worm *Onchocerca volvulus*.

In Africa, the vectors of onchocerciasis belong to the *Simulium damnosum* and *S. neavei* groups of species (Table 11.2) but, in the Zaire River basin, the disease is spread also by *S. albivirgulatum*. The former are two species complexes containing over 30 closely related species, only some of which transmit the disease. Species in the *S. damnosum* complex breed in a variety of

Table 11.2   Summary of the more important pest species of black fly throughout
the world (based on Crosskey, 1990).

Species	Region	Principal effect
Austrosimulium australense & *A. ungulatum*	New Zealand	mass biting of Man
Austrosimulium pestilens	Queensland	mass biting of livestock
Cnephia pecuarum	mid-southern USA	mass biting of Man & livestock (lvsk)
Prosimulium mixtum	Eastern Canada & northeast USA	mass biting of Man
Simulium amazonicum, *S. argentiscutum* & *S. oyapockense*	Brazil, Colombia, (esp. Venezuela)	mass biting of Man, vectors of human onchocerciasis (nematodes)
Simulium arakawae	Japan	mass biting, Man & lvsk
Simulium arcticum	Canada	mass biting of livestock
Simulium cholodkovskii	Siberia	mass biting, Man & lvsk
Simulium chutteri	South Africa	biting of livestock
Simulium colombaschense	Danube Basin	biting, Man & livestock
Simulium damnosum complx. & *S. neavei* group	Tropical Africa	vectors, human onchocerciasis
Simulium equinum	Europe & USSR	biting lvsk, esp. horses
Simulium erythrocephalum	Central Europe	biting, Man & livestock
S. exiguum, *S. metallicum* & *S. ochraceum* complexes	Guatemala, Mexico & Colombia	vectors, human onchocerciasis
Simulium jenningsi complex	Eastern USA	vector, bovine onchocerc.
Simulium luggeri	Canada	mass biting of livestock
Simulium maculatum	USSR	biting, Man & livestock
Simulium meridionale, *S. occidentale*, *S. slossonae* & *S. congareenarum* complex	North America	vect. of poultry leucocyto-zoonosis (protozoan blood parasite)
Simulium ornatum complex	Europe & USSR	vector, bovine onchocerc.
Simulium pusillum	Northern USSR	mass biting, Man & lvsk
Simulium quadrivittatum	Belize to Panama	mass biting of Man
Simulium rugglesi	Canada, N.E. USA	biting, Man, lvsk, poultry
Simulium sanguineum	N.W. South Amer.	mass biting, Man & lvsk
Simulium transiens	USSR	mass biting of Man
Simulium venustum complex	Canada, N.E. USA	mass biting, Man & lvsk

running waters spanning savanna, highlands and lowland forest. In forests, larvae live in the larger, permanent rivers but they occur in smaller rivers in the mountains. In the dry savanna, species breed in large rivers that tend to run dry. The larvae of species belonging to the *S. neavei* group are phoretic, having to pass their early instars attached to lotic species of freshwater crabs (genus *Potamonautes*); they are typically found in turbid, heavily-shaded forest streams (Crosskey, 1990).

In the Americas, species from the *Simulium ochraceum* complex are believed to be the most important vectors (Dalmat, 1955) but other primary vectors include species from the *Simulium exiguum* and *S. metallicum* complexes, together with *S. guianense* (at higher altitudes) and *S. oyapockense* (at lower altitudes). Larvae of *Simulium ochraceum* live in very small forest streams and seeps, and extend over large areas of saturated ground during the rainy season. *Simulium metallicum* larvae occur in medium to large forest streams, especially where there are areas of fast water with rock substrates, in the shade. In contrast, *S. exiguum* breeds in large rainforest rivers (> 5 m wide) where the larvae attach not only to shingle in fast, shallow water, but also to macrophytes in deeper water.

Although onchocerciasis is not directly fatal, it is extremely debilitating as the nematodes cause irritating dermatitis and eye lesions which result in blindness. It is estimated that some 30 million people suffer from onchocerciasis in Africa alone (Nelson, 1981). A person is capable of becoming infected, after being bitten by a black fly carrying the *O. volvulus* parasite, at any time when the larvae are in their infective stage. However, the disease does not appear to take a strong hold until after the victim has received many infective bites over some time. Among other factors, infection depends on Man - fly contact rates. Vector species, like most other black fly species, tend to bite in the open, by day. This contrasts strongly with biting by mosquitoes where people are frequently attacked indoors or while sleeping. Certain outdoor places are more likely to harbour female black flies than others, for example river banks (washing and water drawing areas) and crossings, as well as riparian footpaths. In Latin America, Man - fly contact remains fairly uniform year round whereas, in Africa, particularly in the savannas, contact is highly seasonal (Crosskey, 1990). In Africa, after mating and biting, adult black flies may be dispersed hundreds of kilometres on the wind and thus may reinfect regions previously freed of disease.

The only other filarioid parasite of Man spread by black flies is *Mansonella ozzardi*. This is a New World nematode carried by *Simulium argentiscutum*, *S. oyapockense* and species from the *S. amazonicum* group. Adult worms occur in subcutaneous tissues, while the microfilariae are typically found in the blood but, on occasion, also in the skin; the species is non-pathogenic.

Many species of black fly do not spread disease but represent serious pests because of the sheer numbers of females seeking blood meals from both Man and domestic animals. The more important of these, together with those species that spread parasitic organisms, are listed in Table 11.2. Clearly the impact of black flies is not restricted to tropical regions of the globe.

Tabanidae Horse flies, deer flies, or clegs belong to a family containing more than 3,000 species worldwide. The adults may act as both mechanical and biological vectors. As biological vectors, they are known to spread three species of nematode: *Loa loa,* the cause of Calabar or fugitive swellings in Man in West Africa; *Elaeophora schneideri,* the arterial worm of sheep in North America; and *Dirofilaria roemeri,* a parasite of marsupials. As mechanical vectors, tabanids are among the many transmitters of diseases such as anthrax (caused by *Bacillus anthracis*) and anaplasmosis in cattle, but which sometimes spread to Man (Kettle, 1984). In Africa, tabanids are capable of transmitting sleeping sickness (trypanosomiasis) to humans although the principal vectors are tsetse flies (Glossinidae).

Although larval simuliids primarily occur in permanent running waters, the immature stages of many vector species belonging to the other four families, listed above, live in temporary waters or at the margins of permanent water bodies, particularly in the tropics and subtropics. Intermittent ponds and ditches, irrigation canals, marshes and flooded areas typically support large populations of larvae (Williams, 1987). In addition to breeding in these larger temporary waters, many vector species lay their eggs in phytotelmata - the small, intermittent waterbodies associated with plants (see Chapter 4), and in container habitats, such as cisterns, tin cans and old tyres, associated with human activity. This is especially true of mosquitoes.

Apart from the disease-spreading effect that aquatic insects have on both Man and domestic animals, they are also a considerable nuisance. Bites from mosquitoes, ceratopogonids, black flies and tabanids can be very distracting, and even dangerous if in sufficient numbers, to the point where the normal behaviour of both Man and animal is affected. Cattle grazing in the vicinity of emerging females of haematophilous species, for example, may stampede or, in particularly severe cases, become anaemic through blood loss. The effects on loggers and other field-based workers can influence productivity and hence have financial consequences for industry.

11.4   Aquatic insect fossils as palaeoecological tools

Interpretation of conditions in past environments is based on the study of fossils. Certain parts and whole organisms preserve better than others and, logically, common and well preserved taxa tend to yield the most useful data. Pollen grains have been, for many years, the primary tool of the palaeoecologist because pollinin, the outer structural component of pollen, is extremely inert and remains when the cellulose has rotted (Frey, 1976). There are, nevertheless, problems and disadvantages associated with the use of pollen in interpreting past climatic and edaphic conditions (Wright, 1977). More recently, increasing use has been made of some other organisms whose skeletal materials survive well in sediments. Among these are three insect groups: adults of terrestrial species of Coleoptera, and larval remains of the Chironomidae and Trichoptera. As most fossil-bearing sediments accumulate as a result of deposition by water, it is not surprising to find at least some aquatic insect groups well represented (Williams, N.E., 1981).

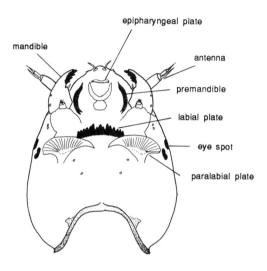

Fig. 11.5   Ventral view of the head capsule of a chironomid larva (Chironomini) showing the major features used in identification, most of which survive in fossil specimens.

Larval chironomids are, as we have noted, particularly abundant in most types of freshwater habitat, and as a group span most of the climates found on the planet. Recent advances in larval taxonomy (e.g. Wiederholm, 1983) enable head capsules, the most commonly surviving fossil remains (Fig. 11.5), to be identified to the level of genus. In addition, chironomid larvae have proved to be good indicators of water quality (e.g. Beck, 1977). These factors make them very valuable tools in interpreting past environments. As an example, Warwick (1980) used fossil chironomids to document Man-induced changes, including eutrophication, in the terrestrial environment bordering the Bay of Quinte, Lake Ontario, Canada, over the last 130 years. Tracking environmental change over thousands of years is also possible using chironomids (Bryce, 1962). One disadvantage in using chironomids, however, is that although species may be good environmental indicators, many genera are widespread and common to a number of habitat types. Environmental interpretation based on such samples therefore must rely heavily on the associations of genera found, rather than on the presence or absence of a single genus or species.

The potential of using caddisfly larval remains in palaeoecology is a more

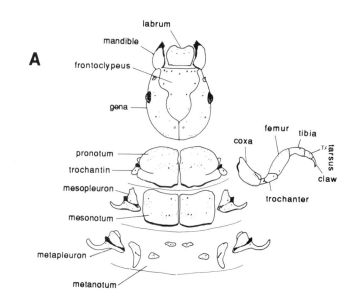

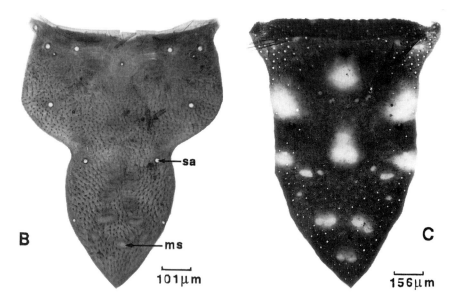

Fig. 11.6   (A) Sclerites of the head and thorax of a limnephilid caddisfly larva;
(B) Fossil frontoclypeus of *Grensia praeterita* (Limnephilidae) from
the Kuskokwim River, Alaska (Boutellier Interstadial, approximately
80,000 yrs B.P.) (sa = setal alveolus; ms = muscle scar); (C) Fossil
frontoclypeus of *Hydropsyche morosa* (Hydropsychidae) from the
Au Sable River, Michigan (approximate age 4,000 yrs B.P.) (from
Williams, N.E., 1989).

recent discovery (Williams and Morgan, 1977). As a consequence, sophistication in interpretation is not as advanced as for some of the other groups of organisms. On the other hand, considerable progress has been made, and already it has been established that larval caddisfly fossils are abundant and identifiable, in many cases to species, and that their assemblages yield considerable information about past habitats and climates (Williams, N.E., 1989). Unlike beetle adults, caddisfly larvae cannot fly into a site, thus their remains are likely to represent local assemblages. Additionally, because caddisfly sclerites are more fragile than those of beetles, they are more likely to be degraded during sediment rebedding processes, making redeposition of older material (contamination) less serious a problem. Although fewer species are usually represented, the nature of the deductions possible from caddisflies is complementary to that possible from diverse beetle assemblages.

Fossil caddisfly larvae are typically found as disarticulated sclerites and the most common of these are the frontoclypeus, pronotum and mesonotum (Fig. 11.6). Identification of these fossil fragments is based on shape and texture together with colour patterns, muscle scar patterns and setal distributions. Two types of information can be derived from fossil caddisfly assemblages: (1) knowledge of present-day caddisfly ecology permits reconstruction of past local aquatic habitats and their surrounding catchments; (2) in combination with distribution records, ecological data can provide indications of past macroclimates (Williams, N.E., 1988). Caddisfly fossils up to at least 500,000 years old can be identified as extant species, making these insects particularly valuable in the study of Quaternary environments. For example, the distinctive head and thoracic sclerites of *Clistoronia* sp. were deposited in Michigan about 14,000 years ago. Today, this cool-adapted genus is restricted to western parts of the U.S.A. and Canada indicating that the present day climate of Michigan is warmer than in the past (Williams, N.E., 1989). Similarly, information from both beetle and caddisfly assemblages suggests that a cold tundra environment existed at Lobsigensee, Switzerland about 13,500 B.P. but that this was replaced by a more temperate climate after 13,000 B.P. (Elias and Wilkinson, 1983). This last example, together with others (e.g. Williams *et al.*, 1981) serves to emphasize the advantages to be gained in co-operative studies of different groups of fossil organisms.

11.5   *Aquatic insects as templets for fishing lures*

Use of artificial baits to catch trout by means of rod and line dates back to the ancient Greeks (Leonard and Leonard, 1962), although the technique was not documented, in detail, until 1496 in "The Treatyse of Fysshynge wyth an Angle." by Dame Juliana Berners. Doubtless, tempting fishes with freshly caught aquatic insects would have preceded this, but the delicate nature of insect bodies would probably soon have led to the manufacture of more robust

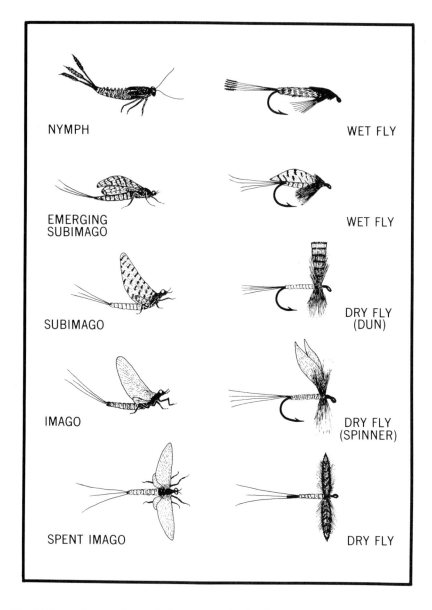

Fig. 11.7   Comparison of the stages in the life cycle of a mayfly and the artificial flies that imitate them (from Williams, 1980b).

imitations from a variety of readily available materials such as fur, feathers and wool. Presumably, the more realistic the artificial insect, the more successful its user would have been. This fact alone probably led to a more careful scrutiny of lake or riverside insects and their habitats, bringing the first aquatic entomologists into being.

A considerable literature on aquatic insect imitation now exists, particularly in the popular press, some of it based on sound entomological principles. The purpose, here, is to summarize some of the basic, good descriptive entomology that has led to the refinement of fly fishing, as illustrated by the Ephemeroptera.

The stages in the life cycle of a typical mayfly and the artificial flies that are meant to imitate them are shown in Fig. 11.7. Aquatic nymphs are mimicked by wet flies fished beneath the water surface. Prominent features of the nymph, such as the cerci, segmented abdomen, legs and darkened, pre-emergence wingpads are all duplicated in the fly. The emerging subimago is also fished as a wet fly, though just below the surface as most mayflies emerge at or near the surface. The characteristic rumpled wings are simulated on the artificial fly by a small portion of feather. The fully-emerged subimago, now a terrestrial stage (see Chapter 2), is represented by a dry fly fished on the water surface. The subimago is similar in appearance to the adult, but has duller colours, the legs and cerci are shorter, and heavy pigmentation along the veins may produce a dark pattern on the wings. The latter rarely persists in the imago but is faithfully copied in the fly, called a dun, by use of a mottled feather. Features of the imago are seen in its counterpart, the spinner, which again is fished on the water surface. The colours in this lure are brighter than in the dun, and the "wings" are made from plain feathers. Segmentation of the abdomen is duplicated by a silk thread binding. Many hundreds of patterns are known for this fly representing the different species of mayfly occurring in local waters. The final stage of the adult mayfly is the prostrate, spent female floating on the water surface. This change in posture is again reflected in the dry fly counterpart. Thus not only is it important for the fly-fisherman to present a bait that looks like the real thing, but he must present it in such a way that it occurs at the appropriate level in the water and "behaves" naturally. Of course many of the other aquatic insect orders are mimicked as tied flies, in particular, the Trichoptera and Plecoptera (Harris, 1952).

Why should the art of deceiving fishes have developed to such a high degree? Availability and not abundance of prey seems to be a primary factor determining which foods are eaten by trout. A certain amount of selectivity in feeding is evident also, and this is best seen in fishes with full stomachs (Allen, 1941; Bryan and Larkin, 1972). In times of high prey availability, many fishes temporarily set their feeding behaviour to take advantage of a transient but plentiful supply of identical organisms (Frost and Brown, 1967). A fish may reset its feeding for successively available aquatic insect species (Williams and Coad, 1979). The successful angler takes advantage of these feeding traits by presenting a close imitation of a particular insect species during the time when the fish's feeding behaviour is "set" for that species; this is the so-called technique of "matching the hatch".

REFERENCES

Aitken, R.B. 1982. Sound production and mating in a waterboatman *Palmacorixa nana* (Heteroptera: Corixidae). Anim. Behav. 30: 54-61.

Alexander, C.P. 1919. The crane-flies of New York. Pt. 1. Distribution and taxonomy of the adult flies. Cornell Univ. Agric. Exp. Stn. Mem. 25: 765-993.

Alexander, R.McN. 1968. Animal Mechanics. Sidgwick & Jackson, London. 346 pp.

Alexander, R.McN. 1982. Locomotion of Animals. Blackie, London. 163 pp.

Allan, J.D. 1978. Trout predation and the size composition of stream drift. Limnol. Oceanogr. 23: 1231-1237.

Allan, J.D. 1982. The effects of reduction in trout density on the invertebrate community of a mountain stream. Ecology 63: 1444-1455.

Allanson, B.R. (ed) 1979. Lake Sibaya. Monogr. Biologicae 36, W. Junk, The Hague. 364 pp.

Allen, K.R. 1941. Studies on the biology of the early stages of the salmon *(Salmo salar)*. 2. Feeding habits. J. Anim. Ecol. 10: 47-76.

Anderson, N.H. 1966. Depressant effect of moonlight on activity of aquatic insects. Nature 209: 319-320.

Anderson, N.H. and A.S. Cargill. 1987. Nutritional ecology of aquatic detritivorous insects. pp. 903-925 *In* Nutritional Ecology of Insects, Mites, Spiders and Related Invertebrates. (F. Slansky and J.G. Rodrigues, eds). John Wiley & Sons, New York. 1016 pp.

Anderson, N.H. and K.W. Cummins. 1979. Influences of diet on the life histories of aquatic insects. J. Fish. Res. Board Can. 36: 335-342.

Anderson, N.H. and E. Grafius. 1975. Utilization and processing of allochthonous material by stream Trichoptera. Verh. Int. Verein. Limnol. 19: 3083-3088.

Angino, E.E., L.M. Magnuson and G.F. Stewart. 1972. Effects of urbanisation on storm water runoff quality: a limited experiment, Naismith Ditch, Lawrence, Kansas. Water Resour. Res. 8: 135-140.

Anholt, B.R. 1990. An experimental separation of interference and exploitative competition in a larval damselfly. Ecology 71: 1483-1493.

Aspey, W.P. and J.E. Blankenship. 1977. Spiders and snails and statistical tails: application of multivariate analyses to diverse ethological data. pp. 75-120 *In* Quantitative Methods in the Study of Behavior. (B.A. Hazlett, ed). Academic Press, New York. 222 pp.

Awachie, J.B.E. 1981. Running water ecology in Africa. pp. 339-366 *In* Perspectives in Running Water Ecology. (M.A. Lock and D.D. Williams, eds). Plenum Publishers, New York. 430 pp.

Axtell, R.C. 1976. Horse flies and deer flies (Diptera: Tabanidae). pp. 415-446 *In* Marine Insects. (L. Cheng, ed). American Elsevier Publ. Co., New York. 581 pp.

Baird, D.J., L.R. Linton and R.W. Davies. 1987. Life-history flexibility as a strategy for survival in a variable environment. Funct. Ecol. 1: 45-48.

Baker, J.H. and L.A. Bradnam. 1976. The role of bacteria in the nutrition of aquatic detritivores. Oecologia 24: 95-104.

Baker, R.L. 1980. Use of space in relation to feeding areas by zygopteran nymphs in captivity. Can. J. Zool. 58: 1060-1065.

Baker, R.L. 1981. Behavioural interactions and use of feeding areas by nymphs of *Coenagrion resolutum* (Coenagrionidae: Odonata). Oecologia 49: 353-358.

Baker, R.L. 1986. Food limitation of larval dragonflies: a field test of spacing behaviour. Can. J. Fish. Aquat. Sci. 43: 1720-1725.

Baker, R.L. 1987. Dispersal of larval damselflies: do larvae exhibit spacing behaviour in the field? J. North Am. Benthol. Soc. 6: 35-45.

Baker, R.L. 1989. Condition and size of damselflies: a field study of limitation. Oecologia 81: 111-119.

Baker, R.L. and B.W. Feltmate. 1987. Development of *Ischnura verticalis* (Coenagrionidae: Odonata): effects of temperature and prey abundance. Can. J. Fish. Aquat. Sci. 44: 1658-1661.

Baker, R.L. and B.W. Feltmate. 1989. Depth selection by larval *Ischnura verticalis* (Odonata: Coenagrionidae): effects of temperature and food. Freshwat. Biol. 22: 169-175.

Bandel, K. 1974. Faecal pellets of *Amphineura* and *Prosobranchia* (Mollusca) from the Caribbean coast of Columbia, South America. Senckenbergiana Marit. 6: 1-32.

Barlocher, F. 1985. The role of fungi in the nutrition of stream invertebrates. Bot. J. Linnean Soc. 91: 83-94.

Barlocher, F. and B. Kendrick. 1973. Fungi in the diet of *Gammarus pseudolimnaeus* (Amphipoda). Oikos 24: 295-300.

Barton, D.R. 1980. Observations on the life histories and biology of Ephemeroptera and Plecoptera in northeastern Alberta. Aquat. Insects 2: 97-111.

Barton, D.R. 1981. Effects of hydrodynamics on the distribution of lake benthos. pp 251-263 *In* Perspectives in Running Water Ecology. (M.A. Lock and D.D. Williams, eds). Plenum Press, New York. 430 pp.

Barton, D.R. and H.B.N. Hynes. 1978. Seasonal study of the fauna of bedrock substrates in the wave zones of lakes Huron and Erie. Can. J. Zool. 56: 48-54.

Barton, D.R. and S.M. Smith. 1984. Insects of extremely small and extremely large aquatic habitats. pp. 456- 483. *In* The Ecology of Aquatic Insects. (V.H. Resh and D.M. Rosenberg, eds). Praeger Scientific, New York. 625 pp.

Barton, D.R. and R.R. Wallace. 1979a. Effects of eroding oil sand and periodic flooding on benthic macroinvertebrate communities in a brown-water stream in northeastern Alberta, Canada. Can. J. Zool. 57: 533-541.

Barton, D.R. and R.R. Wallace. 1979b. The effects of an experimental spillage of oil sands tailings sludge on benthic invertebrates. Environ. Pollut. 18: 305-312.

Bates, M. 1949. The Natural History of Mosquitoes. The Macmillan Co., New York. 379 pp.

Baxter, R.M. 1977. Environmental effects of dams and impoundments. Ann. Rev. Ecol. Syst. 8: 255-283.

Bayly, I.A.E. 1982. Invertebrate fauna and ecology of temporary pools on granite outcrops in southern Western Australia. Aust. J. Mar. Freshwat. Res. 33: 599-606.

Bayly, I.A.E. and W.D. Williams. 1966. Chemical and biological studies on some saline lakes of south-eastern Australia. Aust. J. Mar. Freshwat. Res. 17: 177-228.

Beadle, L.C. 1981. The Inland Waters of Tropical Africa: an Introduction to Tropical Limnology. 2nd ed., Longman Group, London. 475 pp.

Beaver, R.A. 1983. The communities living in *Nepenthes* pitcher plants: fauna and food webs. pp. 129-160 *In* Phytotelmata: Terrestrial Plants as Hosts for Aquatic Insect Communities. (J.H. Frank and L.P. Lounibos, eds). Plexus Pub. Inc., Medford, New Jersey. 292 pp.

Beck, W.M. Jr. 1977. Environmental requirements and pollution tolerance of common freshwater Chironomidae. U.S. Environmental Protection Agency Rept., Cincinnati, EPA/600/4-77/024: 1-261.

Beeton, A.M. 1960. The vertical migration of *Mysis relicta* in lakes Huron and Michigan. J. Fish. Res. Board Can. 17: 517-539.

Bell, H.L. 1971. Effect of low pH on the survival and emergence of aquatic insects. Water Res. 5: 313-319.

Bender, E.A., T.J. Case and M.E. Gilpin. 1984. Perturbation experiments in community ecology: theory and practice. Ecology 65: 1-13.

Benedetto, L. 1970. Observations on the oxygen needs of some species of European Plecoptera. Int. Rev. ges. Hydrobiol. 55: 505-510.

Bergroth, E. 1899. Note on the genus *Aepophilus* Sign. Ent. month. Mag. 35: 282.

Berners, Dame Juliana. 1496. The Treatyse of Fysshynge wyth an Angle. Reprinted (in 1827) from the Book of St. Albans. Pickering, London.

Birks, H.J.B. and H.H. Birks. 1980. Quaternary Palaeoecology. Edward Arnold (Publishers) Ltd., London, England. 289 pp.

Bishop, J.E. 1973. Limnology of a Small Malayan River, Sungai Gombak. Monographiae Biologicae 22. W. Junk, The Hague. 485 pp.

Bishop, J.E. and H.B.N. Hynes. 1969. Downstream drift of the invertebrate fauna in a stream ecosystem. Arch. Hydrobiol. 66: 56-90.

Bohart, G.E. and J.E. Gressitt. 1951. Filth-inhabiting flies of Guam. B.P. Bishop Mus. Bull. 204: 1-152.

Bohle, H.W. 1972. Die Temperaturabhängigkeit der Embryogenese und der embryonalen Diapause von *Ephemerella ignata* (Poda) (Insecta: Ephemeroptera). Oecologia 10: 253-268.

Boon, P.J. 1984. Habitat exploitation by larvae of *Amphipsyche meridiana* (Trichoptera: Hydropsychidae) in a Javanese lake outlet. Freshwat. Biol. 14: 1-12.

Bornhauser, K. 1913. Die Tierwelt der Quellen in der Umgebung Basels. Int. Revue ges. Hydrobiol. Hydrogr. Suppl. 5: 1-90.

Borror, D.J., D.M. DeLong and C.A. Triplehorn. 1981. An Introduction to the Study of Insects. Saunders, Philadelphia. 827 pp.

Bournaud, M. 1963. Le courant, facteur ecologique et ethologique de la vie aquatic. Hydrobiologia 21: 125-165.

Bowden, J. and C.G. Johnson. 1976. Migrating and other terrestrial insects at sea. pp. 97-117 *In* Marine Insects. (L. Cheng, ed). North Holland Pub. Co., Amsterdam. 581 pp.

Bowlby, J.N. and J.C. Roff. 1986. Trophic structure in southern Ontario streams. Ecology 67: 1670-1679.

Bretschko, G. 1982. *Pontomyia* Edwards (Diptera: Chironomidae), a member of the coral reef community at Carrie Bow Cay, Belize. pp. 381-385 *In* The Atlantic Barrier Reef Ecosystem at Carrie Bow Cay, Belize, I. Structure and Communities. (K. Rutzler and I.G. Macintyre, eds). Smithsonian Contrib. Marine Sci. 12., Washington, D.C.

Bretschko, G. 1985. Quantitative sampling of the fauna of gravel streams. (Project Ritrodat - Lung). Verh. Internat. Verein. Limnol. 22: 2049-2052.

Brigham, A.R., W.U. Brigham and A. Gnilka (eds). 1982. Aquatic Insects and Oligochaetes of North and South Carolina. Midwest Aquat. Enterprises, Mahomet, Illinois. 837 pp.

Brittain, J.E. 1977. The effect of temperature on the egg incubation period of *Taeniopteryx nebulosa* (Plecoptera). Oikos 29: 302-305.

Brittain, J.E. 1978. The Ephemeroptera of Øver Heimdalsvatn. Holarct. Ecol. 1: 239-254.

Brittain, J.E. 1982. Biology of mayflies. Ann. Rev. Ent. 27: 119-147.

Brittain, J.E. and T.J. Eikeland. 1988. Invertebrate drift - a review. Hydrobiologia 166: 77-93.

Britton, E.B. 1970. Coleoptera. pp. 495-621 *In* The Insects of Australia. C.S.I.R.O., Melbourne University Press, Canberra. 1029 pp.

Brodsky, K.A. 1980. Mountain Torrent of the Tien Shan. Monographiae Biologicae 39. W. Junk, The Hague. 311 pp.

Brooker, M.P. and R.J. Hemsworth. 1978. The effect of the release of an artificial discharge of water on invertebrate drift in the R. Wye, Wales. Hydrobiologia 59: 155-163.

Brooker, M.P. and D.L. Morris. 1980. A survey of the macro-invertebrate riffle fauna of the River Wye. Freshwat. Biol. 10: 437-458.

Brown, H.P. 1973. Survival records for elmid beetles. Ent. News 84: 278-284.

Brown, J.H. and D.W. Davidson. 1977. Competition between seed eating rodents and ants in desert ecosystems. Science 196: 880-882.

Brundin, L. 1966. Transantarctic relationships and their significance, evidenced by chironomid midges, with a monograph of the subfamilies Podonominae and Aphroteniinae and the austral Heptagyiae. Kgl. Sven. Vetenskapsakad Handl. 11: 1-472.

Brust, R.A. 1967. Weight and development time of different stadia of mosquitoes reared at various constant temperatures. Can. Ent. 99: 986-993.

Brusven, M.A. and K. Prather. 1974. Influence of stream sediments on macrobenthos. J. Entomol. Soc. B.C. 71: 25-32.

Brusven, M.A. and S.T. Rose. 1981. Influence of substrate composition and suspended sediment on insect predation by the torrent sculpin, *Cottus rhotheus*. Can. J. Fish. Aquat. Sci. 38: 1444-1448.

Bryan, E.H. 1972. Quality of stormwater drainage from urban land. Water Resour. Bull. 8: 578-588.

Bryan, J.E. and P.A. Larkin. 1972. Food specialization by individual trout. J. Fish. Res. Board Can. 29: 1615-1624.

Bryce, D. 1962. Chironomidae (Diptera) from freshwater sediments with special reference to Malham Tarn (Yorks). Trans. Soc. Brit. Ent. 15: 41-54.

Burks, B.D. 1953. The Mayflies, or Ephemeroptera, of Illinois. Illinois Nat. Hist. Surv. Div. Bull. 26: 1-216.

Buss, L.W. and J.B.C. Jackson. 1979. Competitive networks: nontransitive competitive relationships in cryptic coral reef environments. Am. Nat. 113: 223-234.

Busvine, J.R. 1979. Arthropod Vectors of Disease. Inst. Biol. Stud. No. 55, Edward Arnold, London. 67 pp.

Butler, M.G. 1982. A 7 year life-cycle for two *Chironomus* species in Arctic Alaskan tundra ponds (Diptera: Chironomidae). Can. J. Zool. 60: 58-70.

Butler, M.G. 1984. Life histories of aquatic insects. pp. 24-55 *In* The Ecology of Aquatic Insects. (V.H. Resh and D.M. Rosenberg, eds). Praeger Scientific, New York. 625 pp.

Buxton, P.A. 1926. The colonization of the sea by insects: with an account of the habits of *Pontomyia*, the only known submarine insect. Proc. Zool. Soc. Lond. 1926: 808-814.

Byers, G.W. 1984. Tipulidae. pp. 491-514 *In* An Introduction to the Aquatic Insects of North America. (R.W. Merritt and K.W. Cummins, eds). Kendal/Hunt Publishers, Dubuque, Iowa. 722 pp.

Cairns, J. Jr. and K.L. Dickson. 1971. A simple method for the biological assessment of the effects of waste discharges on aquatic bottom-dwelling organisms. J. Wat. Poll. Cont. 43: 755-772.

Campbell, I. 1988. Ephemeroptera. pp. 1-22 *In* Zoological Catalogue of Australia, Vol. 6: Ephemeroptera, Megaloptera, Odonata, Plecoptera, Trichoptera. Bureau of Flora & Fauna, Australian Government Publishing Service, Canberra. 316 pp.

Campbell, J.M. 1979. Coleoptera. pp. 357-387 *In* Canada and its Insect Fauna (H.V. Danks, ed). Mem. ent. Soc. Can. 108: 1-573 pp.

Cannings, R.A. and G.G.E. Scudder. 1978. The littoral Chironomidae (Diptera) of saline lakes in central British Columbia. Can. J. Zool. 56: 1144-1155.

Canton, S.P. and J.V. Ward. 1981. Benthos and zooplankton of coal strip mine ponds in the mountains of northwestern Colorado, U.S.A. Hydrobiologia 85: 23-31.

Cantrall, I.J. 1984. Semiaquatic Orthoptera. pp. 177-181 *In* An Introduction to the Aquatic Insects of North America. (R.W. Merritt and K.W. Cummins, eds). Kendal/Hunt Publishers, Dubuque, Iowa. 722 pp.

Cantrell, M.A. and A.J. McLachlan. 1977. Competition and chironomid distribution patterns in a newly flooded lake. Oikos 29: 429-433.

Cargill, A.S., K.W. Cummins, B.J. Hanson and R.R. Lowry. 1985. The role of lipids as feeding stimulants for shredding aquatic insects. Freshwat. Biol. 15: 455-464.

Carlsson, M., L.M. Nilsson, B. Svensson, S. Ulfstrand and R.S. Wotton. 1977. Lacustrine seston and other factors influencing the blackflies (Diptera: Simuliidae) inhabiting lake outlets in Swedish Lapland. Oikos 29: 229-238.

Carpenter, S.R. 1989. Replication and treatment strength in whole-lake experiments. Ecology 70: 453-463.

Carpenter, S.R. 1990. Large-scale perturbations: opportunities for innovation. Ecology 71: 2038-2043.

Carson, R.L. 1962. Silent Spring. Fawcett Publications, New York. 304 pp.

Carter, C.E. 1980. The life cycle of *Chironomus anthracinus* in Lough Neagh. Holarct. Ecol. 3: 214-217.

Carver, M., G.F. Gross and T.E. Woodward. 1991. Hemiptera. pp. 429-509 *In* The Insects of Australia. 2nd Edition, Vol. I. C.S.I.R.O., Melbourne University Press, Canberra. 1137 pp.

Chance, M.M. and D.A. Craig. 1986. Hydrodynamics and behaviour of Simuliidae larvae (Diptera). Can. J. Zool. 64: 1295-1309.

Chandler, H.P. 1971. Aquatic Neuroptera. pp. 234-236 *In* Aquatic Insects of California. (R.L. Usinger, ed). Univ. California Press, Berkeley. 508 pp.

Chapman, R.F. 1975. The Insects: Structure and Function. English Univ. Press, London. 819 pp.

Chapman, R.F. 1982. The Insects: Structure and Function. Harvard University Press, Cambridge, Massachusetts. 919 pp.

Cheng, L. (ed) 1976. Marine Insects. North Holland Publishers, Amsterdam. 581 pp.

Cheng, L. 1985. Biology of *Halobates* (Heteroptera: Gerridae). Ann. Rev. Ent. 30: 111-135.

Chessman, B.C. 1986. Dietary studies of aquatic insects from two Victorian rivers. Aust. J. Mar. Freshwat. Res. 37: 129-146.

Chesson, P.L. and R.R. Werner. 1981. Environmental variability promotes coexistence in lottery competitive systems. Am. Nat. 117: 923-943.

China, W.E. 1955. The evolution of the water bugs. Nat. Inst. Sci. India, Bull. 7: 91-103.

Chutter, F.M. 1963. Hydrobiological studies on the Vaal River in the Vereenigung Area. I. Introduction, water chemistry and biological studies on the fauna of habitats other than muddy bottom sediments. Hydrobiologia 21: 1-65.

Clifford, H.F. 1982. Life cycles of mayflies (Ephemeroptera), with special reference to voltinism. Quaest. Ent. 18: 15-90.

Clifford, H.F. and H. Boerger. 1974. Fecundity of mayflies (Ephemeroptera), with special reference to mayflies of a brown-water stream of Alberta, Canada. Can. Ent. 106: 1111-1119.

Cochran, W.G. 1983. Planning and Analysis of Observational Studies. John Wiley and Sons, New York. 145 pp.

Coffman, W.P. and L.C. Ferrington, Jr. 1984. Chironomidae. pp. 551-710 *In* An Introduction to the Aquatic Insects of North America. (R.W. Merritt and K.W. Cummins, eds). Kendal/Hunt Publishers, Dubuque, Iowa. 722 pp.

Colbo, M.H. 1991. A comparison of the spring-inhabiting genera of Chironomidae from the Holarctic with those from natural and man-made springs in Labrador, Canada. pp. 169-179 *In* Arthropods of Springs, with Particular Reference to Canada (D.D. Williams and H.V. Danks, eds). Mem. ent. Soc. Can. 155. 217 pp.

Colbo, M.H. and G.N. Porter. 1979. Effects of the food supply on the life history of Simuliidae (Diptera). Can. J. Zool. 57: 301-306.

Colbo, M.H. and R.S. Wotton. 1981.Preimaginal blackfly bionomics. pp. 209-226 *In* Blackflies. The Future for Biological Methods in Integrated Control. (M. Laird, ed). Academic Press, London. 399 pp.

Colless, D.H. and D.K. McAlpine. 1970. Diptera. pp. 656-740 *In* The Insects of Australia. C.S.I.R.O., Melbourne University Press, Canberra. 1029 pp.

Collins, N.C., R. Mitchell and R.G. Wiegert. 1976. Functional analysis of a thermal spring ecosystem with an evaluation of the role of consumers. Ecology 57: 1221-1232.

Commonwealth Scientific and Industrial Research Organization. 1970. The Insects of Australia. Melbourne University Press, Canberra. 1029 pp. (2nd edition, 1991)

Connell, J.H. 1980. Diversity and the coevolution of competitors, or the ghost of competition past. Oikos 35: 131-138.

Cook, R.M. and B.J. Cockrell. 1978. Predator ingestion rate and its bearing on feeding time and the theory of optimal diets. J. Anim. Ecol. 47: 529-547.

Cooper, S.D., S.J. Walde and B.L. Peckarsky. 1990. Prey exchange rates and the impact of predators on prey populations in streams. Ecology 71: 1503-1514.

Corbet, P.S. 1957. The life-histories of two spring species of dragonfly (Odonata: Zygoptera). Entomologist's. Gaz. 8: 79-89.

Corbet, P.S. 1963. A Biology of Dragonflies. Quadrangle Books, Chicago. 247 pp.

Corbet, P.S. 1980. Biology of Odonata. Ann. Rev. Ent. 25: 189-217.

Cordery, I. 1976. Some effects of urbanisation on streams. Civil Eng. Trans. 18: 7-11.

Courtemanch, D.L. and K.E. Gibbs. 1980. Short-and long-term effects of forest spraying of carbaryl (Sevin-4-Oil) on stream invertebrates. Can. Ent. 112: 271-276.

Cowell, B.C. and W.C. Carew. 1976. Seasonal and diel periodicity in the drift of aquatic insects in a subtropical Florida stream. Freshwat. Biol. 6: 587-594.

Cranston, P.S., C.D. Ramsdale, K.R. Snow and G.B. White. 1987. Keys to the Adults, Male Hypopygia, Fourth-instar Larvae and Pupae of the British Mosquitoes (Culicidae). Freshwat. Biol. Assoc., U.K. Sci. Pub. 48: 1-152.

Croft, P.S. 1986. A key to the major groups of British freshwater invertebrates. Field Studies 6: 531-579.

Crosskey, R.W. 1981a. Simuliid taxonomy - the contemporary scene. pp. 3-18 *In* Blackflies. The Future for Biological Methods in Integrated Control. (M. Laird, ed). Academic Press, London. 399 pp.

Crosskey, R.W. 1981b. Geographical distribution of Simuliidae. pp. 57-70 *In* Blackflies. The Future for Biological Methods in Integrated Control. (M. Laird, ed). Academic Press, London. 399 pp.

Crosskey, R.W. 1990. The Natural History of Blackflies. John Wiley & Sons, Chichester. 711 pp.

Crowley, P.H., R.M. Nisbet, W.S.C. Gurney and J.H. Lawton. 1987. Population regulation in animals with complex life histories: formulation and analysis of a damselfly model. Adv. Ecol. Res. 17: 1-59.

Crowson, R.A. 1981. The Biology of the Coleoptera. Academic Press, London. 802 pp.

Crowther, R.A. and H.B.N. Hynes. 1977. The effect of road salt on the drift of stream benthos. Environ. Pollut. 14: 113-126.

Culp, J.M. 1986. Experimental evidence that stream macro-invertebrate community structure is unaffected by different densities of coho salmon fry. J. North Am. Benth.

Soc. 5: 140-149.

Cummins, K.W. 1962. An evaluation of some techniques for the collection and analysis of benthic samples with special emphasis on lotic waters. Am. Midl. Nat. 67: 477-504.

Cummins, K.W. 1966. A review of future problems in benthic ecology. pp. 2-51 *In* Organism-substrate relationships in streams. Spec. Publ. No. 4. Pymatuning Laboratory of Ecology, University of Pittsburgh. 145 pp.

Cummins, K.W. 1973. Trophic relations of aquatic insects. Ann. Rev. Entomol. 18: 183-206.

Cummins, K.W. 1975. The ecology of running waters; theory and practice. Proc. Sandusky River Basin, Symp., pp. 277-293, Tiffin, Ohio.

Cummins, K.W. 1977. From headwater streams to rivers. Amer. Biol. Teacher (May): 305-312.

Cummins, K.W. and M.J. Klug. 1979. Feeding ecology of stream invertebrates. Ann. Rev. Ecol. Syst. 10: 147-172.

Cummins, K.W., R.C. Peterson, F.O. Howard, J.C. Wuycheck and V.I. Holt. 1973. The utilization of leaf litter by stream detritivores. Ecology 54: 336-345.

Cushing, C.E. 1963. Filter-feeding insect distribution and planktonic food in the Montreal River. Trans. Am. Fish. Soc. 92: 216-219.

Daborn, G.R. 1971. Survival and mortality of coenagrionid nymphs (Odonata: Zygoptera) from the ice of an aestival pond. Can. J. Zool. 49: 569-571.

Dadd, R.H. 1975. Alkalinity within the midgut of mosquito larvae with alkaline-active digestive enzymes. J. Insect Physiol. 21: 1847-1853.

Dalmat, H.T. 1955. The blackflies (Diptera: Simuliidae) of Guatemala and their role as vectors of onchocerciasis. Smithsonian Misc. Collections 125: 1-425.

Daly, H.V. 1984. General classification and key to the orders of aquatic and semiaquatic insects. pp. 76-81 *In* An Introduction to the Aquatic Insects of North America. (R.W. Merritt and K.W. Cummins, eds). Kendal/Hunt Publishers, Dubuque, Iowa. 722 pp.

Daly, H.V., J.T. Doyen and P.R. Ehrlich. 1978. Introduction to Insect Biology and Diversity. McGraw-Hill, New York. 564 pp.

Danks, H.V. (ed.) 1979. Canada and its Insect Fauna. Mem. ent. Soc. Can. 108: 1-573.

Danks, H.V. 1987. Insect Dormancy: an Ecological Perspective. Biol. Surv. Can. (terr. Arthrop.), Ottawa (Biol. Surv. Can. Monogr. Ser. 1.). 439 pp.

Danks, H.V. and D.R. Oliver. 1972. Diel periodicities of emergence of some high arctic Chironomidae (Diptera). Can. Ent. 104: 903-916.

Danks, H.V. and D.M. Rosenberg. 1987. Aquatic insects of peatlands and marshes in Canada: synthesis of information and identification of needs for research. pp. 163-174 *In* Aquatic Insects of Peatlands and Marshes in Canada (D.M. Rosenberg and H.V. Danks, eds). Mem. ent. Soc. Can. 140: 1-174.

Danks, H.V. and D.D. Williams. 1991. Arthropods of springs, with particular reference to Canada: synthesis and needs for research. pp. 203-217 *In* Arthropods of Springs, with Particular Reference to Canada (D.D. Williams and H.V. Danks, eds). Mem. ent. Soc. Can. 155. 217 pp.

Davies, B.R. 1976. The dispersal of Chironomidae larvae: a review. J. Entomol. Soc. South. Afr. 39: 39-62.

Davies, D.M. 1978. Ecology and behaviour of adult black flies (Simuliidae): a review. Quaest. Ent. 14: 3-12.

Deevey, E.S. 1941. Limnological studies in Connecticut. VI. The quantity and composition of the bottom fauna of thirty-six Connecticut and New York lakes. Ecol. Monogr. 11: 413-455.

Dermott, R.M., J. Kalff, W.C. Leggett and J. Spence. 1977. Production of *Chironomus*, *Procladius*, and *Chaoborus* at different levels of phytoplankton biomass in Lake

Memphremagog, Quebec-Vermont. J. Fish. Res. Board Can. 34: 2001-2007.

Décamps, H. 1967. Écologie des Trichoptères de la vallée d'Aure (Hautes-Pyrénées). Ann. Limnol. 3: 399-577.

Diamond, J.M. 1978. Niche shifts and the rediscovery of interspecific competition. Am. Sci. 66: 322-331.

Digby, P.G.N. and R.A. Kempton. 1987. Multivariate Analyses of Ecological Communities. Chapman and Hall, London. 206 pp.

Dillon, P.J., R.A. Reid and E. de Grosbois. 1987. The rate of acidification of aquatic ecosystems in Ontario, Canada. Nature 329: 45-48.

Dixon, R.D. and R.A. Brust. 1972. Mosquitoes of Manitoba. III. Ecology of larvae in the Winnipeg area. Can. Ent. 104: 961-968.

Dobson, T. 1976. Seaweed flies (Diptera: Coelopidae). pp. 446-464 *In* Marine Insects. (L. Cheng, ed). American Elsevier Publ. Co., New York. 581 pp.

Dobzhansky, T. 1982. Genetics and the Origin of Species. Columbia Univ. Press, New York. 364 pp.

Dodson, S. 1989. Predator-induced reaction norms. Bioscience 39: 447-452.

Donald, D.B. 1980. Deformities in Capniidae (Plecoptera) from Bow River, Alberta. Can. J. Zool. 58: 682-686.

Doyen, J.T. 1976. Marine beetles (Coleoptera: excluding Staphylinidae). pp. 497-520 *In* Marine Insects. (L. Cheng, ed). American Elsevier Publ. Co., New York. 581 pp.

Driscoll, C.T., J.P. Baker, J.J. Bisogni, and C.L. Schofield. 1980. Aluminum speciation and its effects on fish in dilute acidified waters. Nature 284: 161-163.

Dudgeon, D. 1982. Aspects of the microdistribution of insect macrobenthos in a forest stream in Hong Kong. Arch. Hydrobiol. Suppl. 64: 221-239.

Dudgeon, D. 1987. A laboratory study of optimal behaviour and the costs of net construction by *Polycentropus flavomaculatus* (Insecta: Trichoptera: Polycentropodidae). J. Zool., Lond. 211: 121-141.

Dudley, T.L., C.M. D'Antonio and S.D. Cooper. 1990. Mechanisms and consequences of interspecific competition between two stream insects. J. Anim. Ecol. 59: 849-866.

Eberhardt, L.L. 1976. Quantitative ecology and impact assessment. J. Environ. Man. 42: 1-31.

Eberhardt, L.L. and J.M. Thomas. 1991. Designing environmental field studies. Ecol. Monogr. 61: 53-73.

Edmondson, W.T. (ed) 1959. Freshwater Biology. Wiley, New York. 1248 pp.

Edmunds, G.F. Jr. 1972. Biogeography and evolution of Ephemeroptera. Ann. Rev. Ent. 17: 21-42.

Edmunds, G.F. Jr. 1984. Ephemeroptera. pp. 94-125 *In* An Introduction to the Aquatic Insects of North America. (R.W. Merritt and K.W. Cummins, eds). Kendal/Hunt Publishers, Dubuque, Iowa. 722 pp.

Edmunds, G.F., Jr., S.L. Jensen and L. Berner. 1976. The mayflies of North and Central America. Univ. Minnesota Press, Minneapolis, Minnesota. 330 pp.

Edmunds, M. 1972. Defense in animals: a survey of anti-predator defenses. Longman, London, England. 357 pp.

Edwards, F.W. 1931. Chironomidae. pp. 233-324 *In* Diptera of Patagonia and Chile. Pt II: Nematocera. British Museum (Natural History), London. 333 pp.

Eidt, D.C. 1975. The effect of Fenitrothion from large-scale forest spraying on benthos in New Brunswick headwaters streams. Can. Ent. 107: 743-760.

Ekman, S. 1911. Neue apparate zue qualitativen und quantitativen Untersuchung der Bodenfauna der Binnenseen. Int. Revue ges. Hydrobiol. 3: 553-561.

Eldridge, N. 1974. Character displacement in evolutionary time. Am. Zool. 14: 1083-1097.

Elias, S.A. and B. Wilkinson. 1983. Late glacial insect fossil assemblages from

Lobsigensee, Swiss Plateau. Studies in the Late Quaternary of Lobsigensee III. Rev. Paleobiol. 2: 189-204.

Elliott, J.M. 1969. Life history and biology of *Sericostoma personatum* Spence (Trichoptera). Oikos 20: 110-118.

Elliott, J.M. 1970. Methods of sampling invertebrate drift in running water. Ann. Limnol. 6: 133-159.

Elliott, J.M. 1972. Effect of temperature on the time of hatching in *Baetis rhodani* (Ephemeroptera: Baetidae). Oecologia 9: 47-51.

Elliott, J.M. 1977. A Key to the Larvae and Adults of British Freshwater Megaloptera and Neuroptera, with Notes on their Life Cycles and Ecology. Freshwat. Biol. Assoc. U.K., Sci. Pub. No. 35. 52 pp.

Elliott, J.M. 1982. The life cycle and spatial distribution of the aquatic parasitoid *Agriotypus armatus* (Hymenoptera: Agriotypidae) and its caddis host *Silo pallipes* (Trichoptera: Goeridae). J. Anim. Ecol. 51: 923-941.

Elliott, J.M. 1987. Temperature-induced changes in the life cycle of *Leutra nigra* (Plecoptera: Leuctridae) from a Lake District stream. Freshwat. Biol. 18: 177-184.

Elliott, J.M. and U.H. Humpesch. 1980. Eggs of Ephemeroptera. Ann. Rep. Freshwat. Biol. Assoc. 48: 41-52.

Elliott, J.M. and P.A. Tullett. 1978. A bibliography of samples for benthic invertebrates. Freshwat. Biol. Assoc. U.K., Quas. Pub. No. 4. 61 pp.

Ellis, R.A. and J.H. Borden. 1970. Predation by *Notonecta undulata* (Heteroptera: Notonectidae) on larvae of the yellow fever mosquito. Ann. Entomol. Soc. Am. 63: 963-973.

Elton, C. 1927. Animal Ecology. Sidgwick & Jackson, London. 204 pp.

Emlen, J.M. 1966. The role of time and energy in food preference. Am. Nat. 100: 611-617.

Eriksen, C.H., V.H. Resh, S.S. Balling and G.A. Lamberti. 1984. Aquatic insect respiration. pp. 27-37 *In* An Introduction to the Aquatic Insects of North America. (R.W. Merritt and K.W. Cummins, eds). Kendal/Hunt Publishers, Dubuque, Iowa. 722 pp.

Errington, J.R. 1963. Muskrat Populations. Iowa State Univ. Press. 665 pp.

Evans, E.D. and H.H. Neunzig. 1984. Megaloptera and Aquatic Neuroptera. pp. 261-270 *In* An Introduction to the Aquatic Insects of North America. (R.W. Merritt and K.W. Cummins, eds). Kendal/Hunt Publishers, Dubuque, Iowa. 722 pp.

Evans, G. 1977. The Life of Beetles. George Allen & Unwin, Ltd., London. 232 pp.

Evans, J.W. 1958. Insect distribution and continental drift. Continental drift, a symposium. Geol. Dept., Univ. Tasmania, 1956: 134-161.

Evans, M.E.G. 1961. The muscular and reproductive systems of *Atomaria ruficornis*. Trans. R. Soc. Edinb. 64: 297-399.

Extence, C.A. 1978. The effects of motorway construction on an urban stream. Environ. Pollut. 17: 245-252.

Fahy, E. 1973. Observations on the growth of Ephemeroptera in fluctuating and constant temperature conditions. Proc. R. Ir. Acad. (B) 73: 133-149.

Falk, D.L. and W.A. Dunson. 1977. The effects of season and acute sub-lethal exposure on survival times of brook trout at low pH. Water Res. 11: 13-15.

Feltmate, B.W. and D.D. Williams. 1989a. A test of crypsis and predator avoidance in the stonefly *Paragnetina media* (Plecoptera: Perlidae). Anim. Behav 37: 992-999.

Feltmate, B.W. and D.D. Williams. 1989b. Influence of rainbow trout on density and feeding behaviour of a perlid stonefly. Can. J. Fish. Aquat. Sci. 46: 1575-1580.

Feltmate B.W. and D.D. Williams. 1991a. Path and spatial learning in a stonefly nymph. Oikos 60: 64-68.

Feltmate B.W. and D.D. Williams. 1991b. Evaluation of predator-induced stress on field populations of stoneflies (Plecoptera). Ecology 72: 1800-1806.

Feminella, J.W. and V.H. Resh. 1990. Hydrologic influences, disturbance, and intraspecific competition in a stream caddisfly population. Ecology 71: 2083-2094.

Fernando, C.H. 1958. The colonization of small freshwater habitats by aquatic insects. I. General discussion, methods and colonization in the aquatic Coleoptera. Ceylon J. Sci. (Biol. Sci.) 1: 117-154.

Fernando, C.H. and D.F. Galbraith. 1973. Seasonality and dynamics of aquatic insects colonizing small habitats. Verh. Internat. Verein. Limnol. 18: 1564-1575.

Fink, T.J. 1980. A comparison of mayfly (Ephemeroptera) instar determination methods. pp. 367-380 *In* Advances in Ephemeroptera Biology. (J.F. Flannagan and K.E. Marshall, eds). Proc. 3rd Int. Conf. on Ephemeroptera, Winnipeg, Manitoba, 1979. Plenum Pub., New York. 552 pp.

Fish, D. 1983. Phytotelmata: flora and fauna. pp. 1-28 *In* Phytotelmata: Terrestrial Plants as Hosts for Aquatic Insect Communities. (J.H. Frank and L.P. Lounibos, eds). Plexus Pub. Inc., Medford, New Jersey. 292 pp.

Fish, D. and D.W. Hall. 1978. Succession and stratification of aquatic insects inhabiting the leaves of the insectivorous plant, *Sarracenia purpurea*. Am. Midl. Nat. 99:172-183.

Fisher, R.A. and J. Wishart. 1930. The arrangement of field experiments and the statistical reduction of the results. Imperial Bureau of Soil Science (London), Technical Communication Number 10: 1-23.

Fisher, S.G. and G.W. Likens. 1973. Energy flow in Bear Brook, New Hampshire: An integrative approach to stream ecosystem metabolism. Ecol. Monogr. 43: 421-439.

Flannagan, J.F. 1970. Efficiencies of various grabs and corers in sampling freshwater benthos. J. Fish. Res. Board Can. 27: 1691-1700.

Flannagan, J.F. 1979. The burrowing mayflies of Lake Winnipeg, Manitoba, Canada. pp. 103-113 *In* Proc. Second Int. Conf. on Ephemeroptera. (K. Pasternak and R. Sowa, eds). Panstwowe Wydawnietwa Nankowe, Warsaw, Poland. 312 pp.

Folsom, T.C. and N.C. Collins. 1982. An index of food limitation in the field for the larval dragonfly *Anax junius* (Odonata: Aeshnidae). Freshwat. Invert. Biol. 1: 33-40.

Foote, B.A. and W.C. Eastin. 1974. Biology and immature stages of *Discocerina obscurella* (Diptera: Ephydridae). Proc. Entomol. Soc. Wash. 76: 401-408.

Forsyth, D.J. 1978. Benthic macroinvertebrates in seven New Zealand lakes. N. Z. J. Mar. Freshwat. Res. 12: 41-49.

Fox, R.M. and J.W. Fox. 1964. Introduction to Comparative Entomology. Reinhold Pub. Corp., New York. 450 pp.

Fraley, J.J. 1979. Effects of elevated stream temperatures below a shallow reservoir on a cold water macroinvertebrate fauna. pp. 257-272 *In* The Ecology of Regulated Streams. (J.V. Ward and J.A. Stanford, eds.). Plenum Press, New York. 398 pp.

Francissen, F.P.M. and W.M. Mol. 1984. Augerius Clutius and his "De Hemerobio", an Early Work on Ephemeroptera. Basilisken Press, Marburg an der Lahn, West Germany. 128 pp.

Frank, J.H. 1983. Bromeliad phytotelmata and their biota, especially mosquitoes. pp. 101-128 *In* Phytotelmata: Terrestrial Plants as Hosts for Aquatic Insect Communities. (J.H. Frank and L.P. Lounibos, eds). Plexus Pub. Inc., Medford, New Jersey. 292 pp.

Frazier, C.A. 1969. Insect Allergy. W.H. Green, Inc., St. Louis. 493 pp.

Fredeen, F.J.H., J.G. Saha and M.H. Balba. 1975. Residues of methoxychlor and other chlorinated hydrocarbons in water, sand, and selected fauna following injections of methoxychlor blackfly larvicide into the Saskatchewan River, 1972. Pestic. Monit. J. 8: 241-246.

Freeman, B.E. 1968. Studies on the ecology of adult Tipulidae (Diptera) in southern England. J. Anim. Ecol. 37: 339-362.

Fremling, C.R. 1973. Factors influencing the distribution of mayflies along the Mississippi River. pp. 12-25 *In* Proc. 1st Int. Conf. on Ephemeroptera. (W.L. Peters and J.G. Peters, eds), Talahassee, Florida, 1970. Pub. Brill, Leiden. 312 pp.

Fretwell, S.D. 1987. Food chain dynamics: the central theory of ecology? Oikos 50: 291-301.

Frey, D.G. 1976. Interpretation of Quaternary palaeoecology from Cladocera and midges, and prognosis concerning the usability of other organisms. Can. J. Zool. 54: 2208-2226.

Friesen, M.K., T.D. Galloway and J.F. Flannagan. 1983. Toxicity of the insecticide Permethrin in water and sediment to nymphs of the burrowing mayfly *Hexagenia rigida* (Ephemeroptera: Ephemeridae). Can. Ent. 115: 1007-1014.

Frost, W.E. and M.E. Brown. 1967. The Trout. Collins, London.

Fryer, G. 1986. Enemy-free space: a new name for an ancient ecological concept. Biol. J. Linn. Soc. 27: 287-292.

Fuller, R.L. and R.J. Mackay. 1981. Effects of food quality on the growth of three *Hydropsyche* species (Trichoptera: Hydropsychidae). Can. J. Zool. 59: 1133-1140.

Galewski, K. 1971. A study of morphobiotic adaptations of European species of Dytiscidae. Pol. Pismo Ent. 41: 488-702.

Gallepp, G.W. 1977. Responses of caddisfly larvae (*Brachycentrus* sp.) to temperature, food availability and current velocity. Am. Midl. Nat. 98: 59-84.

Garie, H.L. and A. McIntosh. 1986. Distribution of benthic macroinvertebrates in a stream exposed to urban runoff. Water Res. Bull. 22: 447-455.

Gaufin, A.R. 1973. Use of aquatic invertebrates in the assessment of water quality. pp. 96-116 *In* Biological Methods for the Assessment of Water Quality. (J. Cairns, Jr., and K.L. Dickson,eds.), Philadelphia, PA, Am. Soc. Test. Mat. Sp. Publ. 528. 256 pp.

Gaufin, A.R., R. Clubb and R. Newell. 1974. Studies on the tolerance of aquatic insects to low oxygen concentrations. Great Bas. Nat. 34: 45-49.

Gause, G.F. 1934. The Struggle for Existence. Macmillin (Hafner Press), New York. 163 pp.

Geijskes, D.C. 1935. Faunistisch-okologische Untersuchungen an Roerenbach bei Leistal im Basler Tafeljura. Tijdschr. Ent. 78: 249-382.

Geological Society of America. 1983. Geologic time scale (copiled by A.R. Palmer). Geology11: 503-504.

Gerking, S.D. 1957. A method of sampling the littoral macrofauna and its application. Ecology 38: 219-226.

Gilbert, P. and C.J. Hamilton. 1983. Entomology: a Guide to Information Sources. Mansell Publishing Ltd., London. 237 pp.

Giller, P.S. 1980. The control of handling time and its effects on the foraging strategy of a heteropteran predator, *Notonecta* (Hemiptera/ Heteroptera). J. Anim. Ecol. 49: 699-712.

Giller, P.S. 1986. The natural diet of the Notonectidae: field trials using electrophoresis. Ecol. Ent. 11: 163-172.

Gilliam, J.F., D.F. Fraser and A.M. Sabat. 1989. Strong effects of foraging minnows on a stream benthic invertebrate community. Ecology 70: 445-452.

Gilmour, D., D.F. Waterhouse and M.F. Day. 1970. General anatomy and physiology. pp. 29-71 *In* The Insects of Australia. C.S.I.R.O., Melbourne University Press, Canberra. 1029 pp.

Gislason, G.M., U. Halbach and G. Flechtner. 1990. Habitat and life histories of the Trichoptera in Thjorsarver, Central Highlands of Iceland. Fauna Norw. Ser. B. 37: 83-90.

Gooch, J.L. and D.S. Glazier. 1991. Temporal and spatial patterns in mid-Appalachian springs. pp. 29-49 *In* Arthropods of Springs, with Particular Reference to Canada (D.D. Williams and H.V. Danks, eds). Mem. ent. Soc. Can. 155. 217 pp.

Gooley, G.J. 1977. An ecological survey of the Bass River and other streams in the Westernport Bay catchment. B.Sc. Hons. Thesis, Univ. Melbourne, Victoria.

Gore, J.A. 1977. Reservoir manipulations and benthic macroinvertebrates in a prairie river. Hydrobiologia 55: 113-123.

Gore, J.A. 1978. A technique for predicting in-stream flow requirements of benthic macroinvertebrates. Freshwat. Biol. 8: 141-151.

Graesser, A. and P.S. Lake. 1984. Diel changes in the benthos of stones and of drift in a southern Australian upland stream. Hydrobiologia 111: 153-160.

Grafius, E. and N.H. Anderson. 1979. Population dynamics, bioenergetics, and role of *Lepidostoma quercina* Ross (Trichoptera: Lepidostomatidae) in an Oregon coniferous stream. Ecology 60: 433-441.

Gray, L.J. and J.V. Ward. 1982. Effects of sediment releases from a reservoir on stream macroinvertebrates. Hydrobiologia 96: 177-184.

Green, R.H. 1979. Sampling Design and Statistical Methods for Environmental Biologists. John Wiley and Sons, New York. 257 pp.

Greenslade, P.J.M. 1972. Evolution in the staphylinid genus *Priochurus* (Coleoptera). Evolution 26: 203-220.

Greenslade, P.J.M. 1983. Adversity selection and the habitat templet. Am. Nat. 122: 352-365.

Gressitt, J.L. 1974. Insect biogeography. Ann. Rev. Ent. 19: 293-321.

Grime, J.P. 1977. Evidence for the existence of three primary strategies in plants and its relevance to ecological and evolutionary theory. Am. Nat. 111: 1169-1194.

Grossman, G.D., P.B. Moyle and J.O. Whitaker, Jr. 1982. Stochasticity in structural and functional characteristics of an Indiana stream fish assemblage: a test of community theory. Am. Nat. 120: 423-454.

Guthrie, M. 1989. Animals of the Surface Film. Naturalists' Handbooks 12. Richmond Publ. Co., Slough. 87 pp.

Guy, H.P. and D.E. Ferguson. 1962. Sediment in small reservoirs due to urbanisation. Proc. ASCE, J. Hydraulics Div. 88: 27-37.

Hagen, K.S. 1984. Aquatic Hymenoptera. pp. 438-447 *In* An Introduction to the Aquatic Insects of North America. (R.W. Merritt and K.W. Cummins, eds). Kendal/Hunt Publishers, Dubuque, Iowa. 722 pp.

Hall, R.J. and F.P. Ide. 1987. Evidence of acidification effects on stream insect communities in central Ontario between 1937 and 1985. Can. J. Fish. Aquat. Sci. 44: 1652-1657.

Hall, R.J., G.E. Likens, S.B. Fiance and G.R. Hendrey. 1980. Experimental acidification of a stream in the Hubbard Brook experimental forest, New Hampshire. Ecology 61: 976-989.

Hamilton, A.L. 1969. A new type of emergence trap for sampling stream insects. J. Fish. Res. Board Canada 26: 1685-1689.

Hamilton, R.W., J.K. Buttner and R.G. Brunetti. 1975. Lethal levels of sodium chloride and potassium chloride for an Oligochaete, a chironomid midge, and a caddisfly of Lake Michigan. Env. Ent. 4: 1003-1006.

Hammer, O. 1941. Biological and ecological investigations on flies associated with pasturing cattle and their excrement. Vidensk. Medd. dansk. naturh. Foren. Kbh. 105: 1-257.

Hammer, U.T., R.C. Haynes, J.M. Heseltine and S.M. Swanson. 1975. The saline lakes of Saskatchewan. Verh Internat. Verein. Limnol. 19: 589-598.

Hansen, M. 1987. The Hydrophiloidea (Coleoptera) of Fennoscandia and Denmark. Fauna Ent. Scandinavia 18: 1-254.

Hansen, M.H., W.N. Hurwitz and W.G. Madow. 1953. Sample Survey Methods and Theory. Volume 1. Methods and applications. John Wiley and Sons, New York. 476 pp.

Hardin, G. 1960. The competitive exclusion principle. Science 131: 1292-1297.

Harding, J. and M.H. Colbo. 1981. Competition for attachment sites between larvae of Simuliidae. Can Ent. 113: 761-763.

Hardy, A.C. 1924. The herring in relation to its animate environment. Part 1. The food and feeding habits of the herring with special reference to the east coast of England. Ministry of Agriculture and Fisheries, Fisheries Investigations, Series II, 7: 1-45.

Hardy, D.E. 1977. The present and future status of our knowledge of the Diptera faunas of the world. Communicated at 14th Congr. Entomology, Washington, U.S.A.

Harper, P.P. 1973. Emergence, reproduction and growth of setipalpian Plecoptera in Southern Ontario. Oikos 24: 94-107.

Harper, P.P. 1979. Plecoptera. pp. 311-313 *In* Canada and its Insect Fauna (H.V. Danks, ed). Mem. ent. Soc. Can. 108: 1-573.

Harper, P.P. 1981. Ecology of streams at high latitudes. pp. 313-338 *In* Perspectives in Running Water Ecology. (M.A. Lock and D.D. Williams, eds). Plenum, New York. 430 pp.

Harper, P.P. and L. Cloutier. 1986. Spatial structure of the insect community of a small dimictic lake in the Laurentians (Quebec). Int. Revue ges. Hydrobiol. 71: 655-685.

Harper, P.P. and H.B.N. Hynes. 1970. Diapause in the nymphs of Canadian winter stoneflies. Ecology 51: 925-927.

Harris, J.R. 1952. An Angler's Entomology. Collins, London. 268 pp.

Harrison, A.D. 1965. River zonation in Southern Africa. Arch. Hydrobiol. 61: 380-386.

Harrison, A.D. and K.H. Barnard. 1972. The stream fauna of an isolated mountain massif; Table Mountain, Cape Town, South Africa. Trans. roy. Soc. S. Afr. 40: 135-153.

Hart, D.D. 1986. The adaptive significance of territoriality in filter-feeding larval blackflies (Dipetera: Simuliidae). Oikos 46: 88-92.

Hart, D.D. 1987. Experimental studies of exploitative competition in a grazing stream insect. Oecologia 73: 41-47.

Hashimoto, H. 1976. Non-biting midges of marine habitats (Diptera: Chironomidae). pp. 377-414 *In* Marine Insects. (L. Cheng, ed). American Elsevier Publ. Co., New York. 581 pp.

Havas, M. and T.C. Hutchinson. 1982. Aquatic invertebrates from the Smoking Hills, N.W.T.: effect of pH and metals on mortality. Can. J. Fish. Aquat. Sci. 39: 890-903.

Heads, P.A. 1986. The costs of reduced feeding due to predator avoidance: potential effects on growth and fitness in *Ischnura elegans* larvae (Odonata: Zygoptera). Ecol. Entomol. 11: 369-377.

Heckman, C.W. 1979. Rice Field Ecology in North Eastern Thailand. Monogr. Biologicae 34, W. Junk, The Hague. 228 pp.

Heiman, D.R. and A.W. Knight. 1975. The influence of temperature on the bioenergetics of the carnivorous stonefly nymph, *Acroneuria californica* (Banks) (Plecoptera: Perlidae). Ecology 56: 105-116.

Hemphill, N. 1988. Competition between two stream dwelling filter-feeders, *Hydropsyche oslari* and *Simulium virgatum*. Oecologia 77: 72-80.

Hemphill, N. 1991. Disturbance and variation in competition between two stream insects. Ecology 72: 864-872.

Hemsworth, R.J. and M.P. Brooker. 1979. The rate of downstream displacement of macroinvertebrates in the upper Wye, Wales. Holarct. Ecol. 2: 130-136.

Hengeveld, H.G. 1990. Global climate change: implications for air temperature and water supply in Canada. Trans. Am. Fish. Soc. 119: 176-182.

Hennig, W. 1981. Insect Phylogeny (ed. and transl. by A.C. Pont). John Wiley and Sons,

Chichester. 514 pp.

Herricks, E.E. and J. Cairns, Jr. 1974. The recovery of stream macrobenthos from low pH stress. Revista de Biologia 10: 1-11.

Hershey, A.E. and S.I. Dodson. 1987. Predator avoidance by *Cricotopus*: cyclomorphosis and the importance of being big and hairy. Ecology 68: 913-920.

Hess, A.D. 1941. New limnological sampling equipment. Limnol. Soc. Am. Spec. Publ. 6: 1-5.

Hildrew, A.G. and C.R. Townsend. 1980. Aggregation, interference and foraging by larvae of *Plectrocnemia conspersa* (Trichoptera: Polycentropodidae). Anim. Behav. 28: 553-560.

Hildrew, A.G. and C.R. Townsend. 1987. Organization in freshwater benthic communities. pp. 347-372 *In* Organization of Communites Past and Present. (J.H.R. Gee and P.S. Giller, eds). 27th Symp. British Ecol. Soc.; Aberystwyth, 1986. Blackwell Scientific, Oxford.

Hilsenhoff, W.L. 1977. Use of arthropods to evaluate water quality of streams. Tech. Bull. Wisconsin Dept. Nat. Resour. 100. 15 pp.

Hilsenhoff, W.L. 1982. Using a biotic index to evaluate water quality in streams. Tech. Bull. Wisconsin Dept. Nat. Resour. 132. 22 pp.

Hilsenhoff, W.L. 1987. An improved biotic index of organic stream pollution. Great Lakes Entomol. 20: 31-39.

Hilsenhoff, W.L. and R.P. Narf. 1972. Plecoptera (stoneflies). pp. 8-12 *In* Aquatic Insects of the Pine-Popple River, Wisconsin. (W.L. Hilsenhoff, J.L. Longridge, R.P. Narf, K.J. Tennessen and C.P. Walton, eds). Wis. Dept. Nat. Resour. Tech. Bull. 54. 76 pp.

Hinton, H.E. 1955. On the respiratory adaptations, biology, and taxonomy of the Psephenidae, with notes on some related families (Coleoptera). Proc. Zool. Soc. Lond. 125: 543-568.

Hinton, H.E. 1968. Spiracular gills. Adv. Insect Physiol. 5: 65-162.

Hoffman, R.W., C.R. Goldman, S. Paulson and G.R. Winters. 1981. Aquatic impacts of deicing salts in the central Sierra Nevada mountains, California. Wat. Res. Bull. 17: 280-285.

Hogg, I.D. and R.H. Norris. 1991. Effects of run-off from land clearing and urban development on the distribution and abundance of macroinvertebrates in pool areas of a river. Aust. J. Mar. Freshwat. Biol. 42: 507-518.

Holt, R.D. 1984. Spatial heterogeneity, indirect interactions, and the coexistence of prey species. Am. Nat. 124: 377-406.

Horton, R.A. 1945. Erosional development of streams and their drainage basins: hydrophysical approach to quantitative morphology. Geol. Soc. Am. Bull. 56: 275-370.

Hosseinie, S.O. 1976. Effects of the amount of food on duration of stages, mortality rates, and size of individuals in *Tropisternus lateralis nimbatus* (Say) (Coleoptera: Hydrophilidae). Int. Revue ges. Hydrobiol. 61: 383-388.

Howe, R.W. 1967. Temperature effects on embryonic development in insects. Ann. Rev. Ent. 12: 15-42.

Hubault, E. 1927. Contribution a l'etude des invertebres torrenticoles. Bull. biol. Fr. Belg. Suppl. 9. 390 pp.

Humpesch, U.H. 1971. Zur. Faktorenanalyse des Schlupfrhythmus der Flugstadien von *Baetis alpinus* Pict. (Baetidae, Ephemeroptera). Oecologia 7: 328-341.

Humpesch, U.H. 1978. Preliminary notes on the effect of temperature and light condition on the time of hatching in some Heptageniidae (Ephemeroptera). Verh. int. Ver. Limnol. 20: 2605-2611.

Humpesch, U.H. 1980. Effects of temperature on the hatching time of eggs of five

Ecdyonurus spp. (Ephemeroptera) from Austrian streams and English streams, rivers and lakes. J. Anim. Ecol. 49: 317-333.

Humpesch, U.H. and J.M. Elliott. 1980. Effect of temperature on the hatching time of eggs of three *Rhithrogena* spp. (Ephemeroptera) from Austrian streams and an English stream and river. J. Anim. Ecol. 49: 643-661.

Hurlbert, S.H. 1984. Pseudoreplication and the design of ecological field experiments. Ecol. Monogr. 54: 187-211.

Hurlbert, S.H., W. Loayza and T. Moreno. 1986. Fish-flamingo-plankton interactions in the Peruvian Andes. Limnol. Oceanogr. 31: 457-468.

Hutchinson, G.E. 1959. Homage to Santa Rosalia or why are there so many kinds of animals? Am. Nat. 93: 145-159.

Hutchinson, G.E. 1981. Thoughts on aquatic insects. Bioscience 31: 495-500.

Hynes, H.B.N. 1955. Biological notes on some East African aquatic Heteroptera. Proc. R. Ent. Soc. Lond. A. 30: 43-54.

Hynes, H.B.N. 1960. The Biology of Polluted Waters. Liverpool Univ. Press, Liverpool. 202 pp.

Hynes, H.B.N. 1961. The invertebrate fauna of a Welsh mountain stream. Arch. Hydrobiol. 57: 344-388.

Hynes, H.B.N. 1970. The Ecology of Running Waters. Liverpool Univ. Press, Liverpool. 555 pp.

Hynes, H.B.N. 1974. Observations on the adults and eggs of Australian Plecoptera. Aust. J. Zool., Suppl. 29: 37-52.

Hynes, H.B.N. 1976. Biology of Plecoptera. Ann. Rev. Ent. 21: 135-153.

Hynes, H.B.N. 1977. A Key to the Adults and Nymphs of the British Stoneflies. Freshwat. Biol. Assoc. U.K., Sci. Pub. 17. 90 pp.

Hynes, H.B.N. and M.E. Hynes. 1975. The life histories of many of the stoneflies (Plecoptera) of south-eastern mainland Australia. Aust. J. Mar. Freshwat. Res. 26: 113-153.

Hynes, H.B.N. and T.R. Williams. 1962. The effect of DDT on the fauna of a central African stream. Ann. trop. Med. Parasit. 56: 78-91.

Ide, F.P. 1940. Quantitative determination of the insect fauna of rapid water. Univ. Toronto Biol. Ser. 47, Publ. Ont. Fish. Res. Lab. 59: 4-19.

Illies, J. 1965. Phylogeny and zoogeography of the Plecoptera. Ann. Rev. Ent. 10: 117-140.

Illies, J. 1969. Biogeography and ecology of neotropical freshwater insects, especially those from running waters. pp. 685-708 *In* Biogeography and Ecology in South America. (E.J. Fittkau *et al.*, eds). W. Junk, The Hague. 946 pp.

Illies, J. and L. Botosaneanu. 1963. Problemes et methodes de la classification et de la zonation ecologique des eaux courantes considerees surtout du point de vue faunistique. Mitt. int. Verein. theor. angew. Limnol. 12: 1-57.

Imms, A.D. 1990. Insect Natural History. New Naturalist Series, Bloomsbury Books, London. 317 pp.

Intergovernmental Panel on Climate Change (IPCC). 1990. Scientific assessment of climate change (2nd Draft), World Meteorological Organization, Geneva. Meteorological Office, Bracknell, UK. 40 pp.

Iversen, T.M. 1973. Life cycle and growth of *Sericostoma personatum* Spence (Trichoptera: Sericostomatidae) in a Danish spring. Ent. scand. 4: 323-327.

Iversen, T.M. 1976. Life cycle and growth of Trichoptera in a Danish spring. Arch Hydrobiol. 78: 482-493.

Iversen, T.M. 1978. Life cycles and growth of three species of Plecoptera in a Danish spring. Ent. Meddr. 46: 57-62.

Iversen, T.M. 1979. Laboratory energetics of larvae of *Sericostoma personatum*

(Trichoptera). Holarct. Ecol. 2:1-5.

Iversen, T.M. 1980. Densities and energetics of two streamliving larval populations of *Sericostoma personatum* (Trichoptera). Holarct. Ecol. 3: 65-73.

Jaag, O. and H. Ambuhl. 1964. The effects of the current on the composition of bioceonoses in flowing water streams. Int. Conf. Wat. Pollut. Res. Lond., Pergamon Press, Oxford. 31-49 pp.

Jackson, D. 1950. *Noterus clavicornis* and *N. capricornis* in Fife. Ent. mon. Mag. 85: 39-43.

Jackson, D. 1961a. Diapause in an aquatic mymarid. Nature 192: 823-824.

Jackson, D. 1961b. Observations on the biology of *Caraphractus cinctus* Walker (Hymenoptera: Mymaridae), a parasitoid of the eggs of Dytiscidae (Coleoptera). 2. Immature stages and seasonal history with a review of mymarid larvae. Parasitology 51: 269-294.

Jaeger, J. 1988. Developing policies for responding to climatic change: a summary of the discussions and recommendations of the workshops held in Villach (28 September-2 October, 1987) and Bellagio (9-13 November, 1987) under the auspices of the Beijer Institute, Stockholm. World Meteorological Organization, WMO/TD-255, Geneva, Switzerland.

James, H.G. 1969. Immature stages of five diving beetles (Coleoptera: Dytiscidae); notes on their habits and life histories, and a key to aquatic beetles of vernal woodland pools in southern Ontario. Proc. ent. Soc. Ont. 100: 52-97.

James, L.D. 1965. Using a digital computer to estimate the effects of urban development on flood peaks. Water Resour. Res. 1: 223-234.

Jeffries, M.J. and J.H. Lawton. 1984. Enemy free space and the structure of ecological communities. Biol. J. Linn. Soc. 23: 269-286.

Jenkins, M.F. 1960. On the methods by which *Stenus* and *Dianous* (Coleoptera: Staphylinidae) return to the banks of a pool. Trans. R. ent. Soc. Lond. 112: 1-14.

Jewett, S.G. Jr. 1963. A stonefly aquatic in the adult stage. Science 139: 484-485.

Jewett, S.G. Jr. 1971. Plecoptera. pp. 155-181 *In* Aquatic Insects of California. (R.L. Usinger, ed). Univ. California Press, Berkeley. 508 pp.

Johannsen, O.A. 1969. Aquatic Diptera. Cornell Univ. agric. Exp. Stn Mem. 164, 177, 205 and 210. Reprinted by Ent. Reprint Specialist, California.

Johnson, C.G. 1966. A functional system of adaptive dispersal by flight. Ann. Rev. Ent. 11: 233-260.

Johnson, D.M., R.E. Bohanan, C.N. Watson and T.H. Martin. 1984. Coexistence of *Enallagma traviatum* (Zygoptera: Coenagrionidae) in Bays Mountain Lake, Tennessee: an *in situ* enclosure experiment. Adv. Odonatal. 2: 57-70.

Jónasson, P.M. 1972. Ecology and production of the profundal benthos in relation to phytoplankton in Lake Esrom. Oikos Suppl. 14: 1-148.

Jónasson, P.M. 1978. Zoobenthos of lakes. Verh. Internat. Verein. Limnol. 20: 13-37.

Jones, J.R.E. 1950. A further ecological study of the River Rheidol: the food of the common insect of the main-stream. J. Anim. Ecol. 19: 159-174.

Joosse, E.N.G. 1966. Some observations on the biology of *Anurida maritima* (Guerin), (Collembola). S. Morph. Okol. Tiere 57: 320-328.

Junk, W. 1976. Faunal ecological studies in inundated areas and the definition of habitats and ecological niches. Anim. Res. Dev. 4: 47-54.

Kaitala, A. 1987. Dynamic life history strategy of the waterstrider *Gerris thoracicus* as an adaption to food and habitat variation. Oikos 48: 125-131.

Kamento, F.I. and C.J. Goodnight. 1956. The effects of various concentrations of ions on the asexual reproduction of the oligochaete *Aeolosoma hemphrichi*. Trans. Am. Microscop. Soc. 75: 219-228. Kavaliers, M. 1981. Rhythmic thermoregulation in larval cranefly (Diptera: Tipulidae). Can. J. Zool. 59: 555-558.

Kapoor, N.N. 1974. Some studies on the respiration of the stonefly nymph, *Paragnetina media* (Walker). Hydrobiologia 44: 37-41.

Kaushik, N.K. and H.B.N. Hynes. 1971. The fate of dead leaves that fall into streams. Arch. Hydrobiol. 68: 465-515.

Kavaliers, M. 1981. Rhythmic thermoregulation in a larval cranefly (Diptera: Tipulidae). Can. J. Zool. 59: 555-558.

Keast, J.A. 1983. In the steps of Alfred Russel Wallace: Biogeography of the Asian-Australian Interchange Zone. pp. 367-407 *In* Evolution, Time and Space: the Emergence of the Biosphere. (R.W. Sims, J.H. Price and P.E.S. Whalley, eds). Academic Press, London. 492 pp.

Kesler, D.H. 1982. Cellulase activity in four species of aquatic insect larvae in Rhode Island, U.S.A. J. Freshwat. Ecology. 1: 559-562.

Kettle, D.S. 1984. Medical and Veterinary Entomology. Croom Helm, London. 658 pp.

Kevan, D.K.McE. 1979a. Grylloptera and Orthoptera. pp. 318-323 *In* Canada and its Insect Fauna. (H.V. Danks, ed), Mem. ent. Soc. Can. 108: 1-573.

Kevan, D.K.McE. 1979b. Megaloptera and Neuroptera. pp. 351-356 *In* Canada and its Insect Fauna. (H.V. Danks, ed). Mem. ent. Soc. Can. 108: 1-573.

Key, K.H.L. 1970. Orthoptera. pp. 323-347 *In* The Insects of Australia. C.S.I.R.O., Melbourne University Press, Canberra. 1029 pp.

Khoo, S.G. 1964. Studies on the biology of *Capnia bifrons* (Newman) and notes on the diapause in the nymphs of this species. Gewass. Abwass. 34/5: 23-30.

Khoo, S.G. 1968. Experimental studies on diapause in stoneflies. I. Nymphs of *Capnia bifrons* (Newman). Proc. R. ent. Soc. Lond. (A) 43: 40-48.

King, J.M. 1981. The distribution of invertebrate communities in a small South African river. Hydrobiologia 83: 43-65.

Kirk, R.E. 1968. Experimental design: procedures for the behavioral sciences. Brooks/Cole Publ. Co., Belmont, CA. 577 pp.

Kitching, R.L. 1983. Community structure in water-filled treeholes in Europe and Australia - comparisons and speculations. pp. 205-222 *In* Phytotelmata: Terrestrial Plants as Hosts for Aquatic Insect Communities. (J.H. Frank and L.P. Lounibos, eds). Plexus Pub. Inc., Medford, New Jersey. 292 pp.

Kohler, S.L. 1985. Identification of stream drift mechanisms: an experimental and observational approach. Ecology 66: 1749-1761.

Kohler, S.L. and M.A. McPeek. 1989. Predation risk and the foraging behavior of competing stream insects. Ecology 70: 1811-1825.

Kolenati, F.A. 1848. Stettiner entomologische zeitung 9; quoted in: Liebmann, H. 1960. Handbuch der Frischwasser und Abwasserbiologie, Board I and II. Munchen.

Kolkwitz, R. and M. Marsson. 1908. Okologie der pflanzlichen saprobien. Ber. Deutsch. Bot. Ges., 26a: 505-519.

Kolkwitz, R. and M. Marsson. 1909. Ologie der tierischen saprobien. Int. Rev. ges. Hydrobiol. 2: 126-152.

Kondratieff, B.C. and J.R. Voshell. 1981. Influence of a reservoir on the life history of the mayfly *Heterocloen curiosum* (McDunnough) (Ephemeroptera: Baetidae). Can. J. Zool. 59: 305-314.

Konstantinov, A.S. 1958. The effect of temperature on growth rate and development of chironomid larvae. Dokl. Acad. Nauk S.S.S.R. Ser. Biol. 20: 506-509.

Koslucher, D.G. and G.W. Minshall. 1973. Food habits of some benthic invertebrates in a Northern cool-desert stream (Deep Creek, Curlew Valley, Idaho-Utah). Trans. Am. Micros. Soc. 93: 441-452.

Kotler, B.P. and R.D. Holt. 1989. Predation and competition: the interaction of two types of species interactions. Oikos 54: 256-260.

Kozhov, M. 1963. Lake Baikal and its Life. Monogr. Biologicae 11, W. Junk, The

Hague. 344 pp.

Krebs, C.J. 1985. Ecology: the Experimental Analysis of Distribution and Abundance. Harper and Row, New York. 800 pp.

Krebs, C.J. 1988. The experimental approach to rodent population dynamics. Oikos 52: 143-149.

Lacey, L.A. and B.A. Federici. 1979. Pathogenesis and midgut histopathology of *Bacillus thuringiensis* in *Simulium vittatum* (Diptera: Simuliidae). J. Invert. Path. 33: 171-182.

Lacey, L.A. and M.S. Mulla. 1979. Factors affecting feeding rates of black fly larvae. Mosquito News 39: 315-319.

Lackey, J.B. 1949. The biology of sewage treatment. Sewage Wks J. 21: 659 pp.

Lake, P.S., L.A. Barmuta, A.J. Boulton, I.C. Campbell and R.M. St. Clair. 1986. Australian streams and Northern Hemisphere stream ecology: comparisons and problems. Proc. Ecol. Soc. Aust. 14: 61-82.

Lake, P.S., I.A.E. Bayly and D.W. Morton. 1989. The phenology of a temporary pond in western Victoria, Australia, with special reference to invertebrate succession. Arch. Hydrobiol. 115: 171-202.

Lamberti, G.A., J.W. Feminella and V.H. Resh. 1987. Herbivory and intraspecific competition in a stream caddisfly population. Oecologia 73: 75-81.

Landin, J. 1968. Weather and diurnal periodicity of flight by *Helophorus brevipalpis* Bedel (Coleoptera: Hydrophilidae). Opusc. ent. 33: 28-36.

Landin, J. 1976. Seasonal patterns in water beetles. Freshwat. Biol. 6: 89-108.

Landin, J. 1980. Habitats, life histories, migration and dispersal by flight of two water beetles *Helophorus brevipalpis* and *H. strigifrons* (Hydrophilidae). Holarct. Ecol. 3: 190-201.

Lange, W.H. 1984. Aquatic and Semiaquatic Lepidoptera. pp. 348-360 *In* An Introduction to the Aquatic Insects of North America. (R.W. Merritt and K.W. Cummins, eds). Kendal/Hunt Publishers, Dubuque, Iowa. 722 pp.

Langton, P.H., P.S. Cranston and P.D. Armitage. 1988. The parthenogenetic midge of water supply systems, *Paratanytarsus grimmi* (Schneider) (Diptera: Chironomidae). Bull. ent. Res. 778: 317-328.

LaRivers, I. 1971. Aquatic Orthoptera. pp. 154 *In* Aquatic Insects of California. (R.L. Usinger, ed). Univ. California Press, Berkeley. 508 pp.

Larkin, P.A. 1984. A commentary on environmental impact assessment for large projects affecting lakes and streams. Can. J. Fish. Aquat. Sci. 41: 1121-1127.

Lash, T.J.F., D.E.L. Maasland, G. Filteau and P. Larkin. 1974. On doing things differently: an essay on environmental impact assessment of major projects. Issues Can. Sci. Pol. 1: 9-16.

Laurence, B.R. 1954. The larval inhabitants of cow pats. J. Anim. Ecol. 23: 234-260.

Lautenschlager, K.P., N.K. Kaushik and J.B. Robinson. 1978. The peritrophic membrane and faecal pellets of *Gammarus lacustris limnaeus* Smith. Freshwat. Biol. 8: 207-211.

Lawson, D.L., M.J. Klug and R.W. Merritt. 1984. The influence of the physical, chemical, and microbiological characteristics of decomposing leaves on the growth of the detritivore *Tipula abdominalis* (Diptera: Tipulidae). Can. J. Zool. 62: 2339-2343.

Lawton, J.H. and M.P. Hassell. 1981. Asymmetrical competition in insects. Nature 289: 793-795.

Lawton, J.H., B.A. Thompson and D.J. Thompson. 1980. The effects of prey density on survival and growth of damselfly larvae. Ecol. Entomol. 5: 39-51.

Learner, M.A., R. Williams, M. Harcup and B.D. Hughes. 1971. A survey of the macrofauna of the River Cynon, a polluted tributary of the River Taff (South Wales). Freshwat. Biol. 1: 339-367.

Lechleitner, R.A. and B.C. Kondratieff. 1983. The life history of *Pteronarcys dorsata*

(Say) (Plecoptera: Pteronarcyidae) in southwestern Virginia. Can. J. Zool. 61: 1981-1985.

Leggot, M., and G. Pritchard. 1986. Thermal preference and activity thresholds in populations of *Argia vivida* (Odonata: Coenagrionidae) from habitats with different thermal regimes. Hydrobiologia 140: 85-92.

Lehane, M.J. 1991. Biology of Blood-sucking Insects. Harper Collins, London. 310 pp.

Lehmkuhl, D.M. 1971. Stoneflies (Plecoptera: Nemouridae) from temporary lentic habitats in Oregon. Am. Midl. Nat. 85: 514-515.

Lehmkuhl, D.M. 1972. Change in thermal regime as a cause of reduction of benthic fauna downstream of a reservoir. J. Fish. Res. Board Canada 29: 1329-1332.

Lehmkuhl, D.M. 1974. Thermal regime alteration and vital environmental physiological signals in aquatic organisms. pp. 116-222 *In* Thermal Ecology. (J.W. Gibbons and R.R. Sharitz, eds). Nat. Tech. Inf. Serv. Conf. 730505. U.S. Atom. Energ. Comm. 670 pp.

Lehner, P.N. 1979. Handbook of Ethological Methods. Garland STPM Press, New York. 403 pp.

Leivestad, H. and I.P. Muniz. 1976. Fish kill at low pH in a Norwegian river. Nature 259: 391-392.

Leonard, J.W. and F.A. Leonard. 1962. Mayflies of Michigan trout streams. Cranbrook Inst. Sci. Bull. 43: 1-139.

Lepneva, S.G. 1970. Fauna of the U.S.S.R.; Trichoptera, vol 2, no. 1. Larvae and Pupae of Annulipalpia. Zool. Inst. Akad. Nauk S.S.S.R., n.s. 88. Published by the Israel Programme for Scientific Translations.

Lidicker, W.Z., Jr. 1988. The synergistic effects of reductionist and holistic approaches in animal ecology. Oikos 53: 278-281.

Likens, G.E., F.H. Bormann, N.M. Johnson, D.W. Fisher and R.S. Pierce. 1970. Effects of forest cutting and herbicide treatment on nutrient budgets in the Hubbard Brook watershed-ecosystem. Ecol. Monogr. 40: 23-47.

Lindegaard, C., J. Thorup and M. Bahn. 1975. The invertebrate fauna of the moss carpet in the Danish spring Ravnkilde and its seasonal, vertical, and horizontal distribution. Arch. Hydrobiol. 75: 109-139.

Linduska, J.P. 1941. Bottom type as a factor influencing the local distribution of mayfly nymphs. Can. Ent. 74: 26-30.

Linley, J.R. 1976. Biting midges of mangrove swamps and saltmarshes (Diptera: Ceratopogonidae). pp. 335-376. *In* Marine Insects. (L. Cheng, ed). American Elsevier Publ. Co., New York. 581 pp.

Linley, J.R. and J.B. Davies. 1971. Sandflies and tourism in Florida and the Bahamas and Caribbean area. J. Econ. Entomol. 64: 264-278.

Livingstone, D.A., K. Bryan, Jr. and R.C. Leahy. 1958. Effects of an arctic environment on the origin and development of freshwater lakes. Limnol. Oceanogr. 3: 192-214.

Lloyd, M. and R.J. Ghelardi. 1964. A table for calculating the "equitability" component of species diversity. J. Anim. Ecol. 33: 217-225.

Lubchenco, J. 1978. Plant species diversity in a marine intertidal community: importance of herbivore food preference and algal competitive abilities. Am. Nat. 100: 603-609.

Lubchenco, J. 1986. Relative importance of competition and predation: early colonization by seaweeds in New England. pp. 537-555 *In* Community Ecology. (J. Diamond and T.V. Case, eds). Harper and Row, New York. 665 pp.

Lubchenco, J., A.M. Olson, L.B. Brubaker, S.J. Carpenter, M.M. Holland, S.P. Hubbell, S.A. Levin, J.A. MacMahon, P.A. Matson, J.M. Melillo, H.A. Mooney, C.H. Peterson, H.R. Pulliam, L.A. Real, P.J. Regal and P.G. Risser. 1991. The sustainable biosphere initiative: an ecological research agenda. Ecology 72: 371-412.

Lutz, P.E. 1968. Effects of temperature and photoperiod on larval development in *Lestes eurinus* (Odonata: Lestidae). Ecology 49: 637-644.

Macan, T.T. 1938. Evolution of aquatic habitats with special reference to the distribution of the Corixidae. J. Anim. Ecol. 7: 1-19.

Macan, T.T. 1939. Notes on the migration of some aquatic insects. J. Soc. Br. Ent. 2: 1-6.

Macan, T.T. 1962. Ecology of aquatic insects. Ann. Rev. Ent. 7:261-288.

Macan, T.T. 1964. The Odonata of a moorland fishpond. Int. Rev. ges. Hydrobiol. 49: 325-360.

Macan, T.T. 1974. Freshwater Ecology. Longmans, London. 343 pp.

Macan, T.T. 1977. The influence of predation on the compostion of fresh-water animal communities. Biol. Rev. 52: 45-70.

MacArthur, R.H. 1972. Geographical Ecology. Harper and Row, New York. 269 pp.

MacArthur, R.H. and E.O. Wilson. 1967. Theory of Island Biogeography. Princeton Univ. Press. Princeton. 203 pp.

Machado-Allison, D.J., R.R. Berrera and C.G. Cova. 1983. The insect community associated with inflorescences of *Heliconia caribaea* Lamark, in Venezuela. pp. 247-270. *In* Phytotelmata: Terrestrial Plants as Hosts for Aquatic Insect Communities. (J.H. Frank and L.P. Lounibos, eds). Plexus Pub. Inc., Medford, New Jersey. 292 pp.

MacIsaac, H.J. and J.J. Gilbert. 1991. Discrimination between exploitative and interference competition between Cladocera and *Keratella cochlearis*. Ecology 72: 924-937.

Mackay, R.J. and J. Kalff. 1973. Ecology of two related species of caddisfly larvae in the organic substrates of a woodland stream. Ecology 54: 499-511.

Mackay, R.J. and G.B. Wiggins. 1979. Ecological diversity in Trichoptera. Ann. Rev. Ent. 24: 185-208.

Mackerras, I.M. 1970. Composition and distribution of the fauna. pp. 187-203 *In* The Insects of Australia. C.S.I.R.O., Melbourne University Press, Canberra. 1029 pp.

Mackey, A.P. 1977. Growth and development of larval Chironomidae. Oikos 28: 270-275.

Maddux, D.E. 1954. A new species of dobsonfly from California (Megaloptera: Corydalidae). Pan-Pac. Ent. 30: 70-71.

Madsen, B.L. and I Butz. 1976. Population movements of adult *Brachyptera risi* (Plecoptera). Oikos 27: 273-280.

Magnuson, J.J., L.B. Crowder and P. Medvick. 1979. Temperature as an ecological resource. Am. Zool. 19: 331-343.

Malipatil, M.B. and J.D. Blyth. 1982. A qualitative study of the macroinvertebrate fauna of the Thomson River and its major tributaries, Grippsland, Victoria. Rep. Nat. Mus. Vic. 1: 1-95.

Mallet, J., J.T. Longino, D. Murawski and A.S. Gamboa. 1987. Handling effects in *Heliconius*: where do all the butterflies go? J. Anim. Ecol. 56: 377-386.

Manly, B.F.J. 1986. Multivariate Statistical Methods: a Primer. Chapman and Hall, London. 159 pp.

Manton, S.M. 1979. Functional morphology and the evolution of the hexapod classes. pp. 387-465 *In* Arthropod Phylogeny (A.P. Gupta, ed). Van Nostrand Reinhold, New York. 762 pp.

Manton, S.M. and D.T. Anderson. 1979. Polyphyly and the evolution of arthropods. pp. 269-321 *In* The Origin of Major Invertebrate Groups (M.R. House, ed). The Systematics Assn. Spec. Vol. 12, Academic Press, London. 515 pp.

Martin, M.M., J.J. Kukor, J.S. Martin, D.L. Lawson, and R.W. Merritt. 1981a. Digestive enzymes of larvae of three species of caddisflies (Trichoptera). Insect. Biochem. 11: 501-505.

Martin, M.M., J.S. Martin, J.J. Kukor and R.W. Merritt. 1981b. The digestive enzymes of detritus-feeding stonefly nymphs (Plecoptera: Pteronarcyidae). Can. J. Zool. 59: 1947-1951.

Martin, M.M., J.S. Martin, J.J. Kukor and R.W. Merritt. 1980. The digestion of protein and carbohydrate by the stream detritivore, *Tipula abdominalis* (Diptera: Tipulidae). Oecologia 46: 360-364.

Mason, C.F. and R.J. Bryant. 1975. Periphyton production and grazing by chironomids in Alderfren Broad, Norfolk. Freshwat. Biol. 5: 271-277.

Matsuda, R. 1976. Morphology and Evolution of the Insect Abdomen. Pergamon Press, Oxford. 534 pp.

Matter, W.J. and J.J. Ney. 1981. The impact of surface mine reclamation on headwater streams in Southwest Virginia. Hydrobiologia 78: 63-71.

Mattingly, P.F. 1969. The Biology of Mosquito-Borne Disease. George Allen and Unwin Ltd., London. 184 pp.

May, R.M. 1975. Stability and complexity in model ecosystems. Princeton University Press, Princeton, New Jersey. 235 pp.

May, M.L. 1976. Thermoregulation and adaptation to temperature in dragonflies. Ecol. Monogr. 46: 1-32.

May, R.M. 1978. The evolution of ecological systems. Sci. Am. 239: 160-176.

Maynard Smith, J. 1974. The theory of games and the evolution of animal conflicts. J. Theor. Biol. 47: 209-221.

Maynard Smith, J. 1976. Evolution and the theory of games. Am. Sci. 64: 41-45.

McAlpine, J.F. 1979. Diptera. pp. 389-424 *In* Canada and its Insect Fauna. (H.V. Danks, ed). Mem. ent. Soc. Can. 108: 1-573.

McAuliffe, J.R. 1984a. Competition for space, disturbance, and the structure of a benthic stream community. Ecology 65: 894-908.

McAuliffe, J.R. 1984b. Resource depression by a stream herbivore: effects on distributions and abundances of other grazers. Oikos 42: 327-333.

McCafferty, W.P. 1981. Aquatic Entomology. Science Books Internat., Boston, Massachusetts. 448 pp.

McCafferty, W.P. and G.F. Edmunds, Jr. 1979. The higher classification of the Ephemeroptera and its evolutionary basis. Ann. Ent. Soc. Am. 72: 5-12.

McClelland, W.T. and M.A. Brusven. 1980. Effects of sedimentation on the behaviour and distribution of riffle insects in a laboratory stream. Aquatic Insects 2: 161-169.

McCullough, D.A. 1975. The bioenergetics of 3 aquatic insects determined by radioisotope analysis. Battelle Pacific Northwest Labs., B.N.W.L. 1928, special distribution VC-48. 225 pp.

McElhone, M.J. and R.W. Davies. 1983. The influence of rock surface area on the microdistribution and sampling of attached riffle dwelling Trichoptera in Hartley Creek, Alberta. Can. J. Zool. 61: 2300-2304.

McLachlan, A.J. and M.A. Cantrell. 1980. Survival strategies in tropical rain pools. Oecologia 47: 344-351.

McLay, C. 1970. A theory concerning the distance travelled by animals entering the drift of a stream. J. Fish. Res. Board Can. 27: 359-370.

McLellan, I.D. 1975. The freshwater insects. pp. 537-559. *In* Biogeography and Ecology in New Zealand. (G. Kuschel, ed). Monogr. Biologicae 27, W. Junk, The Hague. 689 pp.

Menge, B.A. 1979. Coexistence between the seastars *Asterias vulgaris* and *A. forbesi* in a heterogeneous environment: a non-equilibrium explanation. Oecologia 41: 245-272.

Merritt, R.W. 1976. A review of the food habits of the insect fauna inhabiting cattle droppings in north central California. Pan-Pac. Ent. 52: 13-22.

Merritt, R.W. and K.W. Cummins (eds). 1984. An Introduction to the Aquatic Insects of North America. Kendal/Hunt Publishers, Dubuque, Iowa. 722 pp.

Merritt, R.W. and J.B. Wallace. 1981. Filter-feeding insects. Sci. Am. 244: 132-144.

Michaelis, F.B. 1977. Biological features of Pupu Springs. N.Z.J. mar. Freshwat. Res. 11: 357-373.

Miller, G.T. Jr. 1988. Living in the Environment. Wadsworth Publishing Company, Belmont, CA. 603 pp.

Miller, N.C.E. 1956. The Biology of the Heteroptera. Leonard Hill Ltd., London. 162 pp.

Minckley, W.L. 1963. The ecology of a spring stream, Doe Run, Meade County, Kentucky. Wildlf Monogr. 11: 1-124.

Minshall, G.W. and J.N. Minshall. 1978. Further evidence on the role of chemical factors in determining the distribution of benthic invertebrates in the River Duddon. Arch. Hydrobiol. 83: 324-355.

Minshall, G.W. and R.C. Petersen, Jr. 1985. Towards a theory of macroinvertebrate community structure in stream ecosystems. Arch. Hydrobiol. 104: 49-76.

Mittlebach, G.G. 1984. Predation and resource partitioning in two sunfishes (Centrarchidae). Ecology 65: 499-513.

Mittlebach, G.G. 1988. Competition among refuging sunfishes and effects of fish density on littoral zone invertebrates. Ecology 69: 614-623.

Mittlebach, G.G. and P.L. Chesson. 1987. Predation risk: indirect effects on fish populations. pp. 315-332 *In* Predation: Direct and Indirect Impacts on Aquatic Communities. (W.C. Kerfoot and A. Sih, eds.). Univ. Press of New England, Hanover, New Hampshire. 366 pp.

Mol, A. 1982. The role of the invertebrate fauna in the biological assessment of water quality. Hydrobiol. Bull. 14: 222-223.

Moore, K.A. and D.D. Williams. 1990. Novel strategies in the complex defense repertoire of a stonefly (*Pteronarcys dorsata*) nymph. Oikos 57: 49-56.

Mordukhai-Boltovskoi, Ph.D. 1979. Zoobenthos and other invertebrates living on substrata. pp. 235-268 *In* The River Volga and its Life. (Ph. D. Mordukhai-Boltovskoi, ed). Monogr. Biologicae 33, W. Junk, The Hague. 473 pp.

Morin, P.J., S.P. Lawler and E.A. Johnson. 1988. Competition between aquatic insects and vertebrates: interaction strength and higher order interactions. Ecology 69: 1401-1409.

Morisawa, M. 1968. Streams, their Dynamics and Morphology. McGraw-Hill, New York. 175 pp.

Moum, S.E. and R.L. Baker. 1990. Colour change and substrate selection in larval *Ischnura verticalis* (Coenagrionidae: Odonata). Can. J. Zool. 68: 221-224.

Muirhead-Thomson, R.C. 1973. Laboratory evaluation of pesticide impact on stream invertebrates. Freshwat. Biol. 3: 479-498.

Müller, K. 1954. Investigations on the organic drift in north Swedish streams. Rep. Inst. Freshwat. Res. Drottningholm 35: 133-148.

Müller, K. 1966. Die Tagesperiodik von Fliesswasserorganismen. Z. Morph. Okol. Tierre 56: 93-142.

Mundie, J.H. 1956. Emergence traps for aquatic insects. Mitt. Int. Verh. Limnol. 7: 1-13.

Murphy, G.I. 1968. Patterns in life history and the environment. Am. Nat. 102: 390-404.

Murray, J. 1910. Characteristics of lakes in general and their distribution over the surface of the globe. Bath. Surv. Freshwat. Lochs Scot. 1: 514-618.

Nagell, B. and C.C. Landahl. 1978. Resistance to anoxia of *Chironomus plumosus* and *Chironomus anthracinus* (Diptera) larvae. Hol. Ecol. 1: 333-336.

Nebeker, A.V. 1971. Effect of water temperature on nymphal feeding rate, emergence, and adult longevity of the stonefly *Pteronarcys dorsata* . J. Kans. ent. Soc. 44: 21-26.

Nebeker, A.V. 1973. Temperature requirements and life cycle of the midge *Tanytarsus dissimilis* (Diptera: Chironomidae). J. Kans. ent. Soc. 46: 160-165.

Needham, J.G., J.R. Traver and Y-C. Hsu. 1972. The Biology of Mayflies. Comstock

Pub. Co., New York. (1935), Reprinted by E.W. Classey, Hampton, Middlesex, U.K. 759 pp.

Nelson, G. 1989. Species and taxa: systematics and evolution. pp. 60-81 *In* Speciation and its Consequences (D. Otte and J.A. Endler, eds). Sinauer Assoc. Inc., Massachusetts. 679 pp.

Nelson, G.S. 1978. Mosquito-borne filariasis. pp. 15-25 *In* Medical Entomology Centenary Symp. Proc. (S. Willmott, ed). Roy. Soc. Trop. Med. Hygiene, London.

Nelson, G.S. 1981. Issues in filariasis - a century of enquiry and a century of failure. Acta Tropica 38: 197-204.

Nemenz, H. 1960. On the osmotic regulation of the larvae of *Ephydra cinerea*. J. Insect Physiol. 4: 38-44.

Neves, R.J. 1979. Movement of larval and adult *Pycnopsyche guttifer* (Walker) along Factory Brook, Massachusetts. Am. Midl. Nat. 102: 51-58.

Newbold, J.D., J.W. Elwood, R.V. O'Neill and W. Van Winkle. 1981. Measuring nutrient spiralling in streams. Can. J. Fish. Aquat. Sci. 38: 860-863.

Newbold, J.D., D.C. Erman and K.B. Roby. 1980. Effects of logging on macroinvertebrates in streams with and without buffer strips. Can. J. Fish. Aquat. Sci. 37: 1076-1085.

Norris, K.R. 1991. General Biology. pp. 68-108 *In* The Insects of Australia. 2nd Edition, Vol. I. C.S.I.R.O., Melbourne University Press, Canberra. 1137 pp.

Norris, R.H. 1986. Mine waste pollution of the Molonglo River, New South Wales and the Australian Capital Territory: effectiveness of remedial works at Captains Flat mining area. Aust. J. mar. Freshwat. Res. 37: 147-157.

Norris, R.H., P.S. Lake and R. Swain. 1982. Ecological effects of mine effluents on the South Esk River, North-eastern Tasmania III. Benthic macroinvertebrates. Aust. J. mar. Freshwat. Res. 33: 789-809.

Nursall, J.R. 1952. The early development of a bottom fauna in a new power reservoir in the Rocky Mountains of Alberta. Can. J. Zool. 30: 387-409.

Odum, W.E. and R.T. Prentki. 1978. Analysis of five North American lake ecosystems. IV. Allochthonous carbon inputs. Int. Ver. Theor. Ang. Limnol. Verh. 20: 574-580.

O'Farrell, A.F. 1970. Odonata. pp. 241-261 *In* The Insects of Australia. C.S.I.R.O., Melbourne University Press, Canberra. 1029 pp.

Okland, J. and K.A. Okland. 1980. pH level and food organisms for fish: studies of 1000 lakes in Norway. pp. 326-327 *In* Ecological Impact of Acid Precipitation. (D. Drablos and A. Tollan, eds), Sandefjord, Norway, Proc. Int. Conf., Oslo. 383 pp.

Oliver, D.R. 1968. Adaptations of Arctic Chironomidae. Ann. zool. fenn. 5: 111-118.

Oliver, D.R. 1979. Chironomidae. pp. 402 *In* Canada and its Insect Fauna (H.V. Danks, ed). Mem. ent. Soc. Can. 108: 1-573.

O'Meara, G.F. 1976. Saltmarsh mosquitoes (Diptera: Culicidae). pp. 303-334 *In* Marine Insects. (L. Cheng, ed). American Elsevier Publ. Co., New York. 581 pp.

Orghidan, T. 1959. Ein neuer Lebensraum des unterirdischen Wassers: Der hyporheische Biotop. Archiv fur Hydrobiologie 55: 392-414.

Otto, C. 1974. Growth and energetics in a larval population of *Potamophylax cingulatus* (Steph.) (Trichoptera) in a south Swedish stream. J. Anim. Ecol. 43: 339-361.

Otto, C. 1976. Factors affecting the drift of *Potamophylax cingulatus* (Trichoptera) larvae. Oikos 27: 93-100.

Otto, C. 1984. Adaptive head coloration in case-making caddis larvae. Freshwat. Biol. 14: 317-321.

Otto, C. and J.B. Wallace. 1989. Life cycle variation and habitat longevity in waterlily leaf beetles. Holarct. Ecol. 12: 144-151.

Our Common Future. 1987. World Commission on Environment and Development, Oxford University Press, Oxford 400 pp.

Outridge, P. 1987. Possible causes of high species diversity in tropical Australian freshwater macrobenthic communities. Hydrobiologia 150: 95-107.

Paine, R.T. 1974. Intertidal community structure: experimental studies between a dominant competitor and its principal predator. Oecologia 15: 93-120.

Paine, R.T. 1988. Food webs: road maps of interactions or grist for theoretical development? Ecology 69: 1648-1654.

Pajunen, V. and A. Jansson. 1969. Dispersal of rock-pool corixids *Arctocorixa carinata* (Sahler) and *Callicorixa producta* (Reut) (Heteroptera: Corixidae). Ann. Zool. Fenn. 6: 391-427.

Paterson, C.G. and C.H. Fernando. 1969. Macroinvertebrate colonization of the marginal zone of a small impoundment in Eastern Canada. Can. J. Zool. 47: 1229-1238.

Paterson, C.G. and C.H. Fernando. 1970. Benthic fauna colonization of a new reservoir with particular reference to the Chironomidae. J. Fish. Res. Board Can. 27: 213-232.

Peckarsky, B.L. 1980. Predator-prey interactions between stoneflies and mayflies: behavioral observations. Ecology 61: 932-943.

Peckarsky, B.L. and M.A Penton. 1985. Is predaceous stonefly behavior affected by competition? Ecology 66: 1718-1728.

Pennak, R.W. 1940. Ecology of microscopic Metazoa inhabiting the sandy beaches of some Wisconsin lakes. Ecol. Monogr. 10: 537-615.

Pennak, R.W. 1953. Freshwater Invertebrates of the United States. Ronald Press, New York. 769 pp.

Peterman, R.M. 1990a. The importance of reporting statistical power: the forest decline and acidic deposition example. Ecology 71: 2024-2027.

Peterman, R.M. 1990b. Statistical power analysis can improve fisheries research and management. Can. J. Fish. Aquat. Sci. 47: 2-15.

Petersen, L. B-M. and R.C. Petersen. 1983. Anomalies in hydropsychid capture nets from polluted streams. Freshwat. Biol. 13: 185-191.

Petersen, R.C. and K.W. Cummins. 1974. Leaf processing in a woodland stream. Freshwat. Biol. 4: 343-368.

Peterson, B.V. 1981. Simuliidae. pp. 355-391 *In* Manual of Nearctic Diptera, Vol. 1. (J.F. McAlpine, B.V. Peterson, G.E. Sherwell, H.J. Teskey, J.R. Vockeroth and D.M. Wood, eds). Res. Branch, Agric. Can. Monogr. 27. Ottawa. 674 pp.

Peterson, B.V. 1984. Simuliidae. pp. 534-550 *In* An Introduction to the Aquatic Insects of North America. (R.W. Merritt and K.W. Cummins, eds). Kendal/Hunt Publishers, Dubuque, Iowa. 722 pp.

Pianka, E.R. 1970. On r- and K-selection. Am. Nat. 104: 592-597.

Pianka, E.R. 1975. Niche relations of desert lizards. pp. 292-314 *In* Ecology and Evolution of Communities. (M.L. Cody and J.M. Diamond, eds). Harvard University Press, Cambridge, Mass. 540 pp.

Pierce, C.L. 1988. Predator avoidance, microhabitat shift, and risk-sensitive foraging in larval dragonflies. Oecologia 77: 81-90.

Pimm, S.L. 1982. Food Webs. Chapman and Hall Ltd., London. 219 pp.

Pimm, S.L. and R.L. Kitching. 1988. Food web patterns: trivial flaws or the basis of an active research program. Ecology 69: 1669-1672.

Pimm, S.L. and J.C. Rice. 1987. The dynamics of multispecies, multi-life-stage models of aquatic food webs. Theor. Pop. Biol. 32: 303-325.

Pinder, L.C.V. 1978. A Key to the Adult Males of the British Chironomidae (Diptera). Vol. I, The Key. Freshwat. Biol. Assoc. U.K., Sci. Pub. 37. 169 pp.

Pinder, L.C.V. 1986. Biology of freshwater Chironomidae. Ann. Rev. Ent. 31: 1-23.

Polhemus, J.T. 1976. Shore bugs (Hemiptera: Saldidae, etc.). pp. 225-262 *In* Marine Insects. (L. Cheng, ed). American Elsevier Publ. Co., New York. 581 pp.

Polhemus, J.T. 1984. Aquatic and semiaquatic Hemiptera. pp. 231-260 *In* An

Introduction to the Aquatic Insects of North America. (R.W. Merritt and K.W. Cummins, eds). Kendal/Hunt Publishers, Dubuque, Iowa. 722 pp.

Popham, E.J. 1953. Observations on the migration of corixids (Hemiptera) into a new aquatic habitat. Entomol. mon. Mag. 89: 124-125.

Price, P.W. 1974. Strategies for egg production. Evolution 28: 76-84.

Pritchard, G. 1976. Growth and development of larvae and adults of *Tipula sacra* Alexander (Insecta: Diptera) in a series of abandoned beaver ponds. Can. J. Zool. 54: 266-284.

Pritchard, G. 1983. Biology of Tipulidae. Ann. Rev. Ent. 28: 1-22.

Pritchard, G. 1991. Insects in thermal springs. pp. 89-106 *In* Arthropods of Springs, with Particular Reference to Canada (D.D. Williams and H.V. Danks, eds). Mem. ent. Soc. Can. 155. 217 pp.

Pruthi, H.S. 1932. Colonization of the sea by insects. Nature 130: 312.

Pugsley, C.W. and H.B.N. Hynes. 1983. A modified freeze-core technique to quantify the depth distribution of fauna in stony streambeds. Can. J. Fish. Aquat. Sci. 40: 637-643.

Racovitza, E.G. 1907. Essai sur les problemes biospeleologigues. Biospeologica. I. Arch. de zoologie experimentale et generale 6: 371-488.

Raddum, G.G. and A. Fjellheim. 1984. Acidification and early warning organisms in freshwater in western Norway. Int. Assoc. Theor. Appl. Limnol. 2: 1973-1980.

Raddum, G. and O.A. Saether. 1981. Chironomid communities in Norwegian lakes with different degrees of acidification. Verh. Internat. Verein. Limnol. 21: 399-405.

Redfield, G.W. 1988. Holism and reductionism in community ecology. Oikos 53: 276-278.

Reeve, M.R., T.C. Cosper and M.A. Walter. 1975. Visual observations on the process of digestion and the production of faecal pellets in the chaetognath *Sagitta hipida* Conant. J. Exp. mar. Biol. Ecol. 17: 39-46.

Reice, S.R. and R.L. Edwards. 1986. The effect of vertebrate predation on lotic macroinvertebrate communities in Quebec, Canada. Can. J. Zool. 64: 1930-1936.

Reid, G.K. 1961. Ecology of Inland Waters and Estuaries. Van Nostrand Reinholt, New York. 375 pp.

Resh, V.H., G.A Lamberti and J.R. Wood. 1984. Biological studies of *Helicopsyche borealis* (Hagen) in a coastal California stream. pp. 315-319 *In* Proceedings of the 4th international symposium on Trichoptera. (J.C. Morse, ed.). The Hague, Dr. W. Junk Publishers.

Revill, D.L., K.W. Stewart and H.E. Schlichting. 1967. Passive dispersal of viable algae and Protozoa by certain craneflies and midges. Ecology 48: 1023-1027.

Reznick, D. and J.A Endler. 1982. The impact of predation on life history evolution in Trinidadian guppies (*Poecilia reticulata*). Evolution 36: 160-177.

Richards, W.R. 1979. Collembola. pp. 300-303 *In* Canada and its Insect Fauna (H.V. Danks, ed). Mem. ent. Soc. Can. 108: 1-573.

Richardson, B.A. 1985. The impact of forest road construction on the benthic invertebrate and fish fauna of a coastal stream in southern New South Wales. Aust. Soc. Limnol. Bull. 10: 65-88.

Richardson, J.S. 1991. Seasonal food limitation of detritivores in a Montane stream: an experimental test. Ecology 72: 873-887.

Ricker, W.E. 1934. An ecological classification of certain Ontario streams. Univ. Toronto Stud. Biol. 37: 1-114.

Ricker, W.E. 1964. Distribution of Canadian stoneflies. Gewass. Abwass. 34/35: 50-71.

Riek, E.F. 1970a. Fossil history. pp. 168-186 *In* The Insects of Australia. C.S.I.R.O., Melbourne University Press, Canberra. 1029 pp.

Riek, E.F. 1970b. Megaloptera and Neuroptera. pp. 465-494 *In* The Insects of Australia.

C.S.I.R.O., Melbourne University Press, Canberra. 1029 pp.

Riek, E.F. 1970c. Trichoptera. pp. 741-764 *In* The Insects of Australia. C.S.I.R.O., Melbourne University Press, Canberra. 1029 pp.

Ring, R.A. 1991. The insect fauna and some other characteristics of natural salt springs on Saltspring Island, British Columbia. pp. 51-61 *In* Arthropods of Springs, with Particular Reference to Canada (D.D. Williams and H.V. Danks, eds). Mem. ent. Soc. Can. 155. 217 pp.

Ringler, N.H. 1979. Selective predation by drift-feeding brown trout (*Salmo trutta*). J. Fish. Res. Board Can. 36: 392-403.

Rogers, J.S. 1933. The ecological distribution of the crane-flies of northern Florida. Ecol. Monogr. 3: 1-74.

Rosenberg, D.M. 1986. Resources and development of the Mackenzie system. pp. 517-540 *In* The Ecology of River Systems. (B.R. Davies and K.F. Walker, eds). The Hague, W. Junk Publishers. 432 pp.

Rosenberg, D.M., B. Bilyj and A.P. Wiens. 1984. Chironomidae (Diptera) emerging from the littoral zone of reservoirs, with special reference to Southern Indian Lake, Manitoba. Can. J. Fish. Aquat. Sci. 41: 672-681.

Rosenberg, D.M., H.V. Danks and D.M. Lehmkuhl. 1986. Importance of insects in environmental impact assessment. Environ. Man. 10: 773-783.

Rosenberg, D.M. and A.P. Wiens. 1976. Community and species responses of Chironomidae (Diptera) to contamination of fresh waters by crude oil and petroleum products, with special reference to the Trail River, Northwest Territories. J. Fish. Res. Board Can. 33: 1955-1963.

Rosenberg, D.M. and A.P. Wiens. 1978. Effects of sediment addition on macrobenthic invertebrates in a Northern Canadian river. Water Res. 12: 753-763.

Rosenberg, D.M. and A.P. Wiens. 1980. Response of Chironomidae (Diptera) to short-term experimental sediment additions in the Harris River, Northwest Territories, Canada. Acta Universitatis Carolinae-Biologica 1978: 181-192.

Rosenberg, R. 1976. Benthic faunal dynamics during succession following pollution abatement in a Swedish estuary. Oikos 27: 414-427.

Ross, D.H. and D.A. Craig. 1980. Mechanisms of fine particle capture by larval black flies (Diptera: Simuliidae). Can. J. Zool. 58: 1186-1192.

Ross, D.H. and R.W. Merritt. 1978. The larval instars and population dynamics of five species of black flies (Diptera: Simuliidae) and their responses to selected environmental factors. Can. J. Zool. 56: 1633-1642.

Ross, H.H. 1955. The evolution of insect orders. Ent. News 66: 197-208.

Ross, H.H. 1956. Evolution and Classification of the Mountain Caddisflies. Univ. Illinois Press, Urbana, Illinois. 213 pp.

Ross, H.H. 1963. Stream communities and terrestrial biomes. Arch. Hydrobiol. 59: 235-242.

Ross, H.H. 1965. A Textbook of Entomology. 3rd ed. Wiley, New York. 539 pp.

Ross, H.H. 1967. The evolution and past dispersal of the Trichoptera. Ann. Rev. Ent. 12: 169-206.

Roughgarden, J. and M. Feldman. 1975. Species packing and predation pressure. Ecology 56: 489-492.

Roughley, R.E. and D.J. Larson. 1991. Aquatic Coleoptera of springs in Canada. pp. 125-140 *In* Arthropods of Springs, with Particular Reference to Canada (D.D. Williams and H.V. Danks, eds). Mem. ent. Soc. Can. 155. 217 pp.

Ruttner, F. 1953. Fundamentals of Limnology. Univ. Toronto Press, Toronto. 295 pp.

Rutter, R.P. and T.P. Poe. 1978. Macroinvertebrate drift in two adjoining southeastern Pennsylvania streams. Proc. Pennsylv. Acad. Sci. 52: 24-30.

Saether, O.A. 1975. Nearctic chironomids as indicators of lake typology. Verh. Internat.

Verein. Limnol. 19: 3127-3133.

Saether, O.A. 1977. Taxonomic studies on Chironomidae: *Nanocladius, Pseudochironomus,* and the *Harnischia* complex. Bull. Fish. Res. Board Can. 196: 1-143.

Saether, O.A. 1980. The influence of eutrophication on deep lake benthic invertebrate communities. Prog. Wat. Tech. 12: 161-180.

Sagar, P.M. 1986. The effects of floods on the invertebrate fauna of a large, unstable braided river. New Zealand J. mar. Freshwat. Res. 17: 377-386.

Sattler, W. 1963. Uber den Korperbau und Ethologie der Larvae ind Puppe von *Macronema* Pict. (Hydropsychidae), ein als Larvae sich von "Mikro-Drift" ernahrendes Trichopter aus dem Amazongebiet. Arch. Hydrobiol. 59: 26-60.

Savage, A.A. 1989. Adults of the British Aquatic Hemiptera, Heteroptera: a Key with Ecological Notes. Freshwat. Biol. Assoc., U.K. Sci. Pub. No. 50. 173 pp.

Schaffer, W.M. 1974. Optimal reproductive effort in fluctuating environments. Am. Nat. 108: 783-790.

Schindler, D.W. 1969. Two useful devices for vertical plankton and water sampling. J. Fish. Res. Board Can. 26: 1948-1955.

Schindler, D.W. 1987. Detecting ecosystem responses to anthropogenic stress. Can. J. Fish. Aquat. Sci. 44 (Suppl. 1): 6-25.

Schindler, D.W. 1988. Effects of acid rain on freshwater ecosystems. Science 239: 149-157.

Schlosser, I.J. and K. Ebel. 1989. Effects of flow regime and cyprinid predation on a headwater stream. Ecol. Monogr. 59: 41-57.

Schluter, D. 1986. Character displacement between distantly related taxa? Finches and bees in the Galapagos. Am. Nat. 127: 95-102.

Schneider, R.F. 1962. Seasonal succession of certain invertebrates in a northwestern Florida lake. Quart. J. Fl. Acad. Sci. 25: 127-141.

Schoener, T.W. 1974. Competition and the form of habitat shift. Theoret. Pop. Biol. 6: 265-269.

Schoener, T.W. 1983. Field experiments on interspecific competition. Am. Nat. 122: 240-285.

Schofield, C.L. 1976. Acid precipitation: effects on fish. Ambio 5: 228-230.

Schofield, C.L. 1977. Acid snow-melt effects on water quality and fish survival in the Adirondack Mountains of New York State. Research Program Technical Comprehensive Report Number A-072-NY, Office of Water Research Technology, United States Department of the Interior, Washington, D.C. 122 pp.

Schofield, K., C.R. Townsend and A. G. Hildrew. 1988. Predation and the prey community of a headwater stream. Freshwat. Biol. 20: 85-95.

Schwoerbel, J. 1961. Uber die Lebensbedingungen und die Besiedlung des hyporheische Biotope. Arch. Hydrobiol. Suppl. 25: 182-214.

Scott, D. 1966. The substrate cover-fraction concept. pp. 75-78. *In* Organism-substrate relationships in streams. (K.W. Cummins, C.A. Tryon and R.T. Hartman, eds). Spec. Publ. No. 4. Pymatuning Laboratory of Ecology, University of Pittsburg. 145 pp.

Scott, D.B. Jr. 1971. Aquatic Collembola. pp. 74-78 *In* Aquatic Insects of California (R.L. Usinger, ed). Univ. California Press, Berkeley. 508 pp.

Scriber, J.M. and F. Slansky, Jr. 1981. The nutritional ecology of immature insects. Ann. Rev. Ent. 26: 183-211.

Scudder, G.G.E. 1976. Water-boatmen of saline waters (Hemiptera: Corixidae). pp. 263-290 *In* Marine Insects. (L. Cheng, ed). American Elsevier Publ. Co., New York. 581 pp.

Scudder, G.G.E., D.K. McE. Kevan and E.L. Bousfield. 1979. Higher classification. pp. 235-240 *In* Canada and its Insect Fauna (H.V. Danks, ed). Mem. ent. Soc. Can. 108:

1-573.

Scullion, J. and R.W. Edwards. 1980. The effects of coal industry pollutants on the macroinvertebrate fauna of a small river in the South Wales coalfield. Freshwat. Biol. 10: 141-162.

Serruya, C. (ed) 1978. Lake Kinneret. Monogr. Biologicae 32, W. Junk, The Hague. 501 pp.

Service, M.W. 1986. Blood-sucking Insects: Vectors of Disease. Inst. Biol. Stud. No. 167, Edward Arnold, London. 81 pp.

Shaw, J. 1955. The permeability and structure of the cuticle of the aquatic larvae of *Sialis lutaria*. J. Exp. Biol. 32: 330-352.

Shaw, J. and R.H. Stobbart. 1963. Osmotic and ionic regulation in insects. Adv. Ins. Physiol. 1: 315-399.

Shaw, P-C. and K-K. Mark. 1980. Chironomid farming - a means of recycling farm manure and potentially reducing water pollution in Hong Kong. Aquaculture 21: 155-163.

Sheldon, A.L. 1984. Colonization dynamics of aquatic insects. pp. 401-429 *In* The Ecology of Aquatic Insects. (V.H. Resh and D.M. Rosenberg, eds). Praeger Scientific, New York. 625 pp.

Shepard, R.B. and G.W. Minshall. 1981. Nutritional value of lotic feces compared with allochthonous materials. Arch. Hydrobiol. 90: 467-488.

Short, R.A. and P.E. Maslin. 1977. Processing of leaf litter by a stream detritivore: effect on nutrient availability to collectors. Ecology 58: 935-938.

Sibly, R.M. and P. Calow. 1986. Physiological Ecology of Animals: an Evolutionary Approach. Blackwell Sci. Publs, Oxford. 179 pp.

Sibly, R.M. and P. Calow. 1989. A life-cycle theory of response to stress. Biol. J. Linn. Soc. 37: 101-116.

Siegfried, C.A. and A.W. Knight. 1976. Prey selection by a setipalpian stonefly nymph, *Acroneuria (Calineura) californica* Banks (Plecoptera: Perlidae). Ecology 57: 603-608.

Sih, A. 1980. Optimal behavior: can foragers balance two conflicting demands? Science 210: 1041-1043.

Sih, A. 1986. Antipredator responses and the perception of danger by mosquito larvae. Ecology 67: 434-441.

Sih, A. 1987. Predators and prey lifestyles: an evolutionary and ecological overview. pp. 203-224 *In* Predation: Direct and Indirect Impacts on Aquatic Communities. (W.C. Kerfoot and A. Sih, eds). Univ. Press of New England, Hanover, New Hampshire. 366 pp.

Silsbee, D.G. and G.L. Larson. 1983. A comparison of streams in logged and unlogged areas of Great Smoky Mountains National Park. Hydrobiologia 102: 99-111.

Simpson, K.W. 1976. Shore flies and brine flies (Diptera: Ephydridae). pp. 465-496 *In* Marine Insects. (L Cheng, ed). American Elsevier Publ. Co., New York. 581 pp.

Simpson, K.W. 1980. Abnormalities in the tracheal gills of aquatic insects collected from streams receiving chlorinated or crude oil wastes. Freshwat. Biol. 10: 581-583.

Sjöström, P. 1985. Territoriality in nymphs of *Dinocras cephalotes* (Plecoptera). Oikos 45: 353-357.

Smith, D.C. 1987. Adult recruitment in chorus frogs: effects of size and date at metamorphosis. Ecology 68: 344-350.

Smith, I.M. 1991. Water mites (Acari: Parasitengona: Hydrachnida) of spring habitats in Canada. pp. 141-167 *In* Arthropods of Springs, with Particular Reference to Canada (D.D. Williams and H.V. Danks, eds). Mem. ent. Soc. Can. 155. 217 pp.

Smith, J.A. and A.J. Dartnall. 1980. Boundary layer control by water pennies (Coleoptera: Psephenidae). Aquat. Ins. 2: 65-72.

Smith, R.F. and A.E. Pritchard. 1971. Odonata. pp. 106-153 *In* Aquatic Insects of

California (R.L. Usinger, ed). Univ. California Press, Berkeley. 508 pp.

Smock, L.A., D.L. Stoneburner and D.R. Lenat. 1981. Littoral and profundal macroinvertebrate communities of a coastal brown-water lake. Arch. Hydrobiol. 92: 306-320.

Snellen, R.K. and K.W. Stewart. 1979. The life cycle of *Perlesta placida* (Plecoptera: Perlidae) in an intermittent stream in northern Texas, U.S.A. Ann. ent. Soc. Am. 72: 659-666.

Snodgrass, R.E. 1935. Principles of Insect Morphology. McGraw-Hill, New York. 667 pp.

Sodergren, S. 1976. Ecological effects of heavy metal discharge in a salmon river. Inst. Freshwat. Res. Drottningholm Report 55: 91-131.

Söderström, O. 1988. Effects of temperature and food quality on life-history parameters in *Parameletus chelifer* and *P. minor* (Ephemeroptera): a laboratory study. Freshwat. Biol. 20: 295-303.

Sokal, R.R. and F.J. Rohlf. 1981. Biometry. W.H. Freeman, San Francisco. 859 pp.

Soldan, T. 1979. Struktur und Funktion der Maxillarpalpen von *Arthroplea congener* (Ephemeroptera, Heptageniidae). Acta Entomol. Bohemoslov. 76: 300-307.

Soluk, D.A. 1985. Macroinvertebrate abundance and production of psammophilous Chironomidae in shifting sand areas of a lowland river. Can. J. Fish. Aquat. Sci.. 42: 1296-1302.

Soluk, D.A. and N.C. Collins. 1988. Synergistic interactions between fish and stoneflies: facilitation and interference among stream predators. Oikos 52: 94-100.

Sousa, W.P. 1979. Disturbance in marine intertidal boulder fields: the non-equilibrium maintenance of species diversity. Ecology 60: 1225-1239.

Southwood, T.R.E. 1988. Tactics, strategies and templets. Oikos 52: 3-18.

Spence, J.A. and H.B.N. Hynes. 1971. Differences in benthos upstream and downstream of an impoundment. J. Fish. Res. Board Can. 28: 35-43.

Spence, J.R., D.H. Spence and G.G.E. Scudder. 1980. Submergence behavior in *Gerris*: underwater basking. Am. Midl. Nat. 103: 385-391.

Sprules, W.M. 1940. The effect of a beaver dam on the insect fauna of a trout stream. Trans. Am. Fish. Soc. 70: 236-248.

Standen, L. and T.P. Henry. 1983. Acidification effects on macroinvertebrates and fathead minnows (*Pimephales promelas*) in outdoor experimental channels. Water Res. 17: 47-63.

Statzner, B. and B. Higler. 1986. Stream hydraulics as a major determinant of benthic invertebrate zonation patterns. Freshwat. Biol. 16: 127-139.

Statzner, B. and T. F. Holm. 1982. Morphological adaptations of benthic invertebrates to stream flow - an old question studied by means of a new technique (Laser Doppler Anemometry). Oecologia 53: 290-292.

Statzner, B. and T. F. Holm. 1989. Morphological adaptation of shape to flow: microcurrents around lotic macroinvertebrates with known Reynolds numbers at quasi-natural flow conditions. Oecologia 78: 145-157.

Statzner, B., C. Dejoux and J-M. Elouard. 1984. Field experiments on the relationship between drift and benthic densities of aquatic insects in tropical streams (Ivory Coast). Rev. Hydrobiol. trop. 17: 319-334.

Stearns, S.C. 1976. Life-history tactics: a review of the ideas. Q. Rev. Biol. 51: 3-47.

Stein, R.A. and J.J. Magnuson. 1976. Behavioural response of crayfish to a fish predator. Ecology 57: 571-581.

Steinmann, P. 1907. Die Tierwelt der Gebirgsbache. Eine faunistisch-biologische Studie. Annla Biol. lacustre 2: 30-150.

Stemberger, R.S. 1988. Reproductive costs and hydrodynamic benefits of chemically induced defenses in *Keratella testudo*. Limnol. Oceanogr. 33: 593-606.

Stewart-Oaten, A., W.W. Murdoch and K.R. Parker. 1986. Environmental impact assessment: "pseudoreplication" in time? Ecology 67: 929-940.

Stobbart, R.H. and J. Shaw. 1974. Salt and water balance; excretions. pp. 361-446. *In* The Physiology of Insecta. Vol. 5. (M. Rockstein, ed). Academic Press, New York. 648 pp.

Stocker, Z.S.J. and H.B.N. Hynes. 1976. Studies on the tributaries of Char Lake, Cornwallis Island, Canada. Hydrobiologia 49: 97-102.

Stocker, Z.S.J. and D.D. Williams. 1972. A freezing core method for describing the vertical distribution of sediments in a streambed. Limnol. Oceanogr. 17: 136-138.

Suberkropp, K. and M.J. Klug. 1980. The maceration of deciduous leaf litter aquatic hyphomycetes. Can. J. Bot. 58: 1025-1031.

Sugden, A.M. and R.J. Robins. 1979. Aspects of the ecology of vascular epiphytes in Columbian closed forests I. The distribution of the epiphytic flora. Biotropica 11: 173-188.

Surber, E.W. 1937. Rainbow trout and bottom fauna production in one mile of stream. Trans. Am. Fish. Soc. 66: 193-202.

Sutcliffe, D.W. 1961. Studies on salt and water balance in caddis larvae (Trichoptera) 2. Osmotic and ionic regulation of body fluids in *Limnephilus stigma* Curtis and *Arabolia nervosa* Leach. J. Exp Biol. 38: 501-519.

Sutcliffe, D.W. and T.R. Carrick. 1973. Studies on mountain streams in the English Lake District. I. pH, calcium and the distribution of invertebrates in the River Duddon. Freshwat. Biol. 3: 437-462.

Sweeney, B.W. 1984. Factors influencing life-history patterns of aquatic insects. pp. 56-100 *In* The Ecology of Aquatic Insects. (V.H. Resh and D.M. Rosenberg, eds). Praeger Scientific, New York. 625 pp.

Sweeney, B.W. and J.A. Schnack. 1977. Egg development, growth, and metabolism of *Sigara alternata* (Say) (Hemiptera: Corixidae) in fluctuating thermal environments. Ecology 58: 265-277.

Sweeney, B.W. and R.L. Vannote. 1978. Size variation and the distribution of hemimetabolous aquatic insects: two thermal equilibrium hypotheses. Science 200: 444-446.

Sweeney, B.W. and R.L. Vannote. 1981. *Ephemerella* mayflies of White Clay Creek: Bioenergetic and ecological relationships among six coexisting species. Ecology 62: 1353-1369.

Sweeney, B.W., R.L. Vannote and P.J. Dodds. 1986. The relative importance of temperature and diet to larval development and adult size of the winter stonefly, *Soyedina carolinensis* (Plecoptera: Nemouridae). Freshwat. Biol. 16: 39-48.

Symons, H., S.R. Weibel and G.G. Roebeck. 1964. Influence of impoundments on water quality. Publs U.S. Pub. Hlth Serv. 999-WP-18. 78 pp.

Tallamy, D.W. and R.F. Denno. 1981. Alternative life history patterns in risky environments: an example from lacebugs. pp. 129-147 *In* Insect Life History Patterns: Habitat and Geographic Variation. (R.F. Denno and H. Dingle, eds). Springer-Verlag, New York. 225 pp.

Tavares-Cromar, A.F. 1990. A study of niche overlap, species interactions and the food web of a macroinvertebrate riffle community in Duffin Creek Ontario. M.Sc. Thesis, University of Toronto, Toronto, Canada. 340 pp.

Tavares, A.F. and D.D. Williams. 1990. Life histories, diet and niche overlap of three sympatric species of Elmidae (Coleoptera) in a temperate stream. Can. Ent. 122: 563-577.

Teraguchi, M. and T.G. Northcote. 1966. Vertical distributions and migration of *Chaoborus flavicans* larvae in Corbett Lake, B.C. Limnol. Oceanogr. 11: 164-176.

Thienemann, A. 1925. Die Binnengewässer Mitteleuropas. Eine limnologische

Einführung. Die Binnengewässer 1: 1-125.

Thorpe, W.H. 1931. The biology of the petroleum fly. Science 73: 101-103.

Tillyard, R.J. 1917. The Biology of Dragonflies. Cambridge University Press, Cambridge. 396 pp.

Tillyard, R.J. 1928. Some remarks on the Devonian fossil insects from the Rhynie chert beds, Old Red Sandstone. Trans. ent. Soc. Lond. 76: 65-71.

Tilman, D. 1982. Resource competition and community structure. Princeton University Press, Princeton, New Jersey. 296 pp.

Tones, P. 1978. Osmoregulation in adults and larvae of *Hygrotus salinarius*. Comp. Biochem. Physiol. (A). 60: 247-250.

Towns, D.R. 1979. Composition and zonation of benthic invertebrate communities in a New Zealand kauri forest stream. Freshwat. Biol. 9: 251-262.

Towns, D.R. 1981. Life histories of benthic invertebrates in a kauri forest stream in northern New Zealand. Aust. J. mar. Freshwat. Res. 32: 191-211.

Townsend, C.R. and A.G. Hildrew. 1976. Field experiments on the drifting, colonization and continuous redistribution of stream benthos. J. Anim. Ecol. 45: 759-772.

Tudorancea, C., R.M. Baxter and C.H. Fernando. 1989. A comparative limnological study of zoobenthic associations in lakes of the Ethiopian Rift Valley. Arch. Hydrobiol. Suppl. 83: 121-174.

Turcotte, P. and P.P. Harper. 1982. The macro-invertebrate fauna of a small Andean stream. Freshwat. Biol. 12: 411-419.

Tuxen, S.L. 1944. The hot springs, their animal communities and their zoogeographical significance. The Zoology of Iceland 1: 1-206.

Ulfstrand, S. 1968. Life cycles of benthic insects in Lapland streams (Ephemeroptera, Plecoptera, Trichoptera, Diptera, Simuliidae). Oikos 19: 167-190.

Usinger, R.L. 1957. Marine insects. Geol. Soc. Am. Mem. 67: 1177-1182.

Usinger, R.L. (ed) 1971. Aquatic Insects of California. Univ. California Press. 508 pp.

Van Buskirk, J. 1987. Density-dependent population dynamics in larvae of the dragonfly *Pachydiplax longipennis*: a field experiment. Oecologia 72: 221-225.

Vannote, R.L. 1969. Detrital consumers in natural system. pp. 20-24 *In* AAAS Symp. Tech. Rep. (K.W. Cummins, ed). Michigan State Univ. Inst. Water Res. No. 7.

Vannote, R.L., G.W. Minshall, K.W. Cummins, J.R. Sedell and C.E. Cushing. 1980. The river continuum concept. Can. J. Fish. Aquat. Sci. 37: 130-137.

Vannote, R.L. and B.W. Sweeney. 1980. Geographic analysis of thermal equilibria: a conceptual model for evaluating the effect of natural and modified thermal regimes on aquatic insect communities. Am. Nat. 115: 667-695.

Vepsalainen, K. and M. Nummelin. 1985. Female territoriality in the waterstriders *Gerris najans* and *G. cinereus*. Ann. Zool. Fenn. 22: 433-439.

Verdonschot, P.F.M. and J.A. Schot. 1986. Macrofaunal community types in helocrene springs. Report Res. Inst. Nat. Managt. (Netherlands). pp. 85-103.

Walde, S.J. and R.W. Davies. 1984. The effect of intraspecific interference on *Kogotus nonus* (Plecoptera) foraging behaviour. Can. J. Zool. 62: 2221-2226.

Wallace, A.R. 1876. The Geographical Distribution of Animals: with a Study of the Relations of Living and Extinct Faunas as Elucidating the Past Changes of the Earth's Surface. Vol. I, 503 pp. II, 607 pp. Macmillan, London.

Wallace, J.B. and M.E. Gurtz. 1986. Response of *Baetis* mayflies (Ephemeroptera) to catchment logging. Am. Midl. Nat. 115: 25-41.

Wallace, J.B. and D. Malas. 1976a. The significance of the elongate, rectangular mesh found in capture nets of fine particle filter feeding Trichoptera larvae. Arch. Hydrobiol. 77: 205-212.

Wallace, J.B. and D. Malas. 1976b. The fine structure of capture nets of larval Philopotamidae (Trichoptera), with special emphasis on *Dolophilodes distinctus*.

Can. J. Zool. 54: 1788-1802.

Wallace, M.M.H. and I.M. Mackerras. 1970. The entognathous hexapods. pp. 205-216 *In* The Insects of Australia. C.S.I.R.O., Melbourne University Press, Canberra. 1029 pp.

Wallace, R.R., H.B.N. Hynes and W.R. Merritt. 1976. Laboratory and field experiments with methoxychlor as a larvicide for Simuliidae (Diptera). Environ. Pollut. 10: 251-269.

Walton, O.E. Jr. 1978. Substrate attachment by drifting aquatic insect larvae. Ecology 59: 1023-1030.

Waltz, R.D. and W.P. McCafferty. 1979. Freshwater springtails (Hexapoda: Collembola) of North America. Purdue Univ. Agric. Exp. Sta. Res. Bull. 960. Lafayette, Indiana.

Ward, A.F. and D.D. Williams. 1986. Longitudinal zonation and food of larval chironomids (Insecta: Diptera) along the course of a river in temperate Canada. Holarct. Ecol. 9: 48-57.

Ward, J.V. 1976. Effects of flow patterns below large dams on stream benthos: a review. pp. 235-253 *In* Instream Flow Needs, Solutions to Technical, Legal, and Social Problems Caused by Increasing Competition for Limited Stream Flow. (J.F. Orsborn and G.H. Orsborn, eds). Symposium and specialty conference, Boise, ID, May 1976, Am. Fish. Soc., Bethesda, MD. 1208 pp.

Ward, J.V. 1982. Ecological aspects of stream regulation: responses in downstream lotic reaches. Water Poll. Man. Rev. 2: 1-26.

Ward, J.V. 1986. Altitudinal zonation in a Rocky Mountain stream. Arch. Hydrobiol. Suppl. 74: 133-199.

Ward, J.V. 1989. The four-dimensional nature of lotic ecosystems. J. North Am. Benthol. Soc. 8: 2-8.

Ward, J.V. and R.A. Short. 1978. Macroinvertebrate community structure of four special lotic habitats in Colorado, U.S.A. Internationale Vereinigung fur Theoretische und Angewandte Limnologie Verhandlungen 20: 1382-1387.

Ward, J.V. and J.A. Stanford. 1979. Ecological factors controlling stream zoobenthos with emphasis on thermal modification of regulated streams. pp. 35-56 *In* The Ecology of Regulated Streams. (J.V. Ward and J.A. Stanford, eds). Plenum Press, New York. 398 pp.

Ward, J.V. and J.A. Stanford. 1982. Thermal responses in the evolutionary ecology of aquatic insects. Ann. Rev. Ent. 27: 97-117.

Warren, P.H. 1989. Spatial and temporal variation in the structure of a freshwater food web. Oikos 55: 299-311.

Warwick, W.F. 1980. Paleolimnology of the Bay of Quinte, Lake Ontario: 2800 years of cultural influence. Can. Bull. Fish. Aquat. Sci. 206: 1-117.

Waters, T.F. 1962. Diurnal periodicity in the drift of stream invertebrates. Ecology 43: 316-320.

Waters, T.F. 1972. The drift of stream insects. Ann. Rev. Ent. 17: 253-272.

Waters, T.F. 1979. Benthic life histories: summary and future needs. J. Fish. Res. Board Can. 36: 342-345.

Watson, J.A.L. 1982. Dragonflies in the Australian environment: taxonomy, biology and conservation. Adv. Odonatol. 1: 293-302.

Weatherley, A.H., J.R. Beevers and P.S. Lake. 1967. The ecology of a zinc-polluted river. pp. 252-278 *In* Australian Inland Waters and Their Fauna: Eleven Studies. (A.H. Weatherley, ed.). A.N.U. Press, Canberra. 287 pp.

Weaver, J.S. and J.C. Morse. 1986. Evolution of feeding and case-making behaviour in Trichoptera. J. North Am. Benthol. Soc. 5: 150-158.

Welch, H.E. 1976. Ecology of Chironomidae (Diptera) in a polar lake. J. Fish. Res. Bd Can. 33: 227-247.

Wells, L. 1960. Seasonal abundance and vertical movements of planktonic crustacea in

Lake Michigan. U.S. Fish Wildl. Serv. Fish. Bull. 60: 343-369.

Wentsel, R., A. McIntosh, W.P. McCafferty, G. Atchison and V. Anderson. 1977. Avoidance response of midge larvae (*Chironomus tentans*) to sediments containing heavy metals. Hydrobiologia 55: 171-175.

Wesenberg-Lund, C. 1917. Furesoestudier. Det. Kg. Danske Vedensk.Selk Skrifter, Nat. og. Mathem. Afd., Rakke, III. 1-209.

Westfall, M.J.Jr. 1984. Odonata. pp. 126-176 *In* An Introduction to the Aquatic Insects of North America. (R.W. Merritt and K.W. Cummins, eds). Kendal/Hunt Publishers, Dubuque, Iowa. 722 pp.

Wetzel, R.G. 1983. Limnology. Saunders, New York. 767 pp.

White, D.S., W.U. Brigham and J.T. Doyen. 1984. Aquatic Coleoptera. pp. 361-437 *In* An Introduction to the Aquatic Insects of North America. (R.W. Merritt and K.W. Cummins, eds). Kendal/Hunt Publishers, Dubuque, Iowa. 722 pp.

Whiting, E.R. and H.F. Clifford. 1983. Invertebrates and urban runoff in a small northern stream, Edmonton, Alberta, Canada. Hydrobiologia 102: 73-80.

Whittington, H.B. 1978. The lobopod animal *Aysheaia pedunculata*, middle Cambrian, Burgess Shale, British Columbia. Phil. Trans. R. Soc. B, 284: 165-197.

Wichard, W. 1991. The evolutionary effect of overcoming osmosis in Trichoptera. Proc. 6th Int. Symp. Trichoptera, Lodz, Poland (in press).

Wichard, W. and K. Hauss. 1975. Der Chloridzellenfehlbetrag als Okomorphologischer zeigerwert fur die salinitat von Binnengewassern. Acta Hydrochim. Hydrobiol. 3: 347-356.

Wiederholm, T. (ed) 1983. Chironomidae of the Holarctic region. Keys and diagnoses. Pt. I. Larvae. Ent. Scand. Suppl. 19: 1-457.

Wiegert, R.G. 1988. Holism and reductionism in ecology: hypotheses, scale and systems models. Oikos 53: 267-269.

Wiens, J.A. 1977. On competition and variable environments. Am. Sci. 65: 590-597.

Wiggins, G.B. 1977. Larvae of the North American Caddisfly Genera (Trichoptera). Univ. Toronto Press, Toronto. 401 pp.

Wiggins, G.B. 1982. Trichoptera. pp. 599-612 *In* Synopsis and Classification of Living Organisms (S.P. Parker, ed). McGraw-Hill Book Co., New York.

Wiggins, G.B. 1984. Trichoptera. pp. 271-311 *In* An Introduction to the Aquatic Insects of North America. (R.W. Merritt and K.W. Cummins, eds). Kendal/Hunt Publishers, Dubuque, Iowa. 722 pp.

Wiggins, G.B. and R.J. Mackay. 1978. Some relationships between systematics and trophic ecology in Nearctic aquatic insects, with special reference to Trichoptera. Ecology 59: 1211-1220.

Wiggins, G.B. and W. Wichard. 1989. Phylogeny of pupation in Trichoptera, with proposals on the origin and higher classification of the order. J. North Am. Benthol. Soc. 8: 260-276.

Wigglesworth, V.B. 1965. The Principles of Insect Physiology. Methuen, London. 544 pp.

Wilding, J.L. 1940. A new square-foot aquatic sampler. Limnol. Soc. Am. Spec. Publ. 4: 1-4.

Wiley, M.J. and S.C. Mozley. 1978. Pelagic occurrence of benthic animals near shore in Lake Michigan. Internat. Assoc. Great Lakes Res. 4: 201-205.

Williams, C.J. 1982. The drift of some chironomid egg masses (Diptera: Chironomidae). Freshwat. Biol. 12: 573-578.

Williams, D.D. 1976. Aquatic invertebrates inhabiting agricultural drainage tile systems in southern Ontario. Can. Field-Nat. 90: 193-195.

Williams, D.D. 1979. Aquatic habitats of Canada and their insects. pp. 211-234 *In* Canada and its Insect Fauna (H.V. Danks, ed). Mem. ent. Soc. Can. 108: 1-573.

Williams, D.D. 1980a. Some relationships between stream benthos and substrate heterogeneity. Limnol. Oceanogr. 25: 166-172.

Williams, D.D. 1980b. Applied aspects of mayfly biology. pp. 1-17 *In* Advances in Ephemeroptera Biology. (J.F. Flannagan and K.E. Marshall, eds). Plenum Publ., New York. 552 pp.

Williams, D.D. 1980c. Invertebrate drift lost to the sea during low flow conditions in a small coastal stream in Western Canada. Hydrobiologia 75: 251-254.

Williams, D.D. 1980d. Temporal patterns in recolonization of stream benthos. Arch. Hydrobiol. 90: 56-74.

Williams, D.D. 1984. The hyporheic zone as a habitat for aquatic insects and associated arthropods. pp. 430-455 *In* The Ecology of Aquatic Insects. (V.H. Resh and D.M. Rosenberg, eds). Praeger Scientific, New York. 625 pp.

Williams, D.D. 1987. The Ecology of Temporary Waters. Croom Helm (Routledge, Chapman & Hall Ltd.), London. 205 pp.

Williams, D.D. 1989. Towards a biological and chemical definition of the hyporheic zone in two Canadian rivers. Freshwat. Biol. 22: 189-208.

Williams, D.D. 1990. A field study of the effects of water temperature, discharge and trout on the drift of stream invertebrates. Arch. Hydrobiol. 119: 167-181.

Williams, D.D. 1991. Life history traits of aquatic arthropods in springs. pp. 63-87 *In* Arthropods of Springs, with Particular Reference to Canada (D.D. Williams and H.V. Danks, eds). Mem. ent. Soc. Can. 155. 217 pp.

Williams, D.D. and B.W. Coad. 1979. The ecology of temporary streams III. Temporary stream fishes in southern Ontario, Canada. Int. Rev. ges. Hydrobiol. 64: 501-515.

Williams, D.D. and H.B.N. Hynes. 1977a. The ecology of temporary streams II. General remarks on temporary streams. Int. Revue ges. Hydrobiol. 62: 53-61.

Williams, D.D. and H.B.N. Hynes. 1977b. Benthic community development in a new stream. Can. J. Zool. 55: 1071-1076.

Williams, D.D. and H.B.N. Hynes. 1974. The occurrence of benthos deep in the substratum of a stream. Freshwat. Biol. 4: 233-256.

Williams, D.D. and J.H. Mundie. 1978. Substrate size selection by stream invertebrates, and the influence of sand. Limnol. Oceanogr. 23: 1030-1033.

Williams, D.D., A.F. Tavares and E. Bryant. 1987. Respiratory device or camouflage? - a case for the caddisfly. Oikos 50: 42-52.

Williams, D.D. and N.E. Williams. 1975. A contribution to the biology of *Ironoquia punctatissima* (Trichoptera: Limnephilidae). Can. Ent. 107: 829-832.

Williams, D.D. and N.E. Williams. 1976. Aspects of the ecology of the faunas of some brackishwater pools on the St. Lawrence North Shore. Can. Field-Nat. 90: 410-415.

Williams, D.D. and N.E. Williams. 1987. Trichoptera from cold freshwater springs in Canada: records and comments. Proc. Ent. Soc. Ont. 118: 13-23.

Williams, N.E. 1981. Aquatic organisms and palaeoecology: recent and future trends. pp. 289-303 *In* Perspectives in Running Water Ecology. (M.A. Lock and D.D. Williams, eds). Plenum Publishers, New York. 430 pp.

Williams, N.E. 1988. The use of caddisflies (Trichoptera) in palaeoecology. Palaeogeogr., Palaeoclimatol. Palaeoecol. 62: 493-500.

Williams, N.E. 1989. Factors affecting the interpretation of caddisfly assemblages from Quaternary sediments. J. Paleolimnol. 1: 241-248.

Williams, N.E. 1991. Geographic and environmental patterns in caddisfly (Trichoptera) assemblages from coldwater springs in Canada. pp. 107-124 *In* Arthropods of Springs, with Particular Reference to Canada (D.D. Williams and H.V. Danks, eds). Mem. ent. Soc. Can. 155. 217 pp.

Williams, N.E. and H.B.N. Hynes. 1973. Microdistribution and feeding of the net-spinning caddisflies (Trichoptera) of a Canadian stream. Oikos 24: 73-84.

Williams, N.E. and A.V. Morgan. 1977. Fossil caddisflies (Insecta: Trichoptera) from the Don Formation, Toronto, Ontario, and their use in palaeoecology. Can. J. Zool. 55: 519-527.

Williams, N.E., J.A. Westgate, D.D. Williams, A. Morgan and A.V. Morgan. 1981. Invertebrate fossils (Insecta: Trichoptera, Diptera, Coleoptera) from the Pleistocene Scarborough Formation at Toronto, Ontario, and their palaeoenvironmental significance. Quat. Res. 16: 146-166.

Williams, W.D. 1976. Some problems for Australian limnologists. Search 7: 187-190.

Williams, W.D. 1984. Chemical and biological features of salt lakes on the Eyre Peninsula, South Australia, and an explanation of regional differences in the fauna of Australian salt lakes. Verh. Internat. Verein. Limnol. 22: 1208-1215.

Williams, W.D. 1985. Biotic adaptations in temporary lentic waters with special reference to those in semi-arid regions. pp. 85-110 *In* Perspectives in Southern Hemisphere Limnology. (B.R. Davies and R.D. Walmsley, eds). W. Junk, The Hague.

Williams, W.D. 1988. Limnological imbalances: an antipodean viewpoint. Freshwat. Biol. 20: 407-420.

Wilson, E.O. 1975. Sociobiology. Harvard Univ. Press, Harvard. 697 pp.

Winnell, M.H. and D.J. Jude. 1984. Associations among Chironomidae and sandy substrates in nearshore Lake Michigan. Can. J. Fish. Aquat. Sci. 41: 174-179.

Winner, R.W., M.W. Boesel and M.P. Farrell. 1980. Insect community structure as an index of heavy-metal pollution in lotic ecosystems. Can. J. Fish. Aquat. Sci. 37: 647-655.

Winterbourn, M.J. 1974. The life histories, trophic relations and production of *Stenoperla prasina* (Plecoptera) and *Deleatidium* sp. (Ephemeroptera) in a New Zealand river. Freshwat. Biol. 4: 507-524.

Winterbourn, M.J. 1978. The macroinvertebrate fauna of a New Zealand forest stream. N.Z.J. Zool. 5: 157-169.

Winterbourn, M.J. and N.H. Anderson. 1980. The life history of *Philanisus plebeius* Walker (Trichoptera: Chathamiidae), a caddisfly whose eggs were found in a starfish. Ecol. Ent. 5: 293-303.

Winterbourn, M.J. and S.F. Davis. 1976. Ecological role of *Zelandopsyche ingens* (Trichoptera: Oeconesidae) in a beech forest stream ecosystem. Aust. J. mar. Freshwat. Res. 27: 197-215.

Winterbourn, M.J., J.S. Rounick and B. Cowie. 1981. Are New Zealand stream ecosystems really different? N.Z.J. mar. Freshwat. Res. 15: 321-328.

Wise, E.J. 1980. Seasonal distribution and life histories of Ephemeroptera in a Northumbrian River. Freshwat. Biol. 10: 101-111.

Wissinger, S.A. 1989. Seasonal variation in the intensity of competition and predation among dragonfly larvae. Ecology 70: 1017-1027.

Wood, D.M. 1979. Culicoidea. pp. 398-400 *In* Canada and its Insect Fauna (H.V. Danks, ed). Mem. ent. Soc. Can. 108: 1-573.

Woodward, T.E., J.W. Evans and V.F. Eastop. 1970. Hemiptera. pp. 387-457 *In* The Insects of Australia. C.S.I.R.O., Melbourne University Press, Canberra. 1029 pp.

Wotton, R.S. 1980. Coprophagy as an economic feeding tactic in blackfly larvae. Oikos 34: 282-286.

Wright, H.E. 1977. Quaternary vegetation history - some comparisons between Europe and America. Ann. Rev. Earth Planet. Sci. 5: 123-258.

Wright, J.F., P.D. Armitage, M.T. Furse and D. Moss. 1989. Prediction of invertebrate communities using stream measurements. Regulated Rivers, Res. Management 4: 147-155.

Wrubleski, D.A. 1987. Chironomidae (Diptera) of peatlands and marshes in Canada. pp. 141-161 *In* Aquatic Insects of Peatlands and Marshes in Canada (D.M. Rosenberg and

H.V. Danks, eds). Mem. ent. Soc. Can. 140: 1-174.

Yodzis, P. 1988. The indeterminancy of ecological interactions as perceived through perturbation experiments. Ecology 69: 508-515.

Zar, J.H. 1984. Biostatistical Analysis. Prentice-Hall, Inc., New Jersey. 718 pp.

Zhadin, V.I. and S.V. Gerd. 1963. Fauna and Flora of the Rivers, Lakes and Reservoirs of the USSR. Jerusalem: Israeli Program Sci. Transl. 626 pp.

Zischke, J.A., J.W. Arthur, K.J. Nordlie, R.O. Hermanutz, D.A. Standen and T.P. Henry. 1983. Acidification effects on macroinvertebrates and fathead minnows (*Pimephales promelas*) in outdoor channels. Water Res. 17: 47-63.

Zoltai, S.C. 1980. An outline of the wetland regions of Canada. pp. 1-8 *In* Proc. Workshop on Canadian Wetlands (C.D.A. Rubec and F.C. Pollett, eds). Ecol. Land Classif. Ser. Environ. Can. 12.

Zoltai, S.C. 1987. Peatlands and marshes in the wetland regions of Canada. pp. 5-13 *In* Aquatic Insects of Peatlands and Marshes in Canada (D.M. Rosenberg and H.V. Danks, eds). Mem. ent. Soc. Can. 140: 1-174.

Zwick, P. 1981. Comment, pp. 175. *In* Insect Phylogeny (W. Hennig; A.C. Pont, transl., ed). John Wiley and Sons, Chichester, U.K. 514 pp.

INDEX

A

a posteriori explanations 257
A-selection 161-164, 166-167
Ablabesmyia 122, 133
Acalyptratae 79
Acanthocyclops vernalis 230
Acari 11
Acartia 225
Acentria niveus 98
acid deposition, "rain" 270, 281-283
acid discharge 280
acid stress 283
acid tolerant insects 283
Acridae 42
Acroneuria carolinensis 211
Acroneuria lycorias 246
Aculeata 109
acute toxicity 270
Adephaga 60-61, 66, 69, 73
Adirondack Mountains, U.S.A. 283
aedeagus 2
Aedes 85, 192
Aedes aegypti 153, 247, 290
Aedes africanus 290
Aepophilidae 52
Aepophilus 52
aerial pests 276
aerial spraying 276
aeropneustic insects 185
Aeshna 31, 143
Aeshna juncea 230
Aeshnidae 26, 30, 138
Africa 7, 8, 9, 58, 145, 290-291,
    293-294
Afrotropical Region 8-9, 18, 26, 85,
    90
Agabus bipustulatus 231
Agabus sturmii 231
Agapetus 118, 122, 179
Agarodes 190
aggressive interactions 261
    3rd and 4th order 261
Agnetina capitata 208, 211, 233, 235
agriculture 152

Agrionidae 121, 152
Agriotypidae 109
Agriotypus armatus 109
air-lift sampler 268
alarm substances 241
Alatospora acuminata 219
alderflies 53
algae 15, 226
    filamentous green 227
Algonquin Provincial Park, Canada 92,
    273, 282
alkaline substances 281
alkalinity 198, 280
allelopathy 208
allochthonous input 119, 215, 222, 226
allospecifics 208
Allotrichoma livens 150
aluminium 281-282
Amazon Basin 129, 131
Ameletopsidae 18
Ametropodidae 124
amino acid 222
amoebocytes 1
amphibians 9
Amphipoda 11
Amphipsyche 152
Amphizoidae 60-61, 118
Anabolia 180
anaerobic sediment 277
Anagapetus 118
anal papillae 192
Analysis of Covariance (ANCOVA) 258
Analysis of Variance (ANOVA) 252
analytical sampling 253
anaplasmosis 294
Ancylus fluviatilis 226
anemophilous plants 150
Angara 7
Anisoptera 6, 24-29, 113, 119, 138,
    189-190
Anisozygoptera 6, 24
Annelida 2, 227
Anopheles 85, 288, 290
Anopheles culicifacies 288
Anopheles gambiae 288

M